INTRODUCTION TO GEOMORPHOLOGY

Introduction to Geomorphology

Frank Ahnert

Professor Emeritus of Physical Geography, RWTH Aachen, Germany

ARNOLD

A member of the Hodder Headline Group
LONDON · SYDNEY · AUCKLAND

First published in Great Britain in 1998 by Arnold,
a member of the Hodder Headline Group,
338 Euston Road, London NW1 3BH
http://www.arnoldpublishers.com

© 1996 Eugen Ulmer GmbH & Co
First published under the title *Einführung in die
Geomorphologie*
English translation © Frank Ahnert

Co-published in the US, Central and South America by
John Wiley & Sons, Inc., 605 Third Avenue, New York,
NY10158–0012

British Library Cataloguing in Publication Data
A catalogue record for this book is available from the
British Library

Library of Congress Cataloging-in-Publication Data
A catalog record for this book is available from the
Library of Congress

ISBN 0 340 69260 X (hb)
ISBN 0 340 69259 6 (pb)
ISBN 0 470 23657 4 (hb, Wiley)
ISBN 0 470 23658 2 (pb, Wiley)

Production Editor: James Rabson
Production Controller: Rose James
Cover designer: Terry Griffiths

Typeset in $9\frac{1}{2}/11\frac{1}{2}$ Palatino by Photoprint, Torquay
Printed and bound in Great Britain by The Bath Press, Bath

CONTENTS

Chapter 1	**The purpose and scope of geomorphology**	I
	1.1 Geomorphology	1
	1.2 Landforms as events in time and space	1
	1.3 Landform size and duration	2
	1.4 Methodological components of geomorphology	4
	1.5 Subdivisions of geomorphology by subject matter	9
Chapter 2	**General systems theory in geomorphology**	**10**
	2.1 The concept of systems	10
	2.2 Types of systems	11
	2.3 The role of theoretical models	14
Chapter 3	**The geomorphodynamic system**	**16**
	3.1 Definition	16
	3.2 Structure of the geomorphodynamic system	16
	3.3 The functional relationship between relief and denudation	17
	3.4 Process response models of relief development with no uplift or valley deepening	20
	3.5 Model of relief development with a constant rate of uplift	21
	3.6 Model of relief development with variable rates of uplift	21
	3.7 The maximum possible heights of mountain ranges	22
	3.8 Landforms as an expression of spatial and temporal differentiation of process response systems	23
Chapter 4	**Endogenic process response systems and their geomorphological expression**	**25**
	4.1 Isostasy	25
	4.2 Plate tectonics: the formation and locational change of the continents and oceans	29
	4.3 The large morphostructural units of the continents	32
	4.4 Volcanism and plutonism	36
Chapter 5	**Exogenic factors and systems**	**43**
	5.1 Eustatic changes of sea level	43
	5.2 The morphoclimate	44
	5.3 The major exogenic process response systems	48

Chapter 6   **Rock types and their characteristics**   **51**

6.1 Elements, minerals and rocks   51
6.2 Igneous rocks   51
6.3 Sedimentary rocks   53
6.4 Metamorphic rocks   57

Chapter 7   **The weathering system**   **61**

7.1 The term weathering   61
7.2 Weathering as a process response system   62
7.3 Mechanical weathering processes and their products   69
7.4 Chemical weathering   75
7.5 Soils as products of weathering   79
7.6 The relative share of mechanical and chemical weathering in different
    morphoclimates   86

Chapter 8   **Process response systems of denudation**   **88**

8.1 Introduction   88
8.2 Physical basis of denudative mass movements   89
8.3 Rock fall denudation and landslides   92
8.4 Creep denudation   100
8.5 Periglacial denudation processes   103
8.6 Wash denudation   112
8.7 The aeolian process response system   116
8.8 The determination of denudation rates   123

Chapter 9   **Denudational slope development**   **126**

9.1 Slopes and descriptive slope classification   126
9.2 The mass balance of slope development   127
9.3 Weathering-limited and transport-limited denudation and their
    influence on slope form   129
9.4 Process-specific slope forms   130

Chapter 10   **The hydrological and hydraulic basis of the fluvial system**   **135**

10.1 Global water balance and water budget   135
10.2 Components of the local water budget   135
10.3 Groundwater and springs   137
10.4 Discharge, discharge regimes and fluvial morphoclimate   140
10.5 Fluvial hydraulics   144

Chapter 11   **Stream erosion and stream transport**   **148**

11.1 Types of stream load   148
11.2 Erosion and transport   149
11.3 Discharge and transport rates   152

Chapter 12   **Stream channel formation**   **155**

12.1 The ratio between width and depth   155
12.2 Bedrock channels and gravel stream beds – erosion-limited and
    transport-limited stretches   156
12.3 Gravel bars in the stream channel   156

	12.4 Ripples, dunes and antidunes on sandy channel beds	156
	12.5 Riffles and pools	157
	12.6 Valley floors, natural levees and alluvial loam	159
	12.7 The tendency towards a local dynamic equilibrium in a stream bed	160
Chapter 13	**Stream channel patterns**	**162**
	13.1 Valley form and stream channel pattern	162
	13.2 Stream braiding	162
	13.3 River meanders	165
	13.4 Asymmetry at stream mouths: entrance angles and the deferment of junctions	172
Chapter 14	**The longitudinal stream profile and its development**	**173**
	14.1 The longitudinal profile	173
	14.2 Base level and profile development	174
	14.3 Equilibrium tendency in profile development	176
	14.4 Causes of knick points in longitudinal profiles	178
	14.5 Waterfalls	179
Chapter 15	**River terraces**	**183**
	15.1 Rock-floored terrace	183
	15.2 Accumulation terraces	183
	15.3 Location and preservation of terraces in the stream valley	186
	15.4 The causes of terrace formation	186
	15.5 The diagnostic significance of river terraces	189
Chapter 16	**Alluvial fans and deltas**	**191**
	16.1 Alluvial fans	191
	16.2 Deltas	194
Chapter 17	**Stream and valley networks**	**199**
	17.1 Change and integration of stream networks	199
	17.2 Transversal valleys and water gaps	201
	17.3 Stream order and valley order systems	205
	17.4 Stream and valley network patterns	206
Chapter 18	**Streams and slope development in the fluvial system**	**209**
	18.1 The fluvial process response system	209
	18.2 The linking of processes with differing magnitude–frequencies	212
	18.3 Valley cross-section forms as an expression of the process structure	214
	18.4 Valley head types	218
Chapter 19	**Peneplains, pediments and inselbergs**	**220**
	19.1 The development of surfaces by marine abrasion	220
	19.2 Peneplains as the end stage of the Davis cycle	221
	19.3 The development of denudation surfaces and inselbergs as a result of double planation	222

	19.4 Pedimentation	223
	19.5 Piedmont benchlands, zonal and azonal inselbergs	225
	19.6 Criteria to distinguish peneplains and pseudo-peneplains	227

Chapter 20   **Structurally controlled landforms**   **230**

20.1 Forms determined by joints	230
20.2 Forms determined by fault structures	233
20.3 Forms determined by bedding structures	236

Chapter 21   **Karst forms**   **250**

21.1 Introduction	250
21.2 Karst surface forms	250
21.3 Karst caves	258

Chapter 22   **The glacial system**   **261**

22.1 The formation and characteristics of glacier ice	261
22.2 The mass budget of glaciers	262
22.3 Glacier types	265
22.4 Glacial erosion: process and form	269
22.5 Glacial deposits: materials, processes and forms	272
22.6 Glaciofluvial processes, deposits and landforms	275
22.7 The glacial sequence	278
22.8 The Pleistocene ice age and its morphological significance	279

Chapter 23   **The littoral system**   **286**

23.1 Introduction	286
23.2 Coastal classification	288
23.3 The tides and their geomorphological effect	290
23.4 The surf and its geomorphological effect	297
23.5 Form associations on coasts of unconsolidated material and neutral shorelines	306
23.6 Structurally controlled coastal types	310
23.7 Coastlines determined by climate	311
23.8 Shelf forms and submarine canyons	316

Chapter 24   **Aspects of applied geomorphology**   **317**

24.1 Introduction	317
24.2 Structure of applied geomorphology	317
24.3 Mapping	318
24.4 Functional geomorphological applications	319

Chapter 25   **Brief historical summary of geomorphology**   **325**

Bibliography	**328**
Index	**341**

1

THE PURPOSE AND SCOPE OF GEOMORPHOLOGY

1.1 GEOMORPHOLOGY

Geomorphology is the science that investigates the **landforms** of the earth. Included are the forms on the land surface, the mountains, valleys, slopes, river beds and dunes, for example, and the submarine forms on the sea floor, such as coastal mud flats, coral reefs and submarine canyons.

Landforms change their shape as a result of:

1. movement of the earth's crust: uplift, subsidence, folding and horizontal shifting, all of which change the elevation of the surface and affect the internal structure of the crust;
2. volcanism, which builds up volcanoes;
3. the downwearing or denudation of rocks and soil material from the land surface;
4. the deposition of this material elsewhere.

The **structural landforms** produced by crustal movements and volcanism are modified to become **sculptural landforms** by denudation and deposition. Geomorphology describes the existing landforms, investigates the processes that create them, examines the relationships between landform and process and seeks to explain landform development.

Geomorphology is an earth science and is usually considered to be part of geography and geology. The investigation of the spatial differentiation of the earth's surface and, therefore, also of landforms, is the concern of geography, while geology examines the history of the earth and the development over time of the earth's crust and its surface.

1.2 LANDFORMS AS EVENTS IN TIME AND SPACE

Each landform on the earth's surface occupies a defined space delimited by its location, its outline and its size. No landform exists forever but only within a particular time span in the earth's history. All landforms are created, develop and disappear and other landforms replace them. Where 70 million years ago there was an enormous basin of sediment covered by the sea, today the folded mountain chain of the Alps rises several thousands of metres above sea level. In Europe north of the Alps, older fold mountains, formed in the Palaeozoic era 250 million years ago, have been worn down to lowlands and then covered with sediments for a period of about 200 million years during the Mesozoic era. Later, some of these areas were uplifted as fault blocks and now form the central European uplands into which rivers have incised their valleys. Eventually these uplands will also disappear either by denudation, crustal subsidence, or both, and other landforms develop in their place. All landforms have a beginning, a period of development and an end. When looked at in the framework of earth history they are essentially events in space and time which change during the course of their existence.

1.3 LANDFORM SIZE AND DURATION

There is a clear relationship between the size and the duration of the existence of most landforms. Fig.1.1 shows this relationship for some characteristic form types. The values are not precise but are an indication of the order of magnitude. Size in the diagram is a length measurement such as diameter. The logarithmic estimating equation describes the relationship quantitatively. The durations of individual landforms may of course deviate considerably from the range shown for their type in Fig. 1.1.

At the lower end of the scale is a raindrop impact crater on sandy or silty soil with a diameter of a few millimetres; it can be destroyed again in seconds or minutes by another drop falling in the same place. Once the rain stops the craters may remain for a few days if the weather is dry, before they are destroyed, perhaps by winds (Fig. 1.2). Only if they are buried under a fresh layer of sand or silt will they be preserved, but then they are fossils below the surface and

no longer landforms. At the upper end of the scale are the continental shields with diameters of several thousand kilometres. They have existed since the earliest dated phases of earth history. The Canadian Shield east of the Rockies and the Baltic Shield in northern Europe are the oldest components on their continents and considerably older than the continents themselves. The continents are not shown in Fig 1.1 because they are not landform types but landform aggregates composed of large heterogeneous components.

Geomorphological processes require time for changes to take place and for this reason there is a relationship between landform size and duration: the larger the typical landform units are, the longer they remain preserved and identifiable. In order to change the form of a large individual landform unit, a larger quantity of rock material has to be removed and redeposited, for which a great deal more time is needed than for a small landform. Also, the greater the spatial extent of a landform, the greater are the distances the material has to be transported (Fig. 1.3).

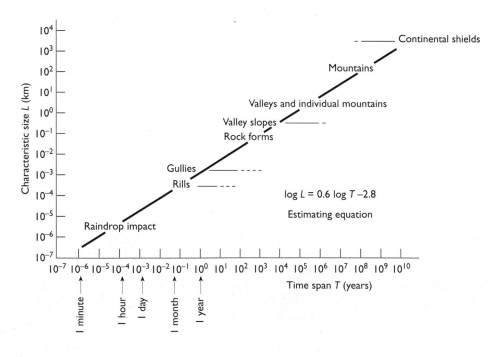

FIGURE 1.1 The relationship between the size and duration of landforms (after Ahnert, 1981b).

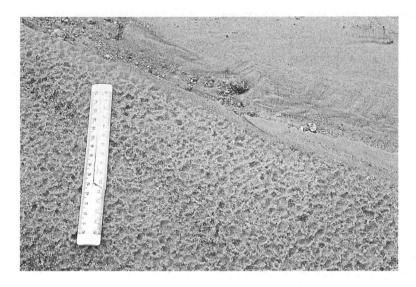

FIGURE 1.2 Raindrop impact craters on sand near Iringa, Tanzania. The smooth surface in the upper part of the photograph has been produced by the runoff of rainwater.

FIGURE 1.3 The snow-covered western Alps, an example of a large complex landform. Photograph by NASA.

Despite its lack of precision, the relationship of landform size and duration is useful because it shows that every form type has a specific spatial and temporal scale and a specific spatial and temporal resolution. For this reason it is important that forms are investigated at the correct scale. To explain erosion rills on a slope, for example, only the process events of a few weeks to a few years need be examined. For a valley this time span is hardly long enough to understand the present state of the processes involved; the genetic explanation of a valley generally includes past events for several hundred thousand years or longer and for mountain ranges, several tens of millions of years are required.

1.4 METHODOLOGICAL COMPONENTS OF GEOMORPHOLOGY

1.4.1. Systematic and regional geomorphology

Geomorphology is usually divided into systematic geomorphology and regional geomorphology. Systematic geomorphology examines the relationships between surface form, rocks and soil material and the processes that act on them, such as the relationship between slope steepness, precipitation and soil denudation by surface runoff. This is the **nomothetic** or law-giving part of geomorphology. Regional geomorphology investigates the landforms of individual regions and aims at a synthesis of all the landforms and processes of significance in the area; it is the **idiographic** part (Greek *idios* = peculiar) of geomorphology because the variations in the combinations of factors that affect landform development from place to place and the particular combination of landforms and processes in an area are unique.

1.4.2 The three research stages of systematic and regional geomorphology

Both systematic and regional geomorphology begin with an inventory based on a description of the present landforms and any traces of processes, which is followed by an investigation of the interrelationships between the components, including observation and, where possible measurement, of the processes that shape the landforms and of their effect. Finally, the long-term development of the present-day landforms is examined in relation to the relevant phases of earth history. The first stage is morphography, the second functional geomorphology and the third historic-genetic morphology (Ahnert, 1978). Every functional relationship has a morphographic inventory as a prerequisite and every historic-genetic investigation is an application of functional geomorphology to landforms and process traces created in the past.

1.4.3 Morphography

The purpose of morphography is to:

1. describe the landforms in terms of their geometric characteristics and dimensions, also their spatial location and arrangement;
2. record the relevant material properties of the land surface, especially the soils and underlying rock;
3. identify specific traces of processes.

This information is the basis for functional and historic-genetic geomorphology. It is also used in other areas, such as geoecology, the search for mineral deposits, agriculture, regional planning or civil engineering.

A description and classification of landforms is based on direct observation in the field, analysis of air photographs, radar and satellite information and topographic, geologic and soil maps.

1.4.4 Functional geomorphology

Functional is used here to describe the relationships, often quantitatively expressed, between two or more geomorphological variables. Functional relationships form the framework of all geomorphological explanation, since explanation is essentially an orderly description of the relationships between the components of a system.

Functional geomorphology describes the present relationships between landforms, rocks and soil materials and processes. It is based on data derived from empirical observation and examination of samples of material in the laboratory from which the relationships between the variables can be established. The determination of local and regional rates of denudation, transport and deposition and their spatial differentiation are particlularly important because from these the current mass balance can be estimated and information about the present development of the landforms obtained. Experiments in the laboratory and in field stations reconstruct and verify observed processes under controlled conditions on the basis of which laws can be derived which make it possible to describe the relationship between forms, materials and processes more precisely. Using these laws, which are based on empirical observation, the development of forms under particular material and process conditions can be reconstructed, usually in quantitative theoretical computer models.

1.4.5 Functional and causal relationships

The simplest functional relationship concerns the locations of different landforms grouped into landform associations, for example, a mountain valley, an erosional landform, and its alluvial fan, a depositional landform consisting of material eroded from the valley which lies at the exit of the valley into the adjacent lowland. The larger the valley the larger, all else being equal, the associated fan. The size of the fan is a function of the size of the associated valley.

The following are types of functional relationships:

1. between properties of different landforms (e.g., valley and alluvial fan);
2. between different properties of the same landform (e.g., slope height and steepness);
3. between properties of different materials (e.g., chemical composition of fresh bedrock and of the soil cover);
4. between different properties of the same material (e.g., grain size and particle shape of sediments);
5. between properties of different processes (e.g., frost frequency and weathering intensity);
6. between different properties of the same process (e.g., stream velocity and turbulence);
7. between properties of landforms and of materials (e.g., gradient of streams and grain size of the gravel in the stream channel);
8. between properties of landforms and of processes (e.g. slope steepness and intensity of downwearing);
9. between properties of materials and of processes (e.g., grain size and perviousness).

All these properties are variables which can be investigated. When a particular functional association between two variables has been identified and one of the variables has been observed in the field, it is possible to estimate the magnitude of the other variable, without reference to the causal connection between the two. It is also unimportant which of the two variables in this functional relationship is used as an independent variable and which as a dependent variable. The reliability of the estimate is based solely on the frequency of the observed association of these values. It is a functional statement, not a causal one. In the functional relationship between the size of the valley and the size of its alluvial fan, for example, neither one is a cause of the other, but the common cause of both is the erosion and sediment transport of the stream.

Only functional relationships containing process characteristics can also be causal relationships. Even then it is possible that the variables may not be directly connected causally but have a causal link in one or more variables not identified in the functional relationship. Simple causal relationships that have an effect in one direction only, from cause to effect, are rare in geomorphology. They tend to be limited to the effect of non-morphological factors on morphological processes, such as the influence of frost frequency on the intensity of weathering of bared rock. Geomorphological process variables and properties of landforms or of material often have **feedbacks** whereby the effect also acts to modify the cause. For example, the steeper the slope the more intensive is the denudation but the more intensive the denudation of the slope, the less steep the slope becomes and the more the denudation is reduced. This cannot be termed a causal relationship. The term **functional process relationship** is more useful and also includes unidirectional causal relationships. Every causal relationship is functional but the reverse is not true. Functional relationships that are based only on association without identified direct process linkages are termed **functional association relationships.**

It is useful to distinguish between **static** and **dynamic** functional relationships. Static functional relationships do not contain process components and cannot, therefore, change over time. Dynamic functional relationships contain process components and can change over time, including any landform and material properties affected by the processes.

1.4.6 Historic-genetic geomorphology

Historic-genetic geomorphology is concerned with morphogenesis, the formation and long term development of landforms. It is often not enough to extrapolate present functional relationships over a long time period because the duration and development of most landforms has lasted from tens of thousands to tens of millions of years, during which time the earth's climate has changed repeatedly. The past two million years have experienced several ice ages, which also resulted in the shifting of the climatic belts in the unglaciated parts of the earth, and the interglacials, when the climate was warmer. The last ice age ended only 10 000 years ago. In addition, the occurrence and intensity of crustal movements has varied greatly. The uplift and folding of the Alps and other young fold mountain ranges, which began over 60 million years ago, has not yet been completed and has included phases of near quiescence as well as phases of very intense activity.

Any investigation of long-term landform development has, therefore, to take climatic and tectonic events into account in so far as traces of their related processes are observable as landform elements, deposits or geologic structures and components of the landscape.

In his principle of **uniformitarianism**, the geologist J. Hutton (1795) stated that past developments in the earth's history were subject to the same laws of nature that are valid today and that they proceeded with the same processes that are observable at the present time, although not necessarily in the same place. The processes that affected northern Europe and North America during the ice ages, for example, can best be understood by observing present-day processes in the Antarctic and in Greenland. A limitation to the principle of uniformitarianism is the progressive evolution of vegetation. Plant covers exist today which protect the land surface and reduce denudation in environmental conditions that would not have allowed plant growth in the geologic past.

1.4.7 Physical time and historical time

There is a major difference in the way time is used in functional and in historic-genetic geomorphology. In functional geomorphology, time is a part of the characterization of processes, in particular of velocities, of process rates and of the effective duration of an individual process. The process rate is a measure of the completed work per time unit:

$$\text{Process rate} = \text{work}/\text{time unit} \qquad (1.1)$$

In physics the process rate is commonly referred to as power. Velocity has the dimension of distance/time and is therefore closely related to the process rate.

The denudation rate of the land surface can also be expressed as a velocity, namely the lowering of the surface, in units of length per time unit. This is usually expressed in mm/1000 years, a measure also known as the **Bubnoff unit** (B) after the geologist Serge von Bubnoff. The Bubnoff unit can also be used as a rate of uplift. The rate of lowering in a catchment can also be expressed as a quantity of weight per time unit, usually tons per year, or as a volume per time unit, m^3/year, which are then related to a unit of the land surface and expressed as tons/(km^2 year) or m^3/(km^2 year). The unit volume denudation rate of 1 m^3/(km^2 year) is equivalent to a mean rate of surface lowering of 1 mm per 1000 years or 1 B.

The total amount of work accomplished by the process during the time period is

$$\text{Total work} = \text{process rate} \times \text{duration} \qquad (1.2)$$

The total denudation for a given time would be obtained by multiplying the denudation rate in Bubnoff units by the duration in units of thousands of years. Similarly, the total distance debris is transported during a time period can be estimated by multiplying the debris movement velocity (cm/year, for example) as process rate and the time period in years.

Time is always **physical time** in functional geomorphology and a necessary component in the measurement of a process rate or of a process result. **Historical time** is used in historic-genetic morphology to order events and periods of geomorphological significance in earth history, such as particular climatic periods or periods of crustal movement and mountain building. The events to which historical time relates and which it orders affect the landforms, materials and

TABLE 1.1 Time units in earth history

Era	Period		Years before present
Cenozoic	Quaternary	Holocene	10 000
		Pleistocene	
			2 million
	Tertiary	Pliocene	
			6 million
		Miocene	
			23 million
		Oligocene	
			37 million
		Eocene	
			52 million
		Palaeocene	
			65 million
Mesozoic	Cretaceous		
			140 million
	Jurassic		
			195 million
	Triassic	Keuper Muschelkalk Bunter Sandstone	
			230 million
Palaeozoic	Permian		
			290 million
	Carboniferous		
			355 million
	Devonian		
			410 million
	Silurian		
			440 million
	Ordovician		
			510 million
	Cambrian		
			570 million
Precambrian			
			4000 million

processes from outside; both the events themselves and the historical time are external to the system of landform development, while physical time is an internal part of the process mechanisms.

The time units of historical time in geomorphology are the subdivisions of geological time (Table 1.1). Most landforms in existence at the present time have been developed in the past 65 million years during the Cenozoic.

1.4.8 Geochronological dating methods

Various chronological dating methods are used to date events and states in earth history, including the age of landforms. **Relative dating** methods estimate age sequences and **absolute dating** methods estimate the actual age in years. Goudie (1981, pp. 277–326) and Wagner and Zöller (1989) have summarized the various methods.

Sediments have long been dated relatively on the basis of the fossils they contain. In geomorphology, sequences of landforms and sedimentary deposits can also be dated relatively in that older forms are generally intersected by younger forms and older deposits covered by younger deposits. Other indications of relative age are the degree to which deposits are weathered and the development stage of soils, especially the occurrence of palaeosoils formed under different climatic conditions and now present as remnants.

Direct absolute dating is possible only if development takes place in countable, usually annual, cycles. **Dendrochronology**, for example, is based on the counting of tree rings. These change annually and comparisons can be made in trees whose life spans overlap and a progressive sequence developed, sometimes over many generations of trees (Worsley in Goudie, 1981, p. 301). Similarly, the annual rhythms of sediment layers of former lakes, known as varves, provide absolute dates for the duration of their deposition (section 22.6.3).

Radiometric dating, in which the age of materials is estimated on the basis of changes in their radioactive elements, reaches farther back. **Radiocarbon dating** is used most frequently to determine ages within the past 50 000 years. The radioactive carbon isotope ^{14}C and the stable

TABLE 1.2 Radioactive substances and their halflife values

Isotope	Halflife (years)	Dating period (years)
Caesium 137	30	≤ 40
Carbon 14	5730	$\leq 50\ 000$
Thorium 230	75 000	$\leq 200\ 000$
Uranium 234	250 000	50 000–100 000
Potassium 40	1 300 000 000	100 000–?
Uranium 238	4 500 000 000	10 million

Source: after Summerfield (1991).

carbon isotope ^{12}C are present in living organisms in constant relative proportions. After the death of the organism, the amount of radioactive carbon decreases at a constant percentage rate. The halflife, the period after which half the original radioactive carbon has decayed, is 5730 years. Based on this decay rate, the age of organic remnants, such as charcoal, bones, shells, peat or humus can be determined in a laboratory by measuring the ratio of the two isotopes ^{14}C and ^{12}C in the sample.

Very recent sediments are dated on the basis of the content of radioactive caesium isotope ^{137}Cs which has a halflife of only 30 years and originates from the atomic tests in the atmosphere during the 1950s and 1960s.

Radioactive isotopes with much longer halflives can be used to date older rocks. Radioactive potassium changes into the inert gas argon, for example, and uranium changes first into thorium and then into lead. Table 1.2 shows the halflives of a number of radioactive isotopes.

Lichenometry is the dating of lichen by calibrating the age and size of an area of lichen growth and is used for dating time spans up to 200 years ago (Winchester, 1984; Worsley in Goudie, 1981, pp. 302–305).

A number of new physical dating methods have been developed in recent years which will gain in importance. They include **fission track analysis** (Brown *et al.*; 1994), **thermoluminescence** dating (TL, Wagner and Zöller, 1989) and **electron spin resonance analysis** (ESR, Radtke and Brückner, 1991).

1.5 SUBDIVISIONS OF GEOMORPHOLOGY BY SUBJECT MATTER

Geomorphology can be subdivided according to a number of criteria, for example, the dominant processes of landform development, the most significant environmental conditions or the rock materials on which the landform is produced. A subdivision on the basis of formation processes includes fluvial morphology, glacial morphology and coastal morphology. **Fluvial morphology** describes landform development determined or controlled directly or indirectly by the work of flowing water. Not only do the streams and rivers themselves, with their channels and their processes of erosion, transport and accumulation, belong to fluvial geomorphology, but also the valleys and valley systems, the valley slopes together with the processes of weathering and downwearing of these slopes and the landforms of fluvial deposition in the lowlands. **Glacial geomorphology** examines the landform development by the movement of glacier ice. It overlaps with fluvial morphology in so far as the glacier usually develops in a previously fluvial landscape which it then modifies. **Coastal geomorphology** is concerned with the landforms on the boundary between land and sea. Here the most effective processes include the surf, currents and tides. They attack and change the landforms above and below sea level and modify the form of the sea floor in the shallow water zone near the shore by erosion, transport and deposition.

Structure and climate are the most important environmental conditions affecting landform development. **Structural geomorphology** investigates the effect of the rock structures on the spatial differentiation of the geomorphological processes and landforms. **Climatic geomorphology** identifies climatic factors such as precipitation intensity, frequency and duration, groundfrost intensity or the strength and direction of the wind and explains the development of landforms under different climatic conditions (Derbyshire, 1976; Büdel, 1977). The historic-genetic branch of climatic geomorphology is **climatogenetic geomorphology**, which seeks to explain the landform development that took place under climatic conditions that differed from the present (Büdel, 1963, 1969; Büdel and Hagedorn, 1975; Bremer, 1977, 1989a).

Rock material is the defining criterion in **karst geomorphology**, which describes landform development in which solution weathering and solution denudation predominate. Solution processes are most effective on rocks with a particular chemical composition, especially on limestones.

Applied geomorphology is an increasingly important branch of geomorphology. It stands more or less by itself and is concerned with the application of the knowledge and working methods of geomorphology to purposes that lie outside of the field. At a time of greater interest in the environment, awareness of the possible applications of geomorphology has increased and there is a very considerable potential for geomorphology to contribute to the solving of many environmental problems.

GENERAL SYSTEMS THEORY IN GEOMORPHOLOGY

2.1 THE CONCEPT OF SYSTEMS

The idea of systems (Greek *systema* = union) is very old. It describes a basic concept present in all sciences. Each system is composed of several components that are linked to one another by functional relationships. Generally, additional relationships exist between the components of the system and its environment, that is, other systems. Scientific explanation is, in effect, a description of the components of a system, their relationships with one another and with other systems. Although each science has its own systems with their own subject matter and their own networks of relationships, the formal characteristics of systems are largely similar for all sciences. The biologist von Bertalanffy (1951, 1962) developed a **general systems theory** in the 1930s.

Chorley (1962; Chorley and Kennedy, 1971) introduced general systems theory into geomorphology. There are two types of systems: a **closed system** receives no supply of energy from outside and transfers no energy outwards. An **open system** receives energy from its surroundings and transfers it out again. The two types differ in the functioning of their inherent, or **ensystemic**, processes.

The energy supply of a closed system is limited; it is progressively used up by the processes functioning within the system, without any additional supply from the outside. The ability of the system to function decreases with the progressive reduction in available energy until the work of the processes ceases altogether and no further change is possible in the system. A mill wheel supplied with water from a non-refillable container would be an example. Once the container is empty the wheel no longer turns. In a truly closed system, the water should be collected below the mill wheel in a second container so that the system remains closed and does not supply any energy to the outside.

In an open system, the energy is continually resupplied from **eksysystemic** sources, that is energy sources from outside. If the example of the mill wheel is extended, the unrefillable container would be replaced by a reservoir fed continuously by a stream so that in the long term there is storage renewal, and energy consumption is equal to the average energy supply.

The identifiable and scientifically understood environment is made up of open systems. These can behave as closed systems temporarily if the energy supply is halted for a period or comes in single thrusts in the form of supply events. If, for example, the stream to the reservoir supplying the mill dries up for a long period, the energy consumption of the mill wheel cannot be compensated for by new energy supply. The reservoir is used up and if the dry period is long enough, the mill wheel is brought to a standstill. Sooner or later the stream probably flows again and the reservoir fills and the mill wheel turns. This occurs because the stream–reservoir–mill system is itself a part of the earth's much larger systems of water circulation and water budget which include condensation, precipitation, runoff and evaporation. The water circulation receives its energy supply from the earth's heat budget, part of the earth's radiation budget which receives its energy supply from the sun's radiation.

Each open system is part of a larger system that receives and gives off energy. The same applies to the supply and delivery of material, the water and water circulation in the mill, the taking up of nutrients and the elimination of waste in biological systems or the supply and removal of gravel in a stream bed.

Material transfers are linked to energy transfers although in the case of radiation energy, there are pure energy transfers from system to system without any supply or removal of material, whether it is in the form of radiation from a stove or the sun's rays across empty space.

The **mass balance** is of major importance for an understanding of the supply and removal of material on the earth's surface. Of the total radiation energy that arrives on the earth and is converted and diffused again into space, only a very small fraction, very difficult to determine, is used for geomorphological processes. Landforms develop and change essentially by the transfer of mass. At different points on the earth's surface, different amounts of rock or soil material are removed by denudation processes or added to by processes of deposition. Removal lowers the land surface and deposition raises the land surface, and the forms change in response to these processes.

2.2. TYPES OF SYSTEMS

The investigation of geomorphological systems is part of functional geomorphology (section 1.4.4). They are made up of the following components:

1. (a) Forms (e.g. stream valleys, fault scarps, barrier beaches);
   (b) form characteristics (e.g. slope angles, curvature of river bends).

2. (a) Types of material (e.g. rocks, gravels, soils);
   (b) material characteristics (e.g. joints in the rock, grain size, perviousness).
3. (a) Processes (e.g. weathering, landslides, stream erosion);
   (b) process characteristics (e.g. runoff frequency, erosion rates, debris velocity).

Systems have been subdivided into **static systems**, **process systems** and **process response systems** (Chorley and Kennedy, 1971).

2.2.1 Static systems

There are three types of static systems.

1. **Form systems** are functional associations of forms or form characteristics, such as coastal cliffs and abrasion platforms, valley meander bends and the asymmetry of valley slope angles, or stream gradients and slope angles.
2. **Material systems** are functional associations of types of materials or material characteristics, such as rock types and joint density, or grain size and perviousness.
3. **Form and material systems** are associations of forms and form characteristics with materials or material characteristics, such as slope angles and soil thickness, or stream gradient and gravel size.

All static systems are **states**, the time factor is disregarded. The observation that there is a relationship between the size of gravel in a stream bed and the gradient is always valid and in itself independent of time. The relationship can only be changed by influences from outside the observed system, such as a change in the climate and, in consequence, the runoff conditions. The functional relationships between the components of static systems are not causal because causality involves the identification of the processes that cause change; instead these systems represent static functional relationships in the sense discussed in section 1.4.5.

2.2.2 Process systems

Process systems consist of associations of processes and/or process characteristics such as debris movement on valley slopes, the transport of material in stream channels at the slope foot or the intensity of precipitation and the rate of soil removal by wash. In many geomorphological process systems material is moved from one process area to the next, usually downslope or down valley from, for example, debris production by weathering on the slope, to the removal and transport by wash of the weathered material from the slope to the stream, and its further transport by the stream to the coast where it is deposited. Chorley and Kennedy (1971) have termed pure process systems 'cascade systems'. However, relationships between processes are not only effective down valley or downslope but also in the opposite direction, sometimes with feedbacks; the accelerating effect of increased stream erosion on the denudation rate of valley slopes is a process association that is directed upslope. The more general term **process system** seems preferable.

Processes are not possible without time, and all process systems always contain the time factor as a system component. This is the most significant difference between static systems and process systems.

2.2.3 Process reponse systems

Process response systems describe the relationships between static components and process components. They are concerned with the response of one or more processes on the form or material characteristics and also with the effects of forms and materials on the processes. A simple example is the interaction between gradient and the vertical erosion rate of a stream that flows into the sea. Initially, the gradient is steep and the erosion rate high; due to erosion the height of the stream above sea level, and therefore its gradient, decreases. The erosion rate also decreases. Gradient and erosion rate interact in such a way that the reduction in the gradient caused by erosion reduces the rate of erosion itself. Erosion and gradient are related in both directions by **negative**, or intensity-reducing **feedbacks**. Most feedbacks in geomorphological process systems are negative (sections 1.4.4 and 1.4.5).

Positive feedback that increases process intensity is much rarer and lasts a much shorter time before it reaches a threshold value which either ends in self-destruction of the system or turns into a negative feedback, as occurs when lava with a high gas content rises in the vent of a volcano. The higher the lava is pushed up by the pressure of its gases, the less is the pressure of the rock above the lava in the vent. More gas is set free in the lava as a result and the augmented gas pressure accelerates the further rise of the lava until an explosive eruption ends the self-reinforcing trend.

Process response systems can be very complex with many different form, material and process components but they are useful for an understanding of landform development. They are complete in the sense that they always contain both static components and the changes that occur to these components. Static systems and process systems do not exist in themselves. They are always part of a process response system and are only separated for analytical purposes. It has to be remembered that the functional relationships between the components of a static system only exist because they are linked by processes, which are not defined when the static system is being observed by itself. This kind of incomplete observation can be useful in applied geomorphology when individual components have to be determined or predicted from partial information and there is not enough time to examine the process response system in full.

2.2.4 Dynamic equilibrium and steady state in process response systems

The components of natural process response systems are normally linked by negative feedbacks. These act as self-regulating mechanisms which steer the system towards a state of equilibrium. Figure 2.1 shows the components of a locally effective process response system in a cross-section of a stream bed.

Initially there is an excess of gravel removal downstream compared to gravel supply, a local deficit in the mass balance of the gravel transport by the stream. As a result, the stream channel bed is lowered at this point and the gradient downstream reduced but increased upstream. The flow velocity of the water downstream is also lower and higher upstream than before. Because gravel transport is a function of flow

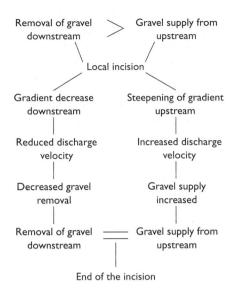

Removal of gravel downstream > Gravel supply from upstream

Local incision

Gradient decrease downstream

Steepening of gradient upstream

Reduced discharge velocity

Increased discharge velocity

Decreased gravel removal

Gravel supply increased

Removal of gravel downstream —— Gravel supply from upstream

End of the incision

Figure 2.1 Functional relationships and negative feedbacks between the components of a process response system in the stream channel (schematic).

velocity, less gravel is removed by the stream but more gravel is supplied than at the beginning of the change. The rates of gravel removal and of gravel supply approach each other until they are equal.

Where there is an excess of gravel supply compared to removal, the rates also become equal eventually. In this case, though, the excess gravel is deposited at the observed stream cross-section and the stream channel is raised with the effect that the gradient is increased downstream and reduced upstream. The supply rate then decreases and the removal rate increases until they equalize. In both cases the characteristic form, gradient, material and the transport capacity of the water are so linked that the initial difference between supply and removal, the cause of the changes, is counteracted. This is the essence of the negative feedback and the self-steering of the system regulated by it. The negative feedbacks lead the system from an initial disequilibrium between two or more process rates, in this example supply and removal rates of the gravel load of the stream, to an equilibrium. Because of the negative feedbacks, there is

an inherent tendency for equilibrium to be re-established after any disturbance to it. The steering of the systems flows from the feedback processes, which themselves are driven by forces, and for this reason the equilibrium is usually termed **dynamic equilibrium** (Greek *dynamis* = force). Characteristic of this equilibrium is the constancy of the process rates, especially the production, or supply, of material and its removal. If equilibrium is reached, then the mass budget of the observed system remains unchanged, as do, quantitatively, the form and material components.

The constancy of form and material related to the process rates of the dynamic equilibrium and, in effect, of the entire process response system, is termed **steady state**. In a steady state the system components do not change although the processes in the system are active. Once the steady state has been reached, there is no further development of the system as long as the energy supply from outside remains constant. In its steady state, the system is **time independent** because time is significant only when changes occur.

It is important to distinguish between dynamic equilibrium, which is related only to process rates and the forces that cause them, that is, to the process systems within the process response system, and the steady state, which relates to

1. the entire process response system whose processes are in dynamic equilibrium;
2. the processes themselves;
3. the static system within the process response system.

There is no equilibrium between form components, between material components or between form and material components of a system. When these components remain constant, they are likely to be in a steady state. This distinction has not always been made by geomorphologists, which has lead to confusion. Also some states have been said to be in equilibrium that are not, such as 'metastable equilibrium' and 'decay equilibrium' (Chorley and Kennedy 1971, p. 202; cf. Ahnert, 1994b; Thorn and Welford, 1994).

In the example shown in Fig. 2.1, it is apparent that to reach equilibrium a **relaxation time** is necessary. The relaxation, the changes in the system components towards equilibrium, takes place rapidly at first then gradually more slowly and in increasingly smaller steps, more or less asymptotically. In nature the relaxation trend is overlain by more or less cyclical fluctuations of the process rates so that it is not possible to determine exactly when the system enters a steady state.

For equilibrium to be established, there must be a sufficiently constant supply of energy. If it remains constant longer than the required relaxation time, a steady state will in fact be reached. Should a change in the energy supply occur earlier, for example a change in the rate of uplift or a relevant change in climatic factors, the previous development trend would be interrupted and the system would then adjust its internal process rates with a new relaxation time in order to reach an equilibrium that corresponded to the altered energy supply. Changes in the material properties can also cause this type of interruption. An incising stream, for example, may cut down into a rock that is more, or less, resistant than the rock in which it formerly cut its bed.

Whether or not a system does reach a steady state depends on its spatial size, its internal complexity and the process mechanisms. When the sea level falls, for example, and the gradient in the area of the stream mouth increases, causing an erosion impulse, then this change in energy supply makes itself felt much more rapidly at the source of a small stream near the coast than at the source of a large river flowing from the interior of a land mass. The relaxation time of the small stream is considerably shorter than that of the large river and the probability that the small stream reaches a steady state before a new sea level change is also much greater.

The practical importance of the equilibrium concept is not whether a particular system has reached a steady state or not but the fact that all geomorphological process response systems that contain negative feedbacks between sufficiently active process components, always move towards a steady state that corresponds to their existing energy supply. The changes in the system are very rapid during the early phases of the relaxation time and slower in the later phases so that, in effect, a state is reached at an early stage

that is near enough, in many respects, to a steady state, for the system to be regarded as being in an approximate steady state after the early phases of rapid development (Ahnert, 1987d).

From this follows that the spatially differentiated mass balance of supply and removal on the land surface, and the **inherent tendency to dynamic equilibrium** of this mass balance, is of fundamental importance in geomorphology.

2.3 THE ROLE OF THEORETICAL MODELS

2.3.1 Model and reality

There are models in many areas of man's existence and there are also many meanings of the word model. In geomorphology and other sciences, the narrower meaning of the term evolves out of the basic structure of scientific thought.

All sciences begin with **perception**. On the basis of this perception a **question** is formulated which provides a goal for further perception and **observation**. The comparison of observed objects or events leads to **abstraction**, that is to a selection of those observed attributes that are considered to be significant and to the omission of others thought to be irrelevant. The attributes form the basis of the **classification** of the observed objects or events. The classification allows the identification of **types**. Types are not segments of the actual, complex reality but **models** of reality abstracted and defined by the investigator. All types are, therefore, models.

Scientific laws are statements about the properties, functional relationships, changes or developments of types of phenomena and, therefore, statements about models. Science moves continually from empirical observation and identification of individual objects and events through abstraction and the development of types to the formulation of general theoretical principles and from there back to empirical observation, in order to test the theory and possibly improve on it.

In geomorphology, the feedback connection of empiricism and theory forms the basis of models: representations of geomorphological systems whose components are linked by functional relationships. They can be either empirical, such as statistically derived functional relationships between observed data, or theoretical, such as a computer model that simulates the development of a valley system. Theoretical models, especially process response models, have an important role in geomorphology because every explanation that is valid beyond a local, individual circumstance is based on a theoretical model concept.

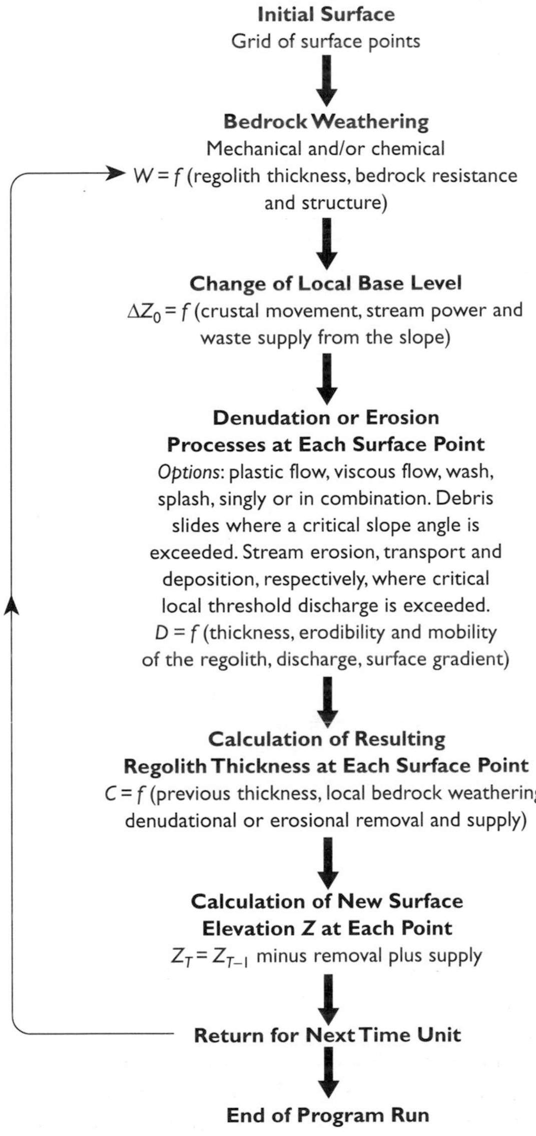

2.3.2 Theoretical process response models

Empirical research can investigate only a very small part of a process response system in the field directly. Representative examples of landforms, materials and processes are investigated on the basis of which conclusions are made about the whole. Significant changes in the landscape take place over a very long period (Fig. 1.1) and can usually be reconstructed only indirectly from landforms and materials left by former processes. Theoretical response models complement empirical research. They describe, qualitatively and, if possible, quantitatively the essential functional relationships of process response systems with their causal linkages and feedbacks, and reproduce the development of their system components. They are based on empirical observations which are used as building blocks for the development of a system that describes as coherently as possible the functional relationships and also the temporal development. The more or less isolated spatial, material and temporal facts gathered in empirical research are linked by estimates which fit empirical experience and general physical laws. Once established, a model system suggests further questions to be answered by empirical investigation, which leads to an improvement and extension of the model. Examples of process response models are discussed in various chapters as a means to explain the successive phases of development of different landform types. All of the models discussed are variants of the model program SLOP3D (Ahnert, 1976b, 1996) and are based, therefore, on the same concepts of weathering, denudation processes and, where applicable, fluvial erosion. Figure 2.2 shows the design structure of SLOP3D as a simplified flow diagram.

Initial Surface
Grid of surface points

↓

Bedrock Weathering
Mechanical and/or chemical
$W = f$ (regolith thickness, bedrock resistance and structure)

↓

Change of Local Base Level
$\Delta Z_0 = f$ (crustal movement, stream power and waste supply from the slope)

↓

Denudation or Erosion Processes at Each Surface Point
Options: plastic flow, viscous flow, wash, splash, singly or in combination. Debris slides where a critical slope angle is exceeded. Stream erosion, transport and deposition, respectively, where critical local threshold discharge is exceeded.
$D = f$ (thickness, erodibility and mobility of the regolith, discharge, surface gradient)

↓

Calculation of Resulting Regolith Thickness at Each Surface Point
$C = f$ (previous thickness, local bedrock weathering, denudational or erosional removal and supply)

↓

Calculation of New Surface Elevation Z at Each Point
$Z_T = Z_{T-1}$ minus removal plus supply

↓

Return for Next Time Unit

↓

End of Program Run

FIGURE 2.2 Simplified design structure of SLOP3D-type landform development models.

3

THE GEOMORPHODYNAMIC SYSTEM

3.1 DEFINITION

Each geomorphological process response system is a part of larger systems to which it is connected by transfers of energy and materials. The larger systems combine in a single **geomorphodynamic system** which includes the entire field of geomorphology, with all the relationships between forms, materials and processes.

3.2 STRUCTURE OF THE GEOMORPHODYNAMIC SYSTEM

Figure 3.1 shows the functional relationships between the major components of the geomorphodynamic system. The static components are in the centre of the diagram: the earth's crust and its surface with its geometric properties (relief, landforms and form characteristics), and its material properties (rocks, soils and vegetation cover). Above and below are two groups of process components which, under the influence of gravity, affect the forms and material.

The **endogenic** (Greek *endon* = within, *genesis* = origin) processes are shown in Fig. 3.1 in the lower part of the diagram. They originate in the earth's interior and include crustal movements (uplift, subsidence and horizontal movements of crustal blocks, folding of strata) and volcanism. With few exceptions, endogenic processes tend to increase relief, or height differences. The uplift of a crustal block or the lowering of a block creates a new height difference between the

block and its surroundings. Folding of formerly horizontal layers of rock, bent upwards into an **anticline** and downwards into a **syncline** also increases height differences. Crustal movement

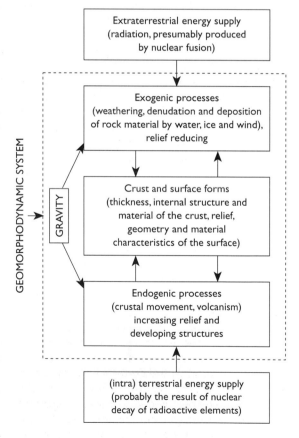

FIGURE 3.1 Structure of the geomorphodynamic system (after Ahnert, 1981b).

and folding can increase relief without an addition of material; volcanism creates height differences by eruptions which transfer material from the earth's interior on to the surface, and by the accumulation of this material in the form of volcanoes.

Exogenic (Greek *exo* = outside, *genesis* = origin) processes, shown on the upper part of Fig. 3.1, are those that affect the earth's surface from the atmosphere and include weathering, denudation, transport and deposition of rock material by water, ice and wind. Because, in general, denudation takes place in areas of higher elevation and deposition in low-lying areas, existing height differences are reduced.

Endogenic and exogenic processes not only affect landforms and surface materials, they are affected by them. The arrows in Fig. 3.1 show the feedbacks that exist, via the static system components in the centre of the diagram, between the endogenic and exogenic processes. For example, if the endogenic uplift of a crustal block increases relief, the gradient and the fluvial erosion also increase and the denudation of the slopes and summits is intensified. The removal of the products of erosion and denudation by the rivers reduces the mass of the uplifted block and, as compensation for this loss of mass, the crustal block is again uplifted (section 4.1). The second uplift of the block is also endogenic because it is the immediate result of a mass displacement in the earth's interior, although it happened because rock material had been removed from the surface by exogenic processes. Through feedbacks, endogenic processes bring about a reaction in the exogenic processes and vice versa. Both processes are subject to the force of the earth's gravity. There are few exceptions to the general rule that endogenic processes increase relief and exogenic processes reduce relief.

The two sources of energy for the process components of the geomorphodynamic system are the sun and the earth's interior. Radiation from the sun reaches the earth and is converted to heat energy both on its surface and in the atmosphere. The heat energy acts as a motor that is in continuous operation and that enables the entire exogenic process response structure on the earth, including all forms of life, to keep functioning. The radiation is generated in the sun by a nuclear fusion process by which four hydrogen nuclei combine to form a helium nucleus, and energy is set free.

The energy for endogenic processes probably originates in the nuclear disintegration processes of radioactive elements in the earth's interior which produce heat in a spatially varying distribution. This uneven distribution leads to the development of pressure and temperature equalizing currents in the earth's mantle more than 100–150 km below the surface. Sections of the lithosphere are moved passively with the currents and pushed, part horizontally, part vertically, against one another. In areas of younger sediments they also cause folding.

All endogenic and exogenic processes on earth can, therefore, be assumed to be driven by nuclear energy. All other energy forms are secondary, produced by terrestrial energy conversions. In Fig. 3.1, the origin of the exogenic and endogenic energy supply is shown at the top and bottom of the diagram but both sources lie outside the boundary of the geomorphodynamic system. The system itself is characterized by the fact that all its components are linked by feedbacks and interact with one another. Both sources of energy for the system are, however, unaffected by these interactions. The earth does radiate its heat energy back into space, but this is of no importance for the energy budget of the sun, and crustal movements appear to have no geomorphologically relevant influence on the generation of energy in the earth's interior.

3.3 THE FUNCTIONAL RELATIONSHIP BETWEEN RELIEF AND DENUDATION

The three basic constituents of the geomorphodynamic system, relief, denudation and uplift, are functionally related to one another. The longer and more intensively a mountain range is uplifted, the higher it is and the more intensively it is denuded; anyone who has walked in hills or mountains will have been aware of the greater intensity of most processes in upland areas. These relationships can be explained quantitatively.

Table 3.1 contains data of the mean relief, mean slope angle and mean denudation rate from 20 large drainage basins or catchments in the middle latitudes of Europe and North America. **Mean relief**, *h*, is here the mean value of the measured height difference between the highest and lowest points, in each of a set of squares 20 × 20 km distributed over the drainage basin. **Mean slope angle** is the mean, expressed as the sine of the slope angle, of all local mean slope angles determined in each square. **The mean denudation rate**, *d*, is calculated by dividing the mean, obtained over a long period, of the volumes of rock material carried out of the drainage basin by the stream, by the surface area of the drainage basin. The volume divided by surface area is equal to the mean lowering of the land surface, that is, the mean denudation rate *d* per time unit; *d* is expressed in m/1000 years.

The denudation rate and the mean slope angle increase, therefore, with increasing relief.

In Fig. 3.2 the functional relationship between the mean denudation rate *d* and the mean relief *h* is shown graphically; the relationship is linear and can be expressed by the regression equation

$$d = 0.0001535\,h - 0.011 (\text{m}/1000 \text{ years}) \quad (3.1)$$

The correlation coefficient $r = 0.98$.

The relief as such, which is only a height difference, is not a denudation determining factor in a causal sense; denudation at any point on the earth's surface in no way depends directly on the size of the height difference in the surrounding 20 × 20 km square area. The denudation rate is directly dependent on the slope angle, more precisely the sine of the slope angle, since the denudation of the land surface results from the physical processes which are influenced by gravity and the laws of inclined planes. For this reason the functional relationship between the mean denudation rate and the mean slope angle

TABLE 3.1 Relief, slope angle and denudation rates in large middle latitude draining basins

Catchment area	Mean relief h(m)	Mean slope angle (sine)	Mean denudation rate d(m/1000 years)
Flint River, Georgia	89	0.033	0.028
Colorado River, Texas	102	0.020	0.016
River Thames, England	159	0.022	0.016
Delaware River, eastern USA	299	0.052	0.042
Canadian River, Texas	353	0.071	0.052
Little Colorado River, Arizona	392	0.056	0.031
Juniata River, Pennsylvania	490	0.089	0.041
Green River, Utah	644	0.113	0.082
Escalante River, Utah	842	0.130	0.135
Dirty Devil River, Utah	912	0.106	0.177
Bighorn River, Wyoming	1004	0.138	0.109
Colorado River, Utah	1040	0.134	0.124
Wind River, Wyoming	1091	0.132	0.115
Animas River, New Mexico	1273	0.150	0.195
Saanen River, Switzerland	1395	0.287	0.210
Alpine Rhine River, Switerland	1994	0.410	0.321
Isère River, above Grenoble, France	2046	n.a.	0.287
Reuss River, above Vierwaldstätter Lake, Switerland	2320	0.345	0.309
Kander River, Switzerland	2428	0.401	0.430
Rhône River in the Valais, Switzerland	2869	0.361	0.418

Source: Ahnert (1970, 247).

of these drainage basins is statistically significant with the regression equation

$$d = 0.967\sin\alpha - 0.007 \text{ (m/1000 years)} \quad (3.2)$$

and $r = 0.95$.

It can be concluded on the basis of this equation that the mean denudation rate, expressed in m/1000 years, is approximately equal to the sine of the mean slope. The steeper the slope, therefore, the more intensively it is eroded.

Because the relief of a 20×20 km square is determined more rapidly than the mean slope angle, it is often used as a substitute. The regression equation for the mean slope and mean relief data in Table 3.1 is:

$$\sin\alpha = 0.005 + 0.00015h(\text{m}) \quad (3.3)$$

with $r = 0.95$ also.

The not immediately obvious relationship between the mean denudation rate and the mean relief exists because the distance between the valleys does not vary greatly even if the relief

varies. Given equal distances between valleys, high relief, for geometric reasons alone, must be associated with steeper slopes than can occur in areas of low relief.

The denudation rate is also influenced by other factors, most importantly climatic, although it would seem from the data shown in Table 3.1 that the influence of climate on the mean denudation rate in these drainage basins is relatively small, despite the fact that the basins lie in the arid southwest and the humid east of the USA, in the oceanic climate of England and in areas of high rainfall and snowfall in the western Alps. Regions with pronounced seasonal contrasts between heavy precipitation and severe aridity, such as the Mediterranean region, monsoonal regions in Asia or the savanna in Africa were not included. The influence of precipitation on the denudation rate would probably be greater in these types of climates.

The mean denudation rates determined from equations (3.1) and (3.2) do not give any direct information about the development and change of the landforms. On an individual slope denudation rates vary between the summit and valley floor. In mountain ranges that are no longer uplifted and whose streams no longer incise, the denudation rate at the slope foot is close to zero and, in any case, less than the mean rate. Elsewhere on the slope the local denudation rate must, therefore, be greater than the mean. If the slope profile from the stream bank to the slope crest is straight, which is the geometrically simplest of all slope forms, and if at the crest the slopes of the adjoining valleys intersect, then the local denudation rate half way between the divide and the slope foot is probably close to the mean denudation rate, and about double the mean rate for the slope at the slope crest or summit (Fig. 3.3a). On a convex slope profile, the summit denudation rate is lower (Fig. 3.3b), and on a concave slope profile (Fig. 3.3c) higher than double the mean rate.

Should stream erosion begin to deepen the valley, these relationships are shifted. If on a straight slope profile the rate of valley deepening, u, is greater than the mean denudation rate, d, then the summit denudation rate, d_s, is smaller than the mean value. If the valley deepening rate is smaller than the mean denudation

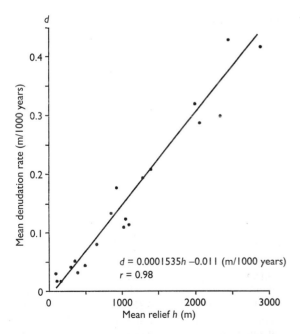

FIGURE 3.2 The functional relationship between the mean denudation rate d (m/1000 years) and the mean relief h (m) for 20 large drainage basins in Europe and North America (*see* Table 3.1; after Ahnert, 1970a).

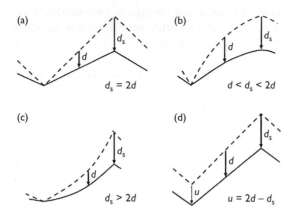

FIGURE 3.3 Relationship between summit denudation rate d_s and mean denudation rate d, (a) on a straight, (b) on a convex and (c) on a concave profile without stream incision at the slope foot; (d) on a straight slope profile with downcutting u at the slope foot.

rate, the summit denudation is larger than the mean but less than double the mean (Fig. 3.3d).

Based on equation (3.1), the mean denudation in 20 drainage basins shown in Table 3.1 in 1000 years is about $0.00015 \times h$ or 0.015 per cent of the existing relief. Included are some basins that are being uplifted at the present time and whose streams are downcutting, and others, like the Thames, where the downcutting rate is probably zero. In most cases the summit denudation rate is probably somewhat higher than the mean. A conservative estimate of summit denudation would be a rate of $0.0002 \times h$ per 1000 years, that is, 0.02 per cent of the existing relief, or $0.2\ h$, or 20 per cent, of the existing relief per million years.

3.4 PROCESS RESPONSE MODELS OF RELIEF DEVELOPMENT WITH NO UPLIFT OR VALLEY DEEPENING

If there is denudation of the summit but no simultaneous uplift, that is, no simultaneous valley deepening, the relief is reduced by about 20 per cent per million years. After 11 million years the relief is, therefore, only 10 per cent and after 22 million years only 1 per cent of the original

relief (Fig. 3.4a). In reality, the total denudation of a mountain range is to some extent compensated by the isostatic uplift that results from the lightening of the earth's crust following denudation (section 4.1). If an allowance is made for

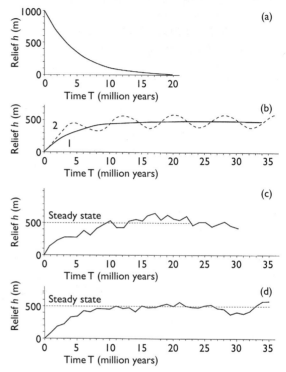

FIGURE 3.4 (a) Relief development without uplift ($u = 0$); initial relief = 1000 m; summit denudation rate $d_s = 0.2h$ m/million years. (b) Relief development curves (1) with constant rate of uplift $u = 100$ m/million years, (2) with cyclically varying rate of uplift between $u = 50$ m/million years and $u = 150$ m/million years, mean value of $u = 100$m/million years; in both cases the initial relief = 0 m and the summit denudation rate $d_s = 0.2h$ m/million years. (c) Relief development with random variation of the rate of uplift around the mean value $u = 100$ m/million years; summit denudation rate $d_s = 0.2h$ m/million years; compare steady state relief with curve (1) in (b). (d) Relief development with random variations of the rate of uplift, similar to (c) and random variations of the summit denudation rate around the mean value $d_s = 0.2h$ m/million years; steady state relief similar to (b) and (c).

this, 10 per cent of the initial relief would be left after 18 million years and 1 per cent after 37 million years (Ahnert, 1970). In general, these estimates agree with what is known about the length of time mountain ranges exist after the end of their main phases of uplift.

3.5 MODEL OF RELIEF DEVELOPMENT WITH A CONSTANT RATE OF UPLIFT

Tectonic uplift increases relief not only because a crustal block is being raised above the surrounding area but also because the streams within the area of uplift deepen their valleys and cause greater height differences locally. A long-term, more or less constant rate of uplift generally results in the streams downcutting at the same rate. In this case, the rate of uplift can be substituted for the rate of valley incision within the area of uplift.

In the initial phases of the development shown in Fig. 3.4b(1) the summit denudation rate, d_s, is lower than the rate of uplift, u, and the relief grows correspondingly. As the relief increases d_s increases too until $d_s = u$. For a summit denudation rate $d_s = 0.2\,h$ (m/million years) this occurs when

$$h = u/0.2 \qquad (3.4)$$

From this point on, a **dynamic equilibrium** exists between uplift and denudation which has been produced by the uplift and is steered by the relief, or more precisely, by the slope angles that are associated with the relief, as expressed in equations (3.1) and (3.2). When equilibrium is reached all process rates in the system are the same:

$$d_s = d = u \qquad (3.5)$$

The landforms, which are now in a **steady state** and being denuded parallel to themselves, have a constant relief whose height h, with a rate of uplift $u = 100$ m/million years according to equation (3.4), is $h = 100/0.2 = 500$m.

Dynamic equilibrium between uplift and denudation is reached only when the rate of uplift is constant for a long period. This rarely seems to occur. Nevertheless, the model is an important key for the interpretation of relief development because:

1. even before equilibrium is reached, the tendency towards development of equilibrium is present and the effects of the processes involved can only be understood in the light of this tendency;
2. the denudation rate approaches the rate of uplift relatively early on so that many aspects of the form development can be interpreted, in the first approximation, as if the system were in dynamic equilibrium.

3.6 MODEL OF RELIEF DEVELOPMENT WITH VARIABLE RATES OF UPLIFT

Constant rates of uplift for many millions of years are unlikely. The assumption that they are constant is, nevertheless, useful for making rough estimates, especially in areas where the total uplift and the time span of the uplift are known but not the variations in the rate of uplift during the relevant period. For example, the Colorado Plateau has been raised approximately 5000 m in the past 40 million years, but little is known about the temporal variation of this uplift. The preliminary assumption of a constant rate of uplift based on the mean value $u = 5000$ m/40 million years $= 125$ m/million years provides, therefore, a useful basis for understanding the development of this region.

Closer to natural relief development is the assumption that the rate of uplift varies. The model shown in Fig. 3.4b(2) varies the rate of uplift periodically between 50 m/million years and 150 m/million years around a mean value of 100 m/million years. The relief fluctuates accordingly between values of 400 m and 600 m, but with a temporal delay of the relief maxima and minima compared to the rate of uplift. The relief increases as long as the rate of uplift is greater than the summit denudation rate and decreases only when the latter is greater than the uplift rate.

Still more realistic are unperiodic variations in the rate of uplift around the long-term mean value of 100 m/million years (Fig. 3.4c). The variations are simulated in the computer with a random number generator. In this case there is a greater variation in relief than when the variations in uplift were periodic.

If, in addition to the endogenic variations in the rate of uplift, there are exogenic climatic variations, shown here as random variations of the summit denudation rate around the long-term mean value $d_s = 0.2h$ million years, then, although the theoretical possible amplitude of the relief variations is still greater, nevertheless, the variations in both rates also often compensate each other and keep the variation of the relief within narrower limits (Fig. 3.4d).

Figure 3.4b–d shows that whether a system is in equilibrium or not depends on the length of the time period considered. Over a relatively short period, up to a few million years, the relief either increases progressively, decreases progressively or even alters the direction of the particular trend. A steady state is not apparent. Over a long period of time, tens of millions of years, these increases and decreases are only variations in a long-term steady state. Clearly time scales are of significance for any interpretation of a geomorphological system (Schumm and Lichty (1965).

3.7 THE MAXIMUM POSSIBLE HEIGHTS OF MOUNTAIN RANGES

The inherent tendency of the endogenic, relief-increasing uplift and the exogenic, relief-decreasing denudation to establish, by means of feedbacks, a dynamic equilibrium between the two, so that the denudation rate at the mountain summit becomes equal to the rate of uplift, has direct consequences for the summit heights of mountain ranges.

In mountain ranges that have been continually uplifted over a long period, such as the Black Forest in Germany and the Alps, the main streams flow without steps in their gradients across the limits of the mountain range on to the adjoining foreland. This indicates that their mean long-term downcutting rate is about the same as the mean long-term rate of uplift of the mountain range and that there is a state at least close to a dynamic equilibrium between the two processes at the margin of the mountain range.

Denudation within the mountain range depends on local relief, that is, on the height difference between summits and valleys. The incision of the valleys in the mountains occurs by **headward erosion** of the stream from the margin of the mountain range. If a stream is to incise a valley at a particular point on its course, the stream requires sufficient gradient. This is created locally by prior vertical erosion in the adjoining stretch immediately downstream. The local vertical erosion of the stream creates, in turn, an increased gradient for the adjoining stretch upstream and leads to incision there. In this way, the erosion impulse triggered by the uplift of the mountain range is progressively transmitted upstream in both the main stream and its tributaries, and eventually reaches the source streams of the drainage network in the area of the crests on the main watershed.

The headward progression of valley downcutting from the margin of the mountain ranges to the main watershed requires time. On the mountain edge, the downcutting commences with the uplift and soon reaches equilibrium with the rate of uplift. At the watershed, equilibrium occurs much later. The denudation rate in the summit region itself depends on the valley downcutting rate and can reach equilibrium with the uplift only after the rate of downcutting has done so.

As long as the denudation rate at the summit remains smaller than the uplift rate, the height of the mountain range increases. This non-equilibrium remains longer at the main watershed than on the margins of the range but here, too, the summit denudation rate eventually catches up with the rate of uplift and there is a state of equilibrium between both rates. From then on, although the uplift continues, the height of the mountains does not increase.

If the long-term rate of uplift is more or less the same for the entire mountain range, the summits on the main watershed are higher than other summits lying closer to the margins of the range because closer to the margins the denudation rate and uplift are in equilibrium sooner.

The broader a mountain range, the more time is required until this state of equilibrium advances to the main watershed, and consequently the higher the summits become there (Ahnert, 1984).

Table 3.2 and Figure 3.5 show that this functional relationship between the relative height of mountain summits above the foreland and the distance of these summits from the edge of the mountains does exist. The 21 denudational, not volcanic, summits shown in Table 3.2 belong to very different mountain ranges, but their relief and distance values can be represented in a single empirical regression equation:

$$H(\mathrm{m}) = 2626.8 \ln L - 23\,524.0(\mathrm{m}) \qquad (3.6)$$

and $r = 0.95$, where H is the relative height of the summit above the foreland of the mountain range and L is the distance of the summit from the margin of the mountain range against the foreland. Equation (3.6) is valid only for values of $L > 8000$ m.

The validity of this functional relationship for so many different mountain ranges would suggest that the long-term (lasting millions of years) rates of uplift of all of these ranges lie in the same order of magnitude. From this it can be concluded that there are upper limits to the long-term rate of uplift because the possible rate of movement of the earth's crust is restricted by the limited mobility of the earth's mantle below (section 4.1).

3.8 LANDFORMS AS AN EXPRESSION OF SPATIAL AND TEMPORAL DIFFERENTIATION OF PROCESS RESPONSE SYSTEMS

The surface of the earth, formed by endogenic and exogenic processes, is spatially differentiated. The forms vary from place to place. Large-scale differences in form between, for example,

TABLE 3.2 Relative height and distance to the foreland from mountain summits

Summits	Height above foreland H(m)	Distance to the foreland margin L(m)
Cross Fell, Pennines, England	770	8 500
Brocken, Harz, Germany	900	9 000
Grand Ballon, Vosges, France	1160	9 600
Snowdon, Wales	1000	9600
Belchen, Black Forest, Germany	1160	15 500
Mount Witney, Sierra Nevada, USA	3200	18 000
Feldberg, Black Forest, Germany	1200	18 500
Pike's Peak, Rocky Mountains, USA	2350	20 000
Zugspitze, Bavarian Alps, Germany	2350	25 000
Ruwenzori, Zaire/Uganda	4200	25 000
Long's Peak, Rocky Mountains, USA	2520	29 000
Gran Paradiso, the Alps	3650	35 000
Pic de Neouvielle, Pyrenees	2700	35 000
Mount St Elias, Alaska	5488	40 000
Monte Rosa, the Alps	4330	50 000
Mont Blanc, the Alps	4400	60 000
Mount McKinley, Alaska	6050	70 000
Dhaulagiri, Himalaya	7880	110 000
Aconcagua, Andes	6958	120 000
Mount Everest, Himalaya	8400	130 000
Nanda Devi, Himalaya	7360	135 000

Source: Ahnert (1984).

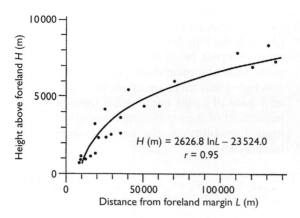

FIGURE 3.5 Functional relationship between the relative height of 21 summits on various mountain ranges and the distance from their forelands (after Ahnert, 1984, 1994b).

high mountains, uplands and lowlands are primarily the result of a large-scale spatial differentiation in the nature and intensity of the endogenic processes, particularly crustal movements , to which the exogenic processes react in a differentiated way.

The nature and intensity of the exogenic processes also varies over much shorter distances.

The diversity of summit, ridge, slope and valley forms develop largely because at different points on the land surface varying quantities of material are removed or deposited. The detailed spatial differentiation of landforms is, therefore, primarily the result of the spatial differentiation of exogenic processes. The intensities of the processes are related by feedbacks to the characteristics of the form that they create. There is, in addition, a spatial differentiation in the composition and, particularly, the resistance of the bedrock and the soil cover.

Forms, material and processes act together as components of geomorphological process response systems (section 2.2.3). Also, at all times, existing forms and material are the products of past processes, which were spatially differentiated, often in other ways than the present day processes and which now further alter forms and material. The **spatial differentiation** of forms, materials and processes is, therefore, superimposed by their **temporal differentiation**. In the following chapters the landforms of the earth and their characteristics are described in terms of the spatial and temporal functional structure of their process response systems, together with their components and their changes.

4

ENDOGENIC PROCESS RESPONSE SYSTEMS AND THEIR GEOMORPHOLOGICAL EXPRESSION

Endogenic processes have an important role in the shaping of the earth's surface and are significant components in the structure of the overall geomorphodynamic system. There are two main groups of endogenic processes:

1. **crustal movements**, which include the displacement of parts of the crust relative to one another, either vertically or horizontally, along **faults**, the tilting of segments of the crust and the deformation of rock structures with the upwarping or folding of rocks in **anticlines** and downwarping in **synclines**;
2. the material transport of molten rock substance, **magma**, from the deeper zones of the earth's interior towards the surface which either causes the deformation of the crust in those areas the magma penetrates or the development of volcanoes if the magma flows out at the surface as **lava**.

Endogenic processes tend to increase the relief (height differences in the landscape) and thereby the potential energy, E_{pot}, available for the exogenic processes of erosion and denudation:

$$E_{pot} = mgh \qquad (4.1)$$

where m is mass, g is gravitational acceleration ($= 9.81 \text{m/s}^2$) and h is available height difference (local relief).

4.1 ISOSTASY

In the area of the continents, the composition of the outer layer of the lithosphere, the earth's crust, is different from that of the ocean floor. The mean thickness of the **continental crust** is about 30 km. It is made primarily of granitic rocks with a mean specific density of 2.7. The **oceanic crust** on the floor of the large ocean basins and in the deep sea trenches, lying at more than 4000 m below sea level, is made of basaltic rocks with a mean density of 3.1 and is only about 6 km thick. The mean surface height of the continents is about 875 m, the mean depth of the ocean basins about 5000 m.

The **hypsographic curve** of the earth's surface is a cumulative frequency diagram of the heights on the surface. It shows five distinct altitude zones (Fig. 4.1a). **High mountains**, over 2000 m, and **deep sea trenches**, below a depth of 6000 m, take up very small shares of the surface. A large proportion of the surface is occupied by the two relatively level zones, the **continental platform** and the **ocean floor**. They appear as separate maxima because they are separated by the intervening less extensive zone of the **continental slope** and the **mid-ocean ridges**. The basic differences in material, structure and morphology between the continental and the oceanic crust are expressed in this bimodality.

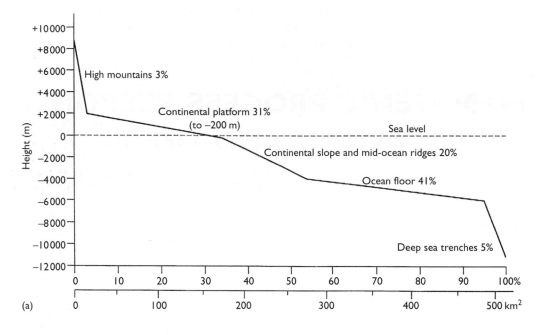

(a)

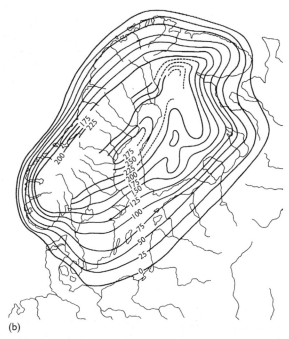

(b)

FIGURE 4.1 (a) the hypsographic curve of the earth's surface; (b) postglacial isostatic uplift of Scandinavia. The lines show the total amount of uplift in m since the melting of the inland ice (after Daly from Fairbridge, 1968b, p. 885).

The shallow sea area extending from the coast down to about 200 m depth, the **continental shelf**, also belongs to the continental platform. The sea level itself is not a structural boundary. It has been subject to considerable fluctuations in the recent geological past and reached its present level only about 6000 years ago.

At the outer edge of the shelf, the sea floor falls relatively steeply to the ocean basins. It is in fact steeper than it appears from the hypsographic curve because the mid-ocean ridges, which are far from the continents, are included in the same section of the curve. The continental slope constitutes the true edge of the continents. In western Europe, for example, it lies west of Ireland. The British Isles therefore belong structurally to the continent of Europe.

The mean surface of the continental crust is about 6 km higher than the surface of the oceanic crust. It has, however, a mean thickness of 30 km, five times the thickness of the ocean crust, and reaches deeper into the internal structure of the 100 km thick upper **earth mantle**. Under continued pressure the mantle yields with plastic flow so that the continental crust has been able to displace parts of the upper mantle, although under sudden stress, the upper mantle reacts by fracturing much like the rocks of the crust. The upper mantle and the crust form the **lithosphere**. Below the upper mantle lies the much more mobile **asthenosphere** (Greek *asthenos* = weak).

Based on the velocities of earthquake waves, a specific density of 3.3 has been determined for the upper mantle, a density that is only slightly higher than that of the oceanic basalt crust but about one-fifth higher than the density of the continental crust. This difference, and the fact that the continental crust reaches deeper into the mantle than the oceanic crust, led in the ninetenth century to the theory of **isostasy** (Greek *isos* = equal, *stasis* = stand) which suggested that the crustal blocks 'float' on the earth's upper mantle. Based on Archimedes' principle, the crust displaces a mass in the mantle equal to its own mass. The depth a crustal block reaches in the mantle is therefore

$$D_k = T_k S_k / S_m \qquad (4.2)$$

where D_k is the depth to which the crust is immersed in the mantle, T_k is the thickness of the crust, S_k is the specific density of the crust, S_m is the specific density of the mantle ($= 3.3$) and the height H_k by which the crust rises above the mantle is

$$H_k = T_k - D_k \qquad (4.3)$$

Because of the very small differences in density between the oceanic crust ($S_k = 3.1$, $T_k = 6$ km) and the mantle, the crust lies almost at the level of the mantle ($H_k = 0.36$ km), while the continental crust ($S_k = 2.7$, $T_k = 30$ km) rises considerably above it ($H_k = 5.5$ km). The actual difference between the mean height of the continents (6 km) and the depth of the large ocean basins (5 km) agrees very well with these estimates.

The isostasy theory provides the key to understanding crustal movements. The 'floating' of the continental crust means that where continental crust rises higher up, in the areas of high mountains and high plateaus, it also reaches deeper into the upper mantle and must, therefore, be thicker (equations (4.2) and (4.3)) than in lowland areas. The pushing together, folding and overthrusting of previously horizontal strata has thickened the crust and caused the isostatic uplift of the crust surface.

Another consequence of isostatic floating is that with additional load, similar to a loaded ship, the crust sinks deeper into the bearing medium, the upper mantle and, with a reduction in load, it rises. In a lowland or shelf area, the accumulation of large masses of sediment leads to isostatic sinking. The lightening by erosion and denudation of the mountain masses leads to isostatic uplift provided that the eroded material is transported away. The amount of the compensatory movements is

$$\Delta H = \pm d S_g / S_m \text{(m)} \qquad (4.4)$$

where ΔH is the the amount of crustal movement (positive when uplift, negative when lowering), d(m) is the mean denudation amount (positive) or sedimentation (negative), S_g is the specific density of the rock material removed or deposited and S_m is the specific density of the mantle ($= 3.3$).

The denudation of rock with a specific density of $S_g = 2.5$ causes an isostatic uplift of

2.5/3.3 = 0.76, that is 76 per cent of the amount of the mean denudation. The specific density of sedimentary material deposited is much less, usually between 1.5 and 2.0 because of the relatively unconsolidated state of the sediment and the high proportion of pore space within it, and the isostatic lowering as a result of sedimentation is between about 1.5/3.3 and 2.0/3.3 or between 45 per cent and 60 per cent of the amount deposited.

Isostatic crustal movement compensates only in part, therefore, for the changes in height of the land surface produced by denudation or sedimentation. Exceptionally, a net denudational isostatic relief increase, that is the net uplift of the highest summits in an area, could occur when the denudation of the summits is considerably smaller than the mean denudation for the entire region. This is possible in a high plateau area where denudation is concentrated in the valleys and on the valley slopes, while in the highest areas, distant from the valleys, little or no denudation takes place.

The additional load on the crust of a large body of water causes **hydro-isostatic subsidence**. Hydro-isostatic uplift occurs when the water body becomes shallower or disappears. During the last Pleistocene ice age, Lake Bonneville in Utah, USA filled a basin of about 50 000 km^2 and was 300 m deep. Today only the Great Salt Lake remains, with a maximum depth of 15 m and an area of 5000–6000 km^2. The depth of Lake Bonneville varied in different parts of its basin, so that the reduction in the load as the body of water became smaller produced different amounts of compensating uplift and left the originally horizontal shorelines of the lake warped. Throughout the earth's history, the long-term raising and lowering of sea level, which has occurred repeatedly, probably led to hydro-isostatic compensating movement of the ocean floors.

The rates and amounts of glacio-isostatic subsidence and uplift caused by the formation and later melting of thick and extensive sheets of inland ice are especially high. Equations (4.2) and (4.3) are also valid for ice if the thickness and density of the ice are substituted for the thickness and density of the crustal rocks. The Greenland ice cap is 3 km thick at its centre

and has a density of 0.9. The weight of the ice has caused the crust to subside about $3000 \times 0.9/3.3$ m $\simeq 800$ m. At the present time the rock surface beneath this ice thickness lies more or less at sea level. The ice-free margins of Greenland are generally between 500 m and 1000 m above sea level so that if the ice melted, the centre would rise about 800 m and then lie at approximately the same height as the margins. The ice-covered Antarctic continent is similarly depressed.

The isostatic equilibrium of the crust is reached only gradually and does not immediately follow the loading or unloading of the ice because the subsidence or uplift of the crust is possible only if material is displaced in the underlying mantle. The mantle's viscosity is, however, very high and reacts to pressure only slowly. For example, the isostatic uplift of Scandinavia, which was covered by the inland ice of the last ice age until about 10 000 years ago, and had subsided accordingly, has not yet been completed. The northern part of the Gulf of Bothnia where the centre of the ice mass lay, is rising at a rate of 9 mm per year at the present time, one of the highest rates of uplift on the earth. The rate of uplift must have been still higher in the past because the total uplift of the area during the past 10 000 years is 275 m (Fig.4.1b). Very similar rates have been measured in the former centre of the inland ice in eastern Canada (Farrand in Fairbridge, 1968b, pp. 884–888).

The formation and melting of ice are, like denudation and sedimentation, exogenic processes. Isostatic movements of the crust are endogenic processes because they are produced directly by the pressure-compensating movements of the mantle material in the earth's interior. Ice formation and melting, loading or unloading of water, denudation or sedimentation are only causes of the movement, not the reason for the mobility.

Epirogeny (Greek *epeiros* = continent, *genesis* = origin) is the term used to describe the widespread vertical movement of large crustal blocks without any significant change to their internal structure. **Orogeny** (Greek *oros* = mountain) describes the building of mountains, and includes internal deformation (folding, upthrusting, normal faulting and overthrusting along fault lines) of the crustal block affected. It is also spatially more confined. Orogeny generally changes the structure, epirogeny does not.

4.2 PLATE TECTONICS: THE FORMATION AND LOCATIONAL CHANGE OF THE CONTINENTS AND OCEANS

The isostatic mobility of the crustal blocks allows both vertical and horizontal movements. At the beginning of this century, this concept received comprehensive treatment in Wegener's (1915) theory of **continental drift**.The theory is based on the fact that the continental margins of the Atlantic are almost parallel, indicating that they might have been connected during an earlier period, and on a great deal of palaeontological and geological evidence, including the continuation of geological structures of the same age and the presence of similar sedimentary rocks and terrestrial fossils on both sides of the ocean. Because the geophysical evidence put forward by Wegener seemed insufficient, his theory remained controversial for a long time. Only with the discovery of sea floor spreading and the development of the theory of plate tectonics was its fundamental correctness confirmed (Dietz and Holden 1970).

Sea floor spreading is the term used to describe the discovery in the North Atlantic near Iceland, and soon after in all other oceans, that the basaltic deep sea floor in the area of the mid-ocean ridges is formed by rising lava which then spreads laterally in both directions. The youngest basalt on the deep ocean floor is, therefore, on the mid-ocean ridges and the greater the distance from the ridge, the older the basalt becomes. The spreading rate, which is determined from the age of the rock, ranges between 1–5 cm/year in the Atlantic to about 10 cm/year in parts of the Pacific.

This horizontal movement of oceanic lithosphere is thought to be caused by convection currents in the mobile earth mantle below, the asthenosphere. The spreading of the sea floor is not accompanied by an increase in the surface of the earth because the lava rising and spreading from the mid-ocean ridges sinks again elsewhere in **subduction zones**, which are largely identical with the deep sea trenches. The subduction zones are also areas of frequent earthquakes and usually associated with the rows of volcanic islands that accompany the deep sea trenches as **island arcs**.

The continental crustal blocks are also carried along by the convection currents. The circulation cells of the currents in the mantle have changed their position repeatedly during the earth's history and with them the currents and the drift direction of the continents. Zones where lava rose and the currents diverged, or where they subsided, have become inactive and zones of activity developed elsewhere.

Plates are areas of the lithosphere that move as a unit. The term **plate tectonics** describes both the mechanism and the sequence of crustal movement during the earth's history. There are eight large plates at the present time: the African, American, Antarctic, Chinese, Eurasian, Indian, Nazca (in front of the west coast of South America) and Pacific plates and a similar number of smaller plates. Within each plate, the movement of the oceanic lithosphere is subdivided again into parallel bands several hundreds of kilometres wide that run in the direction of the movement. The bands move at different rates and are separated by shear zones, the **transform faults**.

The changes in the position of the continents during the past 230 million years are shown in Fig. 4.2 and are based on the reconstruction on the basis of palaeomagnetic evidence of earlier positions of the continents relative to the magnetic poles. From the remanent magnetism in rock minerals it is possible to reconstruct the local direction of the earth's magnetic field lines at the time of the rock's formation and also its geographic position (Tarling, 1971).

During the Permian, the continents of the present day were joined in one large continent, **Pangea**. A large embayment of the sea known as **Tethys** reached deep into this continent on its eastern side. North and northwest of Tethys lay the crustal blocks, still interconnected, that today form North America, Europe and the Asian continent, but not including the Indian subcontinent. These northern interconnected continents are referred to as **Laurasia**. South and southwest of Tethys lay an interconnected southern continent, **Gondwana**, which subsequently broke up to form South America, Africa, India, Australia and the Antarctic.

At the end of the Triassic, about 195 million years ago, Tethys opened westwards and formed a narrow sea passage which completely separated Laurasia from Gondwana. The separation of India from Africa and the Antarctic and of the Antarctic–Australian block from Africa began at the same time. New zones of rising magma in the earth's mantle below the previously connected blocks resulted in the development of diverging currents in the mantle which pulled the continental blocks apart. The lines of separation above the rising magma became gradually widening arms of the ocean. The new oceanic

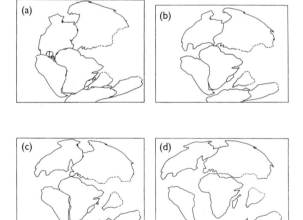

FIGURE 4.2 Plate tectonic changes in the position of the continents in the past 230 million years: (a) Pangea in the Permian about 250 million years ago; (b) division into the northern continent of Laurasia and the southern continent of Gondwana, together with the subdivision of the latter in the upper Triassic, about 200 million years ago; (c) at the transition from the Jurassic to the Cretaceous period, about 140 million years ago; (d) at the beginning of the Tertiary, about 65 million years ago (after Dietz and Holden, 1970). (*see also* Table 1.1 and Figure 4.3.)

crust expanded from both sides of the ridge formed by the zone of magmatic upwelling.

About 140 million years ago at the end of the Jurassic, the separation of the continental blocks had progressed. The triangular Indian block had moved northwards towards southern Asia and the North Atlantic had begun to open between North America on one side and western and southwestern Europe and Africa on the other. North America and Europe remained joined only between Greenland and Norway. In the South Atlantic, a narrow passage ran between South America and southwest Africa.

By the end of the Cretaceous, about 65 million years ago, the southern and middle Atlantic was quite wide. North and South America were moving westwards and were separated by a broad ocean area that connected the Pacific and the middle Atlantic. Greenland and Northern Europe were still attached to one another. The expansion of the South Atlantic shifted Africa's direction of movement towards Europe and western Asia so that the Tethys became narrower. Madagascar separated from the east coast of Africa. The Indian block continued to move north towards southern Asia and the first sea passage began to form between Australia and the Antarctic.

Today, North and South America are joined by a young land bridge formed of volcanic rocks. As these two continents moved progressively westward, the mountain systems of western North America and the Andes in South America developed. The triangular Indian block collided with south Asia and the resulting compression and upthrusting lead to the formation of the high Asian Plateau and the Himalaya. The Alpine fold system was also developed between Africa and Europe and western Asia. The Mediterranean Sea is a remainder of the Tethys. Between southeast Asia and Australia, the belt of island chains and arcs that now belong to Indonesia was formed (Fig. 4.3).

The young mountain ranges of both Americas and the Alpine systems were originally sedimentation basins on the forward, **active** margin of a moving continent. As the North and South American continental blocks advanced towards the subduction zones that today lie on their western margins, they were slowed down with

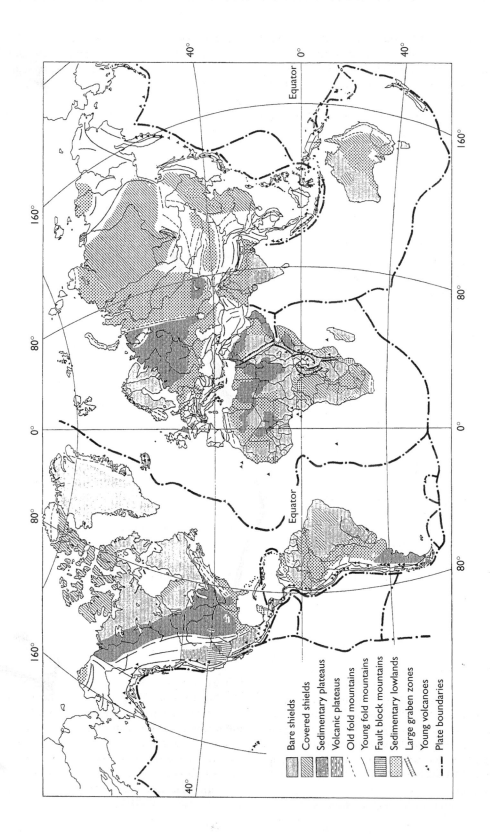

Bare shields
Covered shields
Sedimentary plateaus
Volcanic plateaus
Old fold mountains
Young fold mountains
Fault block mountains
Sedimentary lowlands
Large graben zones
Young volcanoes
Plate boundaries

40°

0°

Equator

40°

160°

80°

0°

80°

160°

Figure 4.3 Morphostructural units of the earth (draft: Ahnert).

the effect that the youngest, easily deformed layers of sediments were compressed and folded on their forward edges and, because of their greater crustal thickness, isostatically uplifted. Compression, folding and overthrusting also took place in the interior of the western part of the advancing continent, particularly in North America. In the Alpine mountain systems, the sediments between the converging continental blocks of the Eurasian and the African plate were pressed together and folding and overthrusting followed. Frequently, the older rocks of the substratum and rising magma were included in the mountain building process.

The sedimentation zones that existed on the **passive** margins of the continents were unaffected by compression and folding. On the eastern coast of the United States, the Atlantic Coastal Plain and the adjacent shelf area, sediments from the denudation of the eastern part of the North American continent have been deposited almost continuously since the plate tectonic opening of the bordering Atlantic Ocean in the Cretaceous. The undisturbed accumulation of sediments is a characteristic of passive continental margins.

Future plate tectonic developments are clearly indicated in several parts of the world. The Red Sea is a young sea passage that has formed as Africa and southwest Asia separate. This rift continues northwards in the Jordan rift valley and southwards in the eastern and central African rift valleys (section 4.3.8). In western Europe, there is a similar but smaller rift zone that can be traced from the mouth of the Rhône River through the Upper Rhine valley as far as southern Norway. On the San Andreas fault in California, two plates, the Pacific and North American, are moving past each other in opposite directions.

These zones of activity are characterized by a high earthquake frequency. The presence of such a dividing line within a continental block does not necessarily mean that a split will occur in this part of the continent. Many divergence lines have never developed beyond an initial stage.

Figure 4.2 shows developments during 230 million years of earth history. The solid earth is at least 4500 million years old. It is quite likely, therefore, that similar sequences of events have taken place before the Permian, probably several times. Evidence is provided by older fold mountains that lie within continental blocks at the present time. The Urals in Russia were folded in the Carboniferous between the even older shields of northern Europe and western Siberia, and the early Palaeozoic mountain systems of eastern North America and western Europe were folded between the north European or Baltic shield and the Laurentian shield which were colliding at that time, evidence that in the early Palaeozoic era North America and Europe converged, before they diverged. The supercontinent Pangea 230 million years ago was not an initial state but a stage in a long-term series of movements, during which there were periods when the continental blocks tended to combine and others when they drifted apart.

4.3 THE LARGE MORPHOSTRUCTURAL UNITS OF THE CONTINENTS

The landforms of the continents can be divided into large units, based primarily on the regional structure of the rocks. The age of the structures and the length of time they have been subject to denudation plays a significant role. Relatively few types of morphostructural units make up each continent but the way in which the units are combined, their extent and their spatial distribution gives each continent its individual geomorphological character. Ten different types of morphostructural units are shown in the world map Fig. 4.3. Transitional areas and spatial superimposition make it difficult to define some boundaries.

4.3.1 Shields

Shields are the oldest parts of the continental crust accessible at the earth's surface. They consist of Precambrian crystalline rocks and originally made up the basement of Precambrian fold mountains that have been almost continuously eroded since. The removal and transport of material from the crust caused slow, long-term isostatic upift and the rocks that formed at great depth are now at the surface. The **exposed shields** generally have a very low relief with only minor adjustments for variations in rock resistance. Elevations often coincide with resistant rock, and valleys with zones of less resistant rock or lines of structural weakening by jointing or faulting. On the **covered shield**, the Precambrian crystalline rock is covered by a relatively thin layer of younger sedimentary rocks or lava sheets. These usually lie almost horizontal because the rigidity of the old crystalline base does not allow any further folding. At most there is a local vertical movement on faults. Otherwise, only epirogenetic movement over wide areas is possible, such as the current glacial isostatic subsidence of the shield area of Greenland or the postglacial uplift of the Baltic Shield in Scandanavia and the Laurentian Shield in eastern Canada.

The Laurentian Shield, the Baltic Shield and the covered Angara Shield in northern Siberia are parts of the old core area of Laurasia. The shield areas of South America, Africa, the Arabian Peninsula, India, Australia and Antarctica formed the core of Gondwana. Many shields are subdivided, particular in Africa whose large shield consists of higher areas, such as the Lunda Rise, with exposed crystalline rock, and basins with a sedimentary cover.

4.3.2 Sedimentary plateaus, tablelands and cuestas

A **plateau** is a more or less horizontal land surface that is higher than its surroundings. Because of their altitude, plateaus are often incised by streams. **Sedimentary plateaus** are made up of nearly horizontal sedimentary strata (sandstones, claystones, limestones). The term **tableland** includes plateaus but it is also applied to extensive areas of sedimentary rock that are not higher than their surroundings; the Russian Tableland between the Baltic Sea and the Ural Mountains is an example.

Many sedimentary tablelands have a crystalline basement that impedes small-scale deformation. The sedimentary deposits on tablelands are generally thicker than those on covered shields. It is only in higher areas that streams cut their valleys through these thick sediment covers into the underlying crystalline rocks, as has taken place in the Grand Canyon of the Colorado River in Arizona, USA.

If the layers are slightly warped and therefore a little inclined, the varying resistance of the exposed rocks results in the formation of cuesta scarps or, if the rocks dip more steeply, of hogbacks (section 20.3). Well developed cuesta scarps dominate the landscape in northern France, southern England, southern Germany, parts of the Sahara and on the Colorado Plateau of the USA.

4.3.3 Volcanic plateaus

Volcanic plateaus are areas of large flood eruptions. The lava in the eruptions is of very low viscosity, usually basaltic, that has flowed from fissures in the ground and spread over the land surface. Where volcanic activity lasted long enough for hundreds of metres or more of lava to accumulate, the lava sheets form volcanic plateaus. Unlike sedimentary plateaus, the lava layers have a composition and resistance that is essentially uniform and interrupted only by occasional deposits of volcanic ashes; their surface forms hardly vary. Most large volcanic plateaus were formed in the Tertiary, especially the Eocene, Oligocene and Miocene (Table 1.1). Three of the largest volcanic plateaus are the Columbia Plateau in the northwestern USA, the Highland of Ethiopia and the Deccan Plateau in India. The basalt plateau in the Parana region of Brazil was formed in the late Jurassic and early Cretaceous.

The flood eruptions in Ethiopia and the Deccan lie on a crystalline basement. The Columbia Plateau covers complex fault and fold structures which, in part, continue southwards in the fault block region of the Great Basin and eastwards to the Rocky Mountains. Several uplifted blocks, such as the Blue Mountains in Oregon, rise above the lava.

4.3.4 Old fold mountains

The old fold mountains are made up of rocks folded in the Palaeozoic, either in the **Caledonian orogeny** which lasted from the upper Cambrian to the beginning of the Devonian

(Table 1.1) or the **Variscan orogeny** which began in the Devonian and ended in the Permian. Folding and mountain building within both these orogenies was concentrated in different regions over varying shorter periods of time.

The mountains formed during the Palaeozoic had been worn down to lowland by the beginning of the Mesozoic. Their fold structures were capped by denudation, but preserved below the land surface. In the Mesozoic this folded basement was covered in many areas by horizontally bedded younger sedimentary rocks. The boundary plane between the capped, folded basement rocks and the cover rocks is a **unconformity** which indicates the sequence that occurred: folding of the basement rocks, denudation, deposition of the cover rocks.

In central Europe, parts of the Palaeozoic fold mountains were uplifted along fault lines as crustal blocks during the Tertiary and Quaternary, a long-range effect of plate tectonic pressure that accompanied the folding and uplift of the Alps taking place at this time (section 4.2). The Mesozoic cover was removed in part or entirely from the uplifted blocks, leaving the basins between covered with Mesozoic sediment. Most of the central uplands of Europe are uplifted blocks of this type. The old fold mountains of Europe are now spatially divided into fault blocks. They could have been designated fault block mountains in Fig. 4.3, since they are both. Other old fold mountains such as the Variscan Urals, the Caledonian Highlands of Scotland and the Appalachians, which are part Variscan and part Caledonian, can only be defined as old fold mountains. The Variscan Australian Alps are today more of a scarp than a mountain range over long stretches on the edge of the continent towards the Pacific. The mountains of Central Asia, the Tien Shan, Kun Lun Shan and others are Palaeozoic in origin but their present mountainous character has resulted from more recent crustal movements.

Old fold mountains are more varied in form than the other morphostructural types. Originally they were large mountain systems that stretched across the Laurasian and Gondwana continents, similar to the Alpine system that extends from the western Mediterranean to southeast Asia. The widely scattered occurrence of old fold mountains today is a consequence of the breaking up of the original continents by plate tectonics and, in central Europe, by younger fault tectonic activity.

4.3.5 Young fold mountains

As a morphostructural unit, the young fold mountains were formed in the **Alpine orogeny** which began in the late Mesozoic but took place largely in the Tertiary. The orogeny itself, with folding and uplift, was preceded by a long phase of sedimentation of the strata that were later folded in **geosynclines**. The fold mountains were created by the plate tectonic movement of continental lithospheric blocks described in section 4.2. The Alpine orogeny is still continuing at the present time, particularly the uplift phase.

There are two belts of young fold mountains. The **Alpine mountain belt** stretches from the Atlas Mountains in North Africa, through the Sierra Nevada and Pyrenees in Spain, the Alps, the Apennines, Carpathians, Dinaric Alps, Balkans and marginal mountains in Asia Minor, the Caucasus, Elburz and Zagros to the Hindu Kush, the Pamir and the Himalaya into southeast Asia where it joins the **circum Pacific belt**. This mountain belt surrounds the Pacific, partly in the form of island chains and partly as mountain ranges, in east Asia and North and South America. It then extends from southern Patagonia, over the island arc of the South Sandwich Islands to the Palmer Peninsula (Graham Land) and into Antarctica. The New Zealand Alps are an isolated but particularly active part of the chain that has a high rate of uplift at the present time.

In some areas the young mountain chains subdivide and then come together again, sometimes to form high mountain nodes such as the Pamir and the Armenian highland. Some of the chains divide because the folds are made up of easily deformed rock strata lying on either side of rigid, older masses which remain at a lower position. The large arc of the Carpathians and the Dinaric Alps on either side of the Hungarian Plain, and the Taurus and Pontic Mountains in Turkey are of this type.

4.3.6 Fault block mountains

In Fig. 4.3, only the Great Basin in the western United States belongs to this morphostructural type. It lies within the young fold mountain system of North America and was once part of it. After folding in the Cretaceous and early Tertiary, the area broke up along fault lines into numerous blocks in the late Tertiary and early Quaternary. Some blocks are uplifted as horsts bordered on both sides by faults, some as tilted blocks. Others have been tectonically lowered. The area now consists of a large number of small ranges separated by basins which receive the debris removed by denudation from the mountains. Some old fold mountains can also be classified as fault block mountains, particularly those in central Europe (section 4.3.4).

4.3.7 Sedimentary plains

The sedimentary plains of the earth are heterogenous in origin. Some, like the Kalahari Basin in South Africa and the west Siberian lowland, are relatively low-lying areas of shields covered with young, in part already consolidated, sediments. Others, like the Ganges lowland in India, the much smaller Alpine foreland in Germany and the Po Basin in Italy, are sediment-filled troughs in front of young fold mountains. Still others are coastal lowlands whose sediments can cover a variety of structures below. Common to all is a low relief and a surface made up of fluvial, sometimes also glacial, aeolian, marine or lacustrine sediments that were deposited in the Quaternary, Tertiary or, in some areas, in the Cretaceous.

4.3.8 Large graben zones

A **graben** develops when one or more blocks are lowered between two or more roughly parallel running faults. Only two large graben zones are shown in Fig. 4.3. The African–Near Eastern graben system, which is by far the largest in the world, begins south of Lake Malawi, divides northwards into two branches, the **east African graben**, in the northern part of which lies Lake Turkana, and the **central African graben** in which Lake Tanganyika, Lake Kiwu and Lake

Edward lie. Between the two grabens is the large, but relatively shallow depression of Lake Victoria. In southern Ethiopia the east African graben turns northeastward, reaching the southern exit of the Red Sea at the Afar triangle. It then follows the Red Sea northwards and divides on both sides of the Sinai Peninsula into the Gulf of Suez and the Gulf of Aqaba. The latter branch continues north to the Aravah Valley, the Dead Sea and the Jordan Valley via the Sea of Galilee into Lebanon. The entire fault system covers more than 50° of latitude (5600 km) and seems to be the youngest plate tectonic dividing line in the progressive breaking up of the old Gondwana continent.

The other major graben zone extends in a north–south direction through western Europe from the mouth of the Rhône to Lyon, then northwards along the Saône to the Burgundy Gap between the Vosges and the Jura. From there it follows the most well-defined section of the system, the approximately 300 km long Rhine graben between Basel and Mulhouse in the south and the edge of the Taunus hills near Wiesbaden and Frankfurt in the north. The graben zone then turns northeastwards into the Wetterau and continues in the lowland of Hessen and the graben of the Leine near Göttingen. In north Germany the system is covered with young sediments, but it appears again in southern Norway as the Oslo graben and ends at the Mjösa Lake. The entire system has been named the Mediterranean–Mjösa graben zone. The various sections of the system seem to be of different ages. Its development and geotectonic significance have not yet been fully explained.

4.3.9 Large young volcanoes and areas of volcanoes

Not even the largest single volcanoes constitute morphostructural units. A number of young volcanoes and volcanic areas are shown in Fig. 4.3, mainly to indicate the distribution of volcanic activity at the present time. Around the Pacific, they are clearly related to the subduction zones of the circum Pacific mountain belt but they also occur in the Alpine mountain belt in the Mediterranean, along the African graben system and rise as volcanic islands from the ocean floor forming, for example, the Islands of Hawaii, Iceland, the Azores and Ascension.

4.3.10 Form generations and types of morphostructure

The shields, the old fold mountains and the young fold mountains have been formed by essentially the same process sequence of folding and uplift followed by denudation. They are distinguished by the age of their orogenies and, therefore, also the duration of their denudation. The rocks and structures of the shields were formed in the Precambrian between 600 million and 4 billion years ago. Since then they have been subjected repeatedly to renewed pressure and tension from subsequent orogenies. They could not react by further folding, at most only by movement of their crustal blocks because their rocks were too massive and rigid from previous folding.

The shields have also been worn down over a very long period, causing a long-term isostatic upift which was, nevertheless, smaller than the amount of material removed because the specific density of the material removed was lower than that of the mantle (section 4.1). For this reason the shields have become progressively lower. The existence of younger sedimentary covers on some shield areas are a sign that at those locations the uplift ceased, at least for a period, and was perhaps even replaced by a temporary subsidence.

The old fold mountains are Palaeozoic and were formed between 600 and 230 million years ago. Their later structural development did not include further folding but was due to fault tectonics. The original folded and uplifted mountains were usually worn down after a few tens of millions of years, although even today they have not been worn down nearly as far into the deep basement structures as the shields. Where the old fold mountains have a mountainous relief today, it is due largely to young fault block uplift and the erosive and denudative differentiation of particularly resistant rocks which lie as monadnocks above the surrounding areas. In the Appalachian mountains, the resistant strata of the old fold mountains today form long ridges or hogbacks and the weaker strata longitudinal valleys (section 20.3.10).

Young fold mountains are usually less than 140 million years old and originated in the plate tectonic activity that lead to the present distribution of the continents. In most cases, uplift has not yet ceased, relief is high and denudation is proceeding rapidly.

At some time in the Precambrian era the structure, rocks and landforms of the shields resembled those of the young fold mountains. The same can be said of the old fold mountains in the Palaeozoic era. The young fold mountains can be regarded as an early stage, the old fold mountains as a middle stage and the shields as a later, perhaps the last, stage of a more or less continuous morphostructural development. These three types of large morphostructural units represent therefore the three essential major landform generations of the earth's surface.

4.4 VOLCANISM AND PLUTONISM

Volcanism and plutonism are the only endogenic processes that increase relief on the earth's surface by transferring molten rock upwards in the crust or to the land surface. If the molten mass, the magma (Greek = kneaded mass) solidifies several kilometres below the surface, it becomes a batholith. Intrusion is the process of penetration by the magma into the rock of the upper crust. If the rising magma reaches the surface, it becomes lava. Magma and lava differ in that lava gives off gases into the atmosphere when it reaches the surface and is, therefore, poorer in substances than the original magma.

Intrusions usually displace the rocks upwards and outwards, causing uplift and upwarping within the rocks and at the surface. Volcanic eruptions increase relief directly by bringing lava to the surface. The high intensity of material output from volcanoes can lead to local rates of relief increase that greatly exceed any possible rates of relief increase by uplift of the crust.

There are two main types of volcanic eruption: explosive ash or cinder eruptions accompanied by a large output of gas, and effusive eruptions of lava. The contrast is determined not only by the amount of gas present that is released but also by the viscosity of the magma which is dependent on its chemical and mineral composition. Basic magmas, poor in silica, are more fluid than acid, silica–rich magmas. Basic magmas generally reach the surface as effusive lava eruptions, while acid magmas often erupt explosively. The boundary between the two types is not sharp and in many volcanoes both types of eruptions take place. Explosive **phreatic eruptions** (Greek *phrear* = well) may occur in any magma if water enters the magma chamber and changes into steam, thereby increasing the gas pressure.

Ash or **cinder eruptions** generally come from a **volcanic vent** that reaches deep into the magma. The high energy release of the exploding gases produces the funnel- or bowl-shaped **crater** of the volcano and throws the molten material into the air. During flight it is torn into pieces, cools and solidifies. Large pieces of lava may rotate and be rounded into **volcanic bombs**. Particles between 4 mm and 32 mm in size are termed **lapilli** and particles less than 4 mm **volcanic ash**. Ash eruptions also throw out rock fragments from the walls of the volcanic neck. All the unconsolidated products of explosive

volcanic eruptions are combined under the terms **tephra** or cinder. When hardened into solid rock, they are **tuff**. Both are **pyroclastic sediments**.

Lava eruptions can be relatively quiet with the lava flowing out of fissures rather than craters. Mauna Loa and other Hawaiian volcanoes erupt from fissures. But a lava eruption can also follow an ash eruption if the lava fills the crater created by an ash eruption and eventually flows over its edge. Often tephra is torn loose from the crater rim which is lowered where the lava flows over, giving it a horseshoe shape.

Various types of eruptions and their associated landforms are described in the following sections (*see also* Fig. 4.4).

4.4.1 Maars

Volcanic eruptions that consist almost entirely of gas explosions force away the rock cover on the volcanic vent but bring little material to the surface. Instead of a volcanic mountain, a funnel-shaped explosive crater is produced. If its rim rises above its surroundings at all, it is formed of debris from the rock cover. There are a large number of these craters in the southern Eifel upland in Germany, some of which contain lakes known locally as **maars**. In geomorphology the term maar is applied to all craters of this type, whether they contain a lake or not. Most have been formed by a single, probably phreatic eruption. Continuous activity would normally bring enough material to the surface to make a cinder volcano.

4.4.2 Cinder cones

During an explosive eruption the coarse particles fall back close to the eruption area. If there is sufficient material and the eruption lasts long enough, a cone-shaped **cinder volcano**, also termed ash volcano, is built up. At its summit is a crater whose central axis is formed by the pipe or **vent** through which the eruptions occurred. Cinder volcanoes have solidified lava only in the vent or neck; the rest of the volcano is formed of unconsolidated material. Initially the angle of inclination of the cone's sides corresponds to the angle of internal friction for unconsolidated material, about 30–35°. Denudation may reduce

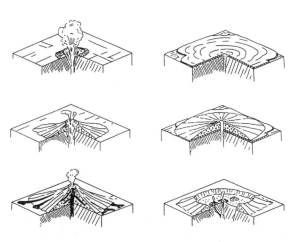

FIGURE 4.4 Types of volcano. Left, from above: maar, cinder volcano, strato-volcano; right, from above: plateau lava, shield volcano, caldera with small young cinder volcanoes (after Cloos from Wagner, 1960, p. 278).

FIGURE 4.5 Ship Rock, New Mexico, USA, a volcanic neck.

the inclination later if no further eruptions occur.

Cinder volcanoes can grow very rapidly. Paricutin in Mexico erupted in a maize field on 25 February 1943. By the following morning it had built a cone 40 m in height and within a year was more than 300 m high (Francis, 1976, p. 215). The cinder eruptions were accompanied by lava eruptions which, however, came out of fissures at the foot of the cone, not from the crater.

Pure cinder volcanoes are not active for long and do not remain as landforms for a long period. The unconsolidated material of a cinder volcano is removed rapidly by denudation after the end of activity. All cinder volcanoes that have been preserved originated in the Quaternary, within the last 1–2 million years; most are less than 150 000 years old. Only volcanic necks of resistant tuff and lava remain from pre-Quaternary cinder volcanoes (Fig. 4.5).

4.4.3 Strato-volcanoes

Strato-volcanoes, as their name suggests, are built up of alternating layers of pyroclastics and lava which slope down from the central axis of the volcano in all directions. None of the lava layers extends around the entire cone. Single lava flows have flowed down on different sides of the volcano at different times, and have been covered later by loose material and new lava flows.

The solidified lava gives the volcano a firmer structure and a higher resistance to denudation than a cinder volcano. Ash and lava eruptions may take place for a long time, so that some of these volcanoes reach considerable heights, forming cones whose slopes rise from a broad base and become steeper towards the summit. Mayon (2421 m) on the Philippine island of Luzon has an almost perfect cone shape (Fig. 4.6a) and Fujiyama (3776 m) in Japan comes close to this ideal form (Fig. 4.6b).

Vesuvius in Italy is also a strato-volcano but consists of several generations of volcanoes that have been superimposed on one another. Before the eruption of 79 AD, which buried Pompei and Herculaneum with ash, Vesuvius was thought to be extinct. The phase of activity begun at that time has continued until the present day. The eruption of 79 AD blasted the summit of the mountain away and left a crater 4 km in diameter which was created mainly by the explosion itself, but to some extent also by the collapse of parts of the crater walls. Subsequent eruptions built a new strato-volcano in the southern part of this huge crater which now forms the main summit (1277 m). With each eruption the shape of the summit changes. The last major eruptions occurred in 1872, 1906 and 1944. Part of the rim of the 79 AD crater forms a semicircular ridge on the northeast flank, Monte Somma. With a height

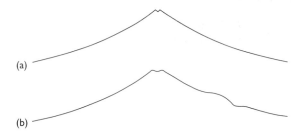

FIGURE 4.6 Profiles of (a) the volcano Mayon and (b) Fujiyama (after Francis, 1976, p. 225).

of 1132 m, it is only a little lower than the main summit (Fig. 4.7). A comparable explosion occurred in the State of Washington in the USA in 1980 when Mt St Helens erupted. The summit and north flank of the mountain were removed in the explosion and the height of the mountain was reduced from 2950 m to 2551 m. The eruption created a large horseshoe crater, open to the north, in which a new dome of lava is rising.

Large strato-volcanoes usually reach their height of a few thousand metres after a few million years. Even if there are some periods of relatively rapid growth, the mean long-term net rate of relief increase remains in the order of 1m/1000 years. Both Vesuvius and Mt St Helens show that the height of active volcanoes can fluctuate considerably depending on whether the summit has been blasted off or built up. Exogenic denudation has hardly any influence on the summit height during an active period. Only when volcanic activity has ceased does exogenic denudation progressively wear the mountain down, in the same way that denudation affects a non-volcanic mountain after an uplift phase has finished (section 3.3).

FIGURE 4.7 Vesuvius, Italy, a complex strato-volcano.

FIGURE 4.8 The shield volcanoes of Mauna Kea (in front) and Mauna Loa (behind), Hawaii, seen from the north. Note the asymmetrical distribution of the trade wind clouds.

4.4.4 Shield volcanoes

If the outlet for low viscosity lava is a point or a short fissure and new lava flows repeatedly from the same place so that radially diverging lava streams become superimposed on one another, a **shield volcano** develops. They are broad at the base and, because of the low viscosity of the lava and the virtual absence of cinders have a much lower slope angle than strato-volcanoes, generally 6–10° (Fig. 4.4). Mauna Kea (4210 m) and Mauna Loa (4168 m) on Hawaii are the largest shield volcanoes in the world (Fig. 4.8). The base of their lava lies 4000–5000 m below sea level so that the total thickness of the lava flows is about 9000 m. The entire island of Hawaii above sea level covers 10 400 km². Together with the base of the island below sea level, Hawaii forms an enormous single mass of lava. Iceland also has a number of shield volcanoes. In the geological past several regions of the world have been covered by extensive sheet eruptions of highly mobile, low-viscosity lava (flood basalt) which today form lava plateaus (section 4.3.3).

4.4.5 Calderas

The term **caldera** describes a volcanic hollow form produced either by a large explosive eruption, an **explosion caldera**, that leaves a crater with a diameter that far exceeds the measurements of ordinary craters (Fig. 4.9), or by the collapse of a volcanic summit into a magma chamber below the surface which has been emptied after an eruption, a **collapse caldera**. Both processes have been present in the formation of many calderas and it is not always possible to distinguish which has played the dominating role. The crater formed on Vesuvius in 79 AD was a caldera caused mainly by the explosive eruption, but apparently with some additional collapse. The Laach Lake in the Eifel in Germany resulted from the combined effect of an explosion and collapse. Its last eruption occurred

11 000 years ago. The product of the explosion, a light-coloured pumice, provides an index horizon in Quaternary deposits all over Germany (Meyer, 1986, pp. 431–435).

The caldera of Crater Lake in the State of Oregon in the USA has a diameter of 10 km. Eight thousand years ago a 3000 m high stratovolcano erupted. Part of the mountain was blasted away and the remainder collapsed into the empty magma chamber. Wizard Island in Crater Lake is a small new cone that has developed since.

4.4.6 Subvolcanic structures

Subvolcanoes are lava masses that have solidified underground but close to the surface of the earth. As the lava rises, it fills out hollow spaces in the surrounding rock. Lava that fills a rock fissure or joint is termed a **dyke**; lava that penetrates the bedding plane between two horizontal layers forms a **sill**. A **subvolcanic boss**
develops when lava rises in a neck and upwarps but does not break through the cover rock. The summit of the Drachenfels in the Siebengebirge near Bonn in Germany, is a subvolcanic boss of trachyte from which the cover has been eroded. Highly viscous lava with little gas content sometimes also reaches the surface and solidifies without spreading, forming a rounded knoll or **volcanic dome**. The Puy de Dome, Puy de Clersoux and Grand Sarcouy are all volcanic domes among the more numerous, well-preserved cinder volcanoes in the Auvergne in France (Fig. 4.10).

A **laccolith** is formed when lava flows along the bedding planes and pushes up the rock layers above the intrusion while the layers below remain unaltered. The cross-section of a laccolith resembles a plano-convex lens, with the convex side upward. On its margins the intrusion thins out laterally along the bedding planes. Denudation has generally removed the cover rock of older laccoliths. On the Colorado Plateau in the

FIGURE 4.9 Halemaumau crater, Hawaii, an explosion caldera.

FIGURE 4.10 Volcanoes in the Auvergne, France: left, the cinder volcano Puy de Pariou (1209 m); to the right, with rounded summits, several volcanic domes; in the background, the Puy de Dome (1464 m), a higher volcanic dome.

USA, there are a large number of laccoliths, including the Henry Mountains, the La Sal Mountains, Abajo Peak and Navajo Mountain.

4.4.7 Plutons

The masses of molten material that have solidified into plutonic rocks to form **batholiths** differ from subvolcanoes in several respects. Their magma has solidified at depths of more than 1 km so that the molten mass cooled much more slowly than lava near the surface. This meant that the time available for crystallization was greater and that the crystals are, therefore, much larger than those in volcanic or subvolcanic rocks. The spatial extent of batholiths is also much greater than subvolcanoes and, unlike laccoliths, they have no lower boundary but increase in width downwards. Small batholiths are termed **stocks**. The Brocken in the Harz Mountains in Germany is a granite stock from which the cover rock has been removed. The more than 4000 m high Sierra Nevada in California is a batholith that has been uncovered over an area of about 80 000 km^2. About half that size is the Idaho batholith which forms the Salmon River Mountains in Idaho. The intrusion of magma uplifted the cover rock and produced enough of an increase in denudation, in combination with isostatic uplift, for batholiths which had solidified at considerable depths, eventually to be exposed at the surface. The present height of many batholiths is due though to later tectonic uplift, unrelated to the original development of the batholith.

5

EXOGENIC FACTORS AND SYSTEMS

In contrast to the endogenic processes, which have their origin in the earth's interior, exogenic processes originate outside the solid earth in the water bodies, the atmosphere and indirectly in space. The main exogenic factors are the position of the sea level, the climates of the earth and climatic changes. In conjunction with the endogenic factors, they steer the exogenic process response systems that govern landform evolution.

5.1 EUSTATIC CHANGES OF SEA LEVEL

The **base level** of erosion is the lowest level to which a land surface can be worn down. In regions where all rivers ultimately drain into the sea, sea level is the base level. The position of the sea relative to the land is, therefore, of fundamental importance in determining the possible magnitude and intensity of denudation of the land surface. On many coasts the position of sea level relative to the land changes as a result of local endogenic crustal movements (Chapter 4). **Eustatic** changes of sea level are long-term changes in the height of the sea itself. The changes are by the same amount world-wide and take place simultaneously in all the oceans and in the marginal and shelf seas that are joined to the oceans by straits or channels. Eustatic changes have two causes:

1. changes in the hollow volume of the ocean basins;

2. changes in the amount of water present in the oceans and seas.

The volume of the ocean basins can change in two ways: **depositional eustasy** and **tectonic eustasy**. Depositional eustasy takes place when large amounts of sediment are transported from the continents and deposited in the oceans, thereby reducing the volume of the basins and raising sea level. This type of eustasy usually continues over a long period of geological time but at so low a rate that its influence is almost negligible. Tectonic eustasy occurs in conjunction with crustal movements of continental blocks, often associated with plate tectonics (section 4.2) which cause the opening, widening, narrowing or closing of ocean basins and also vertical movement of the ocean crust. The widespread occurrences of marine deposits on present-day land areas of the continents indicate that the total capacity of the ocean basins was smaller in the past and the sea level higher. Such **thalassocratic periods** (Greek *thalassa* = sea), during which the low-lying regions of the continental shelves were inundated more extensively than today, have occurred repeatedly during the earth's history. In the **geocratic periods**, the hollow volume of the ocean basins was greater and the sea level lower. Since the middle of the Tertiary, about 30 million years ago, there has been an increase in the volume of the ocean basins and a progressive lowering of sea level, although with some significant oscillations.

The geomorphologically most important variations in sea level have occurred during the two million years of the Quaternary period as a result of **glacio-eustatic changes** in the amount of water in the ocean basins, that were brought about by the repeated formation and subsequent melting of large masses of glacier ice on the land areas of the earth. At the present time, about 2 per cent of the earth's available water is contained in glacier ice, an ice volume of 26 million km³. During the Quaternary ice ages, the ice volume was about three times as great. If all the present-day glaciers melted completely, including the huge ice masses of the inland ice of Antarctica and Greenland, there would be a glacio-eustatic rise in sea level of about 60 m. At the maximum extension of the last ice age, which ended about 10 000 years ago, the sea level was about 100 m lower than at present. The glacio-eustatic changes were relatively rapid and had a major influence on the landform development in all coastal regions. When sea level was lower and the gradient of the newly exposed sea floor was steeper than the older land surface, the streams incised their courses more deeply near the coast; during the interglacials, when sea level was high, the lower sections of stream valleys became embayments and estuaries in which the streams deposited their sediments. The former high sea levels left beaches and shore platforms which are now **glacio-eustatic terraces** at various elevations above the present sea level.

5.2 THE MORPHOCLIMATE

All exogenic geomorphological process systems obtain their energy supply, directly or indirectly, from the sun's radiation. The spatial differentiation of the process systems is, therefore, closely related to the spatial differentiation of the climates. The subdivision of the world's climate into climatic zones based on the mean annual and monthly values of precipitation and temperature, is not very useful for geomorphology. Most geomorphological processes are made up of discontinuous process events that have little relationship to the totals and means of climatic data. A **morphoclimatology** oriented to geomorphology still has to be developed (Ahnert, 1987a; De Ploey *et al.* 1991).

The **morphoclimate** of a location or region is made up of the totality of those climatic characteristics that influence the type, frequency, duration and intensity of geomorphological processes at that location or in that region. Included in the climatic characteristics are the regimes of precipitation, temperature and wind.

5.2.1 Magnitude–frequency analysis of the precipitation regime

The pecipitation governs the water budget in and on the soil and affects the surface runoff, infiltration, soil moisture, groundwater recharge and spring discharge. In addition, the impact of raindrops on bare earth in a heavy rain can directly cause soil particles to move.

Precipitation amounts measured over months, seasons, or years primarily influence changes in the water budget of the soil in terms of soil moisture stored as a result of infiltration and in the groundwater budget. The moisture present in the soil and in the underlying rock are of significance for chemical weathering.

An assessment of the effectiveness of geomorphological processes such as stream erosion or the removal of soil particles by surface runoff depends, however, much more on the size, frequency and intensity of single precipitation events than mean values. **Magnitude–frequency analysis** is one method that identifies the morphoclimatic importance of these events quantitatively, especially the frequency of precipitation events of various sizes. It was originally used mainly to estimate the frequency of floods (Chow, 1964,pp. 8–29; Morisawa, 1968, p.4; Dury; 1969, pp.322–325) but can be applied to other climatic and hydrological events (Wolman and Miller, 1960; Ahnert, 1987a, 1988a).

The data used are discrete precipitation values of uniform, short time periods, usually daily rainfall. Because rainfall is generally not continuous for 24 hours and not of constant intensity all day, the daily precipitation data do not describe the actual duration and intensity of precipitation during the course of the day. They have, however, the advantage that daily rainfall is measured at numerous weather stations throughout the world and published in meteorological year books and are generally used when more detailed data are not available. Precipitation measurements for shorter time intervals that also allow a direct determination of intensity are limited in most countries to a few stations which are often far from the areas in which the data are required.

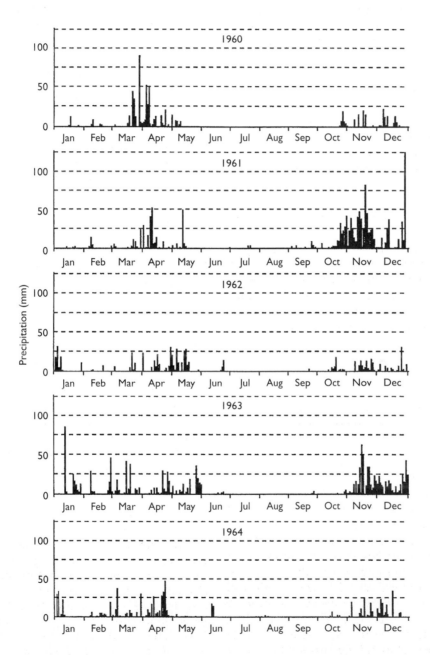

Figure 5.1 Diurnal precipitation in Machakos, Kenya, 1960–1964 (Ahnert, 1982).

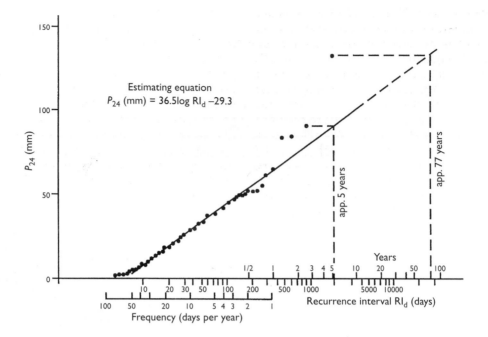

FIGURE 5.2 Magnitude frequency diagram of the diurnal precipitation in Machakos, Kenya 1960–1964 (Ahnert, 1982).

In Machakos, Kenya, data for the distribution of daily precipitation during the 5 years 1960–1964 are plotted in Fig. 5.1, and a magnitude–frequency analysis diagram is shown in Fig. 5.2 (Ahnert,1982). The data were ranked with $r = 1$ the highest value in the series. The recurrence interval RI for each value, that is, the time span after which an event of similar or greater size as the observed event was to be expected is:

$$RI = (N+1)/r \qquad (5.1)$$

where N is the total number of time units, in this case 5 years or 1826 days (days without rain are included) and r is the rank of the daily precipitation value.

Because this method only provides a rough estimate it is usually sufficient to give N in years, although the number of days also can be used instead. The difference in the result is small, particularly if the measurement period covers many years.

The precipitation values are plotted in a diagram as the function of the decadic logarithm of their respective recurrence intervals. They form

an almost straight line from which, generally, only the highest and lowest values may deviate (Fig. 5.2). The highest value deviates because the maximum value can occur in the measurement period by chance although, in reality, it may have a much larger recurrence interval. In Figure 5.2, the maximum value lies high above the rest of the points at the recurrence interval of 1827 days, about 5 years. If this highest rainfall had not occurred during the measurement period, then the present second highest value would have the recurrence interval of 1827 days and would lie on the straight line linking the other data points. At the lower end of the data series the values deviate from the straight line because they approach zero.

The straight line representing the data points for values from $P_{24} > 10$ mm can be quantified using a regression equation of the type

$$P_{24} = a + b \log_{10}(\text{RI}) \text{ (mm)} \qquad (5.2)$$

P_{24} is the precipitation (mm) per 24 hours. If RI is expressed in days, then the constant for Machakos is $a = -29.3$ and the coefficient $b = 36.5$; if the recurrence interval RI is represented in years, then $a = 64.2$ and the value of b

remains unchanged. The use of RI in years and of the decadic logarithms rather than natural logarithms has the practical advantage that the constant a in equation (5.2) corresponds directly to the precipitation of the occurrence interval RI = 1 year, for Machakos P_{24} 64.2 mm, because $\log_{10} 1.0 = 0.0$, and that the sum of $a + b$ indicates directly the precipitation with a recurrence interval RI = 10, namely, $P_{24} = 64.2 + 36.5 = 100.7$ mm, since $\log_{10} 10.0 = 1.0$. The two values a and b characterize, therefore, not only the entire regression line but also its one-year and ten-year values. Together they form a useful **magnitude–frequency index** (MFI) (Ahnert, 1987a). The index is written in brackets and consists of the regression constant a and the regression coefficient b of equation (5.2); the two numbers are separated by a semicolon. For Machakos the MFI is (64.2;36.5).

The index makes it possible to estimate the mean time intervals in which precipitation of a particular magnitude is expected to recur, or at which mean time intervals particular daily rainfall values would probably be exceeded, for example, the threshold value for the onset of soil erosion. The method offers a useful way to identify and predict the geomorphological capability of a precipitation regime. In the example used here the estimates are valid only for recurrence intervals of one year or more because the MFI in

equation (5.2) indicates nothing about the seasonal distribution of rainfall during the year. In Machakos, with its two relatively short rain periods (Fig. 5.1), this is particularly clear. For periods of less than a year it is more correct to use instead of the recurrence interval of precipitation of a particular magnitude, the frequency per year, which corresponds to the reciprocal of RI. Seasonal fluctuations can also be covered if separate magnitude–frequency analyses are carried out for the seasons or individual months.

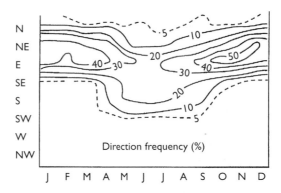

FIGURE 5.4 Seasonal distribution of wind directions, Nairobi, Kenya (Ahnert, 1982).

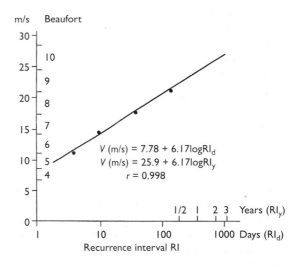

FIGURE 5.5 Magnitude–frequency diagram of wind force on the island of Norderney, Germany (Ahnert, 1987a).

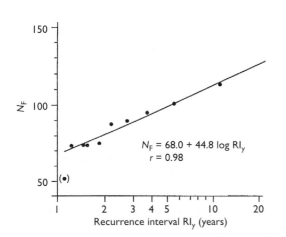

FIGURE 5.3 Annual frequency of the occurrence of freeze–thaw cycles in Munich (Ahnert, 1987a).

The monthly magnitude frequencies of stream discharges are shown in Fig.10.8, for example.

5.2.2 Temperature and wind regimes

Mean annual temperatures have often been used to distinguish morphological regions but they are not very suitable for this purpose. Mean monthly data are not much more useful. They allow only a rough differentiation, such as the approximate spatial distribution of permafrost or an estimate of the annual period during which the water temperature at the earth's surface remains above the freezing point. The latter is of significance for runoff and for chemical weathering which cannot take place without water in a fluid state.

Of greater importance are the mean monthly temperature ranges, the extreme temperatures reached and the frequency with which certain threshold temperatures are transgressed, especially the freezing point. The degree of frost and the frequency and duration of severe frost are also relevant. Magnitude–frequency analysis can be used to characterize these components of morphoclimate (Fig. 5.3).

The significance of wind depends on the temporal frequency of different wind directions and different wind speeds. Wind directions are often shown by a wind rose. Seasonal variation are more usefully expressed in a diagram showing the proportional share of different wind directions per month (Fig. 5.4). Magnitude–frequency analysis can be used to show the frequency of wind speeds above a particular threshold. (Fig.5.5).

5.3 THE MAJOR EXOGENIC PROCESS RESPONSE SYSTEMS

5.3.1 The systems of weathering, denudation and slope development

Exogenic processes shape the land surface by removing rock material from one location and depositing it in another. Before this can happen the hard bedrock must first be in a state in which it is removable. Weathering is the process that fragments, decomposes or dissolves parts of the rock so that it can be removed.

Weathering takes place at every point on the land surface. The type and intensity in a particular area depends in a complex way on the environmental conditions, the type of rock and other variables that are produced by the processes themselves, such as soil or debris cover on the bedrock. The totality of these components forms the **weathering process response system**.

Denudation is the process of **areal downwearing** of the surface and transport of the products of weathering, usually downslope in the direction of the steepest gradient. The intensity of denudation varies temporally and spatially. Spatial variations change the form of the slopes but are, at the same time, through feedbacks, dependent on the local form characteristics.

Denudation is in itself a process system and is, like weathering, a subsystem of the larger complex of processes, material, form characteristics and environmental conditions that make up the **system of slope development**. The slope development system is, in turn, a subsystem of the systems that are regionally effective: the **fluvial system**, dominated by streams, the **glacial system**, regulated by glaciers and the **littoral system**, influenced by the surf, coastal currents and tides.

The **aeolian system**, which is a largely independent exogenic system dominated by the wind, is an exception. The wind is areally effective and is, therefore, denudational but it cannot, like other denudation processes, be wholly included in slope development because it produces independent forms of its own, such as dunes, and is also able to transport sand and dust against the direction of gravity.

5.3.2 The fluvial system

By far the largest part of the earth's land surface is occupied by the catchment areas of streams. In reaction to the relief-increasing endogenic processes, the streams have incised their channels into the land and form the low lines of the surface. The gradients of the channels are continuously unidirectional, from the small creeks to the large rivers, until they reach the sea or an interior basin.

The crest lines on the watersheds are the lines of divergence and the stream channels are the lines of convergence of the surface gradients. Together they give the land surface a hierarchical spatial order as small streams flow into larger streams and small watersheds combine to form large watersheds.

The geomorphological ordering function of streams grows out of their process function. They are not only lines of incision, but also lines of collection and transport paths for the rock material prepared by weathering on the slopes and transported by denudation downslope. The streams continue to transport this material along their channels, together with the material eroded by their own downcutting. Where the gradient is insufficient for it to be carried further, the material is deposited.

The entire surface of the land in each catchment area can be understood as an interconnected process response system that is made up of the process components of:

1. debris production by the weathering of the bedrock;
2. the removal of debris or downwearing by denudative transport processes such as wash, landslides and slow mass movement that bring the debris to the stream at the foot of the slope;
3. the erosion, transport and sedimentation work of the stream itself.

Closely related to these process components by numerous causal connections and feedbacks is the nature of the material that is produced and transported and the development of the landforms on which all of this occurs. The forms themselves are a direct result of the process events and, in turn, have a considerable influence on the spatial differentiation of process types and their intensity.

Because of the dominating role of flowing water in this complex structure of processes, material and landforms, this process response system is termed the **fluvial system**.

5.3.3 The glacial system

Where the earth's surface lies beneath a thick and extensive layer of ice, as in the Antarctic and Greenland, the ice and its slow movement is almost the only factor changing the shape of the surface. Temperature fluctuations are generally very small at the base of the thick ice masses. Rock blocks are detached from the bedrock surface primarily because of local pressure differences in the weight of the moving ice, rather than by normal weathering processes.

On the margins of the inland ice, the ice flows as outlet glaciers in linear ice streams with speeds of up to several tens of metres per day. Frequently they follow the course of preglacial valleys originally formed by stream erosion. The erosive effect of the ice deepens and widens the valley, changing the stream valley form into a glacier trough with a U-shaped cross-section. The same is true for the valley glaciers of the high mountains, except that they are being fed by small plateau glaciers, ice fields and cirque glaciers lying above the snowline, rather than by an inland ice mass. Above the ice surface, on the flanks of outlet glaciers and valley glaciers, rise slopes on which weathering and denudation take place in the same way as on the slopes of non-glaciated valleys. The glaciers transport the debris received by denudation from the slopes down valley in much the same way as a conveyor belt.

In most areas where they have been active or are active now, **glacial process response systems** have existed for only relatively short periods of time, as the sequence of glacials and interglacials of the last ice age shows. This explains the widespread use of former river valleys by glaciers. Nevertheless, despite the short period of their effectiveness, glacial erosion, transport and deposition have made a considerable impact on the landscape.

5.3.4 The littoral system

A littoral is defined as the shore of a body of water, particularly of the seas and oceans. The **littoral process response system** is dominated by the effect of the waves, currents and tides upon the form development of the shore zone. Weathering and denudation do also take place on a sea cliff, for example, and in this respect a cliff is not very different from a rock wall in the mountains, except that the control of the undercutting of the cliff and the transport of the material is by the movement of sea water.

The fundamental difference between the littoral and the fluvial systems is that the effectiveness of the sea is related to the position of sea level. A stream erodes downwards if the land is uplifted, or deposits if the surface is lowered, without changing its horizontal position. On the shore, every vertical movement of sea level or of the land means that the processes of the littoral zone also shift their location horizontally: landwards if the sea level rises, seawards if it sinks. With every relative change in sea level, the formation of the littoral zone begins again at another location.

The present-day coastal landforms of the earth are at an early stage of development because the last eustatic rise in sea level ended only 6000 years ago.

5.3.5 The aeolian system

In the **aeolian process response system** the wind is the dominating medium for the erosion, transport and deposition of unconsolidated material. The wind's effectiveness depends not only on its strength but on the quantity and grain size of the unconsolidated material and the amount of vegetation cover. Deserts and ocean beaches are, therefore, areas of particularly effective aeolian erosion, transport and deposition. Wind erosion produces small forms mostly, such as blown out hollows and wind abrasion on rock faces and pebbles. Sand grains are moved along near the surface and form dunes. Finer dust is carried up high into the air and often transported over great distances before it falls to the ground.

The effectiveness of the aeolian system is less than either the fluvial or glacial systems. Its influence on the landforms in regions where the vegetation cover is sparse and where there are neither streams or glaciers is, nevertheless, considerable.

6

ROCK TYPES AND THEIR CHARACTERISTICS

6.1 ELEMENTS, MINERALS AND ROCKS

Rocks are the material that make up the earth's crust. The composition of the crust is dominated by only a few chemical elements. Nine elements account for 98 per cent of the crust, ranging from oxygen with 49 per cent to hydrogen with 1 per cent (Table 6.1). The first five elements in the list (O, Si, Al, Fe, Ca) add up to 86 per cent of the rock substance.

These and other rarer elements combine to form the solid **minerals**, usually in a crystalline form such as potassium feldspar ($KAlSi_3O_8$), quartz (SiO_2) and biotite or black mica $K(MgFe)_3$ $(OH)_2/AlSi_3O_{10}$). **Rocks** consist of minerals. Potassium feldspar, quartz and biotite, for example, are components of granite, a rock that occurs frequently.

The rocks themselves are the material product of endogenic and exogenic processes in the past. Different rocks weather differently under the same environmental conditions. For this reason, knowledge of rock types is a prerequisite for an understanding of weathering. There are three groups of rocks:

1. igneous rocks;
2. sedimentary rocks;
3. metamorphic rocks.

6.2 IGNEOUS ROCKS

6.2.1 Types

Igneous rocks form when a molten mass, either magma underground or lava on the surface, cools and solidifies. Lava usually contains fewer substances than magma because the gaseous components in the lava are released into the

TABLE 6.1 Composition of the upper 16 km of the earth's crust

Element	Mean content in the rocks (%)
Oxygen (O)	49.13
Silicon (Si)	26.00
Aluminium (Al)	7.45
Iron (Fe)	4.20
Calcium (Ca)	3.25
Sodium (Na)	2.40
Potassium (K)	2.35
Magnesium (Mg)	2.35
Hydrogen (H)	1.00
Titanium (Ti)	0.61
Carbon (C)	0.35
Chlorine (Cl)	0.20
Phosphorus (P)	0.12
Sulphur (S)	0.10
Manganese (Mn)	0.10
Others	0.39
	100.00

Source: Pirsson and Knopf (1958, p. 13).

atmosphere when it reaches the surface. In a magma below the surface, the gases remain in the molten material and become components of the minerals when the magma solidifies. Rocks formed at depth are termed **plutonites** and those produced by the solidification of lava, **volcanic** rocks.

Magma is well insulated below the surface and cools only slowly. The molecular movement during its liquid phase of cooling continues for a long time so that a long period is available for the growth of crystals. Plutonites tend to have a **phaneritic** (Greek *phaneros* = visible) texture with crystals large enough to be seen with the naked eye. **Granite** is the most common plutonic rock (Fig. 6.1); others occurring frequently are **syenite**, **diorite** and **gabbro**.

Volcanic rocks solidify much more rapidly. Their texture is therefore mostly **aphanitic** (Greek *aphanes* = not visible), made up of crystals that are too small to be seen without a magnifying glass. The most common volcanic rock is **basalt** (Fig. 6.2); others are **rhyolite**, **trachyte**, **phonolite** and **andesite**. If cooling is too rapid for recrystallization to takes place, the lava solidifies to a **volcanic glass**. **Obsidian** forms in lava that contains little or no gas and **pumice** where there is a large amount of gas present. Like manufactured glass, volcanic glasses are amorphous.

Porphyries occupy an intermediate position between phaneritic and aphanitic rocks. The name comes from their colour, which is usually reddish, (Greek *porphyra* = purple). They are

FIGURE 6.2 Basaltic lava with flow structure in Hawaii.

characterized by coarse crystals in a matrix of fine crystals and develop in a two-stage solidification process in which the large crystals are formed before the rest of the molten mass becomes solid.

6.2.2 Chemical and mineral composition

Based on their chemical composition, igneous rocks can be placed on a scale from acid to basic. Acid igneous rocks contain a large share of minerals with a high SiO_2 content, particularly orthoclase or potassium feldspar ($KAlSi_3 O_8$) and quartz (crystalline SiO_2). These minerals are light in colour and have a low specific gravity (potassium feldspar, 2.5–2.6; quartz, 2.65).

Basic rocks are dominated by heavy minerals containing iron and magnesium which have a low SiO_2 content and high specific gravity. Pyroxene $Ca(Mg,Fe)(SiO_3)_2$, for example, has a specific gravity of 3.0–3.6, hornblende $Ca(Mg,Fe)_3 (SiO_3)_4$, 2.9–3.4 and olivine $(Mg,Fe)_2SiO_4$, 3.2–3.6. They are usually dark coloured.

Each column in Table 6.2 shows rocks, from plutonites to aphanitic rocks, with the same or a very similar chemical and mineral composition. The aphanitic rock rhyolite is in many respects the volcanic equivalent of the phaneritic granite. Similarly, trachyte corresponds to syenite, andesite to diorite and basalt to gabbro.

Porphyries are named after the texture of the matrix in which the crystals are embedded. If the

FIGURE 6.1 Granite in Tanzania.

matrix is phaneritic then the porphyry is named after the chemically and mineralogically similar plutonite, granite porphyry, for example. If the matrix is aphanitic, the porphyry is named after the volcanic rock with a similar composition, such as rhyolite porphyry.

In Table 6.2, the rocks are arranged in sequence from acid rocks on the left to basic rocks on the right, with the exception of amorphous rocks and tephra. **Ultrabasic rocks** (peridotite, pyroxenite and hornblendite) occur only rarely at the earth's suface and are not included.

Unconsolidated volcanic material, **tephra** (section 4.4) is composed chemically and mineralogically of lava but is a deposit and therefore a pyroclastic (Greek *pyr* = fire, *klastikos* = broken) sediment. When hardened it becomes **tuff**.

6.3 SEDIMENTARY ROCKS

6.3.1 Sediment

Sediments are anorganic and organic material, deposited for the most part under water but also under glacier ice or, in the case of dune sand, on dry land. Most sediments are the products of weathering or the remnants of rocks that have been weathered and transported to a lower-lying area. The sediments become hardened into sedimentary rocks by processes termed collectively

diagenesis. The hardening process takes place either as a result of pressure produced by the layers of sediment lying on top of one another or through the cementing of the deposited grains by various bonding agents.

6.3.2 Clastic sedimentary rocks

6.3.2.1 Formation and composition

Clastic sediments consist of fragments or hard particles remaining after rocks have been weathered. Depending on their grain sizes, the sediments are defined as **clay**, **silt**, **sand**, **gravel** or **coarse debris** (Table 6.3). The names of the clastic sedimentary rocks reflect the sediments they are formed of: **claystone**, of which **shale** is a platy version, **siltstone** and **sandstone**. The latter is composed largely of quartz grains. **Conglomerates** are sedimentary rocks made of rounded coarse material; if the coarse component is angular, the rocks are **breccia**.

Arkose and **greywacke** are special types of sedimentary rocks. In grain size they are similar to sandstone but their chemical composition differs. Arkose has a high proportion of chemically unweathered feldspar fragments. Because feldspar weathers easily to clay in humid climates, arkoses are an indication that dry conditions dominated when they were formed. Greywacke generally contains, besides quartz grains, more than 50 per cent other mineral grains and rock

TABLE 6.2 Most common igneous rocks

Colour:	Acid Light			Basic Black
SIO$_2$ content:	>70%	~60%	~55%	<50%
Specific weight:	2.6–2.75	2.6–2.8	2.8–2.9	2.9–3.3
Plutonite, phaneritic	Granite	Syenite	Diorite	Gabbro
Porphyry:				
(a) Phaneritic	Granite	Syenite	Diorite	Gabbro
matrix	porphyry	porphyry	porphyry	porphyry
(b) Aphanitic	Rhyolite	Trachyte	Andesite	Basalt
matrix	porphyry	porphyry	porphyry	porphyry
Vulcanite, aphanitic	Rhyolite	Trachyte	Andesite	Basalt
Amorphous	Pumice, volcanic glass, e.g. obsidian			
Tephra, pyroclastic		Unconsolidated: ash, lapilli, bombs		
		Consolidated: tuff		

TABLE 6.3 Clastic sediments and sedimentary rocks

Characteristic grain size (mm)	Sediment	Sedimentary rock
>0.002	Clay	Claystone
		Shale
0.002–0.063	Silt	Siltstone
0.063–2.0	Sand	Sandstone
		Arkose (with feldspar)
		Greywacke
		(with various minerals and rock fragments)
>2.0	Gravel, pebbles (rounded)	Conglomerates
	Coarse fragments (angular)	Breccia

fragments. As its name implies, it is grey in colour.

6.3.2.2 Cementing agents

The most common cementing agents in sedimentary rocks are clay particles, **quartz** (SiO_2), **iron oxide** (Fe_2O_3) and **calcium carbonate** ($CaCO_3$). They are in part already present during sedimentation, in part added later, having been precipitated out of water circulating in the pores of the sediment or produced by the weathering of minerals deposited in the sediments. The cementing agents vary in their vulnerability to attack by weathering. Also, the more cementing agent a rock contains, for example sandstone, the smaller the pore space remaining between the quartz grains and the lower its perviousness. For this reason, the type and quantity of cementing substance has a major influence on the weathering and resistance of the rock.

6.3.2.3 Colour and bedding

Most clastic rocks, particularly sandstone, conglomerates and breccias, are deposited on land surfaces, in shallow water areas of lakes, in rivers and in shallow coastal waters. If deposition took place in water, it was usually oxygen rich because of currents and turbulence and the iron compounds in the sediments were oxidized. This is the cause of the yellow, reddish and brown colouring of many sandstones, conglomerates and breccias. Subsequent seepage of water containing iron oxide into the pores and joints contributed to this colouring.

As the currents in which the sediments are being deposited shift, the grain size of the sediments being deposited changes. In areas of shallow still water, fine sediment is laid down but in a neighbouring area of faster-flowing water, only coarse particles are deposited. Subsequent shifting of the line of highest current velocity can result in coarse material being deposited on top of fine material and fine material on coarse deposits. Lateral shifting of currents in streams, erosion and redeposition on gravel bars and sand banks produces bedding locally inclined at varying angles and in varying directions; this kind of stratification is known as **cross-bedding**.

Cross-bedding also occurs in dunes and aeolian **sandstones** where sand transported by the wind is sorted according to the dominating wind force and, if the wind changes direction, may be deposited in alternating directions (Fig. 6.3). In general, the layers in aeolian sandstones dip predominantly to the lee side so that the wind conditions at the time of deposition can be estimated from this type of rock. Unlike sandstones deposited in water, aeolian sandstones have no clay and gravel components.

Clay particles sink to the bottom in stagnant water, although in large quantities only at some distance from the shore where the conditions for sedimentation remain constant for long periods over large areas.There is no cross-bedding in clays and claystones (Fig. 6.4) and layering only if clay and silt are deposited alternately. Clays are usually grey. Red and yellow colouring is less frequent than in sandstones.

6.3.3 Limestone, marl and dolomite

6.3.3.1 Formation and composition

Limestone is made up predominantly of calcium carbonate ($CaCO_3$). It is formed either by precipitation from solution, partly in conjunction with the metabolism of organisms, or by the deposition of the calcareous remains of skeletons and shells, particularly of sea animals such as mussels, snails, sea lilies, corals and foraminifera. Former coral reefs become **reef limestone**.

Marl is a mixture of limestone and 35–65 per cent clay. Rocks with a content of 65–75 per cent limestone and 25 per cent or more clay are called marly limestone. Clayey marl has a clay content of 65–75 per cent and marl clay more than 75 per cent clay.

Dolomite is a calcium magnesium carbonate ($MgCa(CO_3))_2$ that is similar to limestone but generally more resistant. It has usually developed from a limestone that has been altered by solutions containing magnesium.

6.3.3.2 Colouring, bedding and hardness

Limestones and dolomites range in colour from white, chalk for example, to dark grey (Fig. 6.5). Yellow and red tones that would indicate the presence of iron oxide occur only rarely. Most limestones were deposited in the sea under conditions that were the same over large areas and are usually evenly stratified so that it is possible to follow layers containing particular fossils over great distances. Very thick, uniform beds of limestone have formed where conditions were constant for a very long period. In shallow seas, near

FIGURE 6.3 Aeolian sandstone (Triassic Fort Wingate Sandstone) near Fort Wingate, New Mexico, USA.

FIGURE 6.4 Claystone in the Painted Desert, Arizona, USA.

land, deposition conditions alternated frequently with the result that thinner limestone layers are often interbedded with clay or marl. The clay particles have been washed into the sea from the land (Fig. 6.6). Reef limestones are unlayered and are generally more resistant than stratified limestones.

It would seem that the hardening process in limestones progresses through geological time, long after their original formation. Limestones formed in the Palaeozoic seas 230 millon years ago are generally harder and more resistant than the limestones formed in the Cretaceous, 145–65 million years ago, and considerably harder than the Tertiary limestones formed 65–2 millon years ago.

6.3.3.3 Difference between limestone and dolomite in the field

Limestone and dolomite are easily confused in the field because of their similar appearance. A simple way to distinguish them is the application of hydrochloric acid (HCl), diluted to 10 per cent. If the acid is dripped on limestone it fizzes very noticeably, as calcium chloride ($CaCl_2$), carbon dioxide (CO_2) and water (H_2O) form:

$$CaCO_3 + 2HCl \rightarrow CaCl_2 + CO_2 + H_2O$$

By contrast, dolomite fizzes in this test only when the rock surface has been ground to a fine powder with a hammer or knife, and then only weakly.

6.3.4 Other sedimentary rocks

Organic and biochemical deposits such as coal and strata containing oil and gas are of little geomorphological significance.

Evaporites are salts formed by the evaporation of solutions, usually the evaporation of sea water, but also in arid regions following the drying out of lakes and the capillary rise of solutions in the soil. They include **anhydrite** ($CaSO_4$), **gypsum** ($CaSO_4 \cdot 2H_2O$), **halite** (NaCl) and **potash** (KCl).

These salts become plastic under pressure within the earth's crust. Originally deposited in horizontal layers, they have been pressed into zones of weakness in the layers above following tectonic compression in the region, and form **diapirs**, also known as **salt domes** or **salt stocks**. Diapirs are highly soluble in water and are seldom encountered at the surface in humid climates; in extremely dry climates they persist for a long time as surface forms and in some cases rise to considerable heights.

6.4 METAMORPHIC ROCKS

Metamorphic rocks are former igneous or sedimentary rocks that have been altered by high pressure and/or high temperatures. For example, tectonic subsidence may cause rocks that were formerly near the surface to be brought to depths in the crust where the temperatures are very high. As the rocks subside, they are covered by other deposits which cause an increase in the pressure. Pressure and heat together result in the development of parallel pressure surfaces or foliation in the rock. If the heat becomes great enough, the rocks may melt partially or completely and new minerals form by recrystallization.

Depending on the original rock and conditions of temperature and pressure, different types of metamorphic rocks develop. **Foliated metamorphic rocks** were formerly composed of a variety of mineral components or formed new minerals during metamorphism; their minerals are arranged in parallel layers, foliation planes, which are oriented at right angles to the direction of pressure.

Unfoliated metamorphic rocks either consist essentially of a single type of mineral which was not suitable for a foliated structure, or the metamorphism took place without a significant pressure component and only under the influence of

FIGURE 6.5 Chalk cliff with marine stacks, Ballard Point, Dorset, England.

high temperatures and the addition of solutions and gases from a cooling mass of magma nearby. The term **contact metamorphism** refers to a metamorphism which is limited to the contact zone with a magma, to distinguish it from the more extensive **regional metamorphism**.

6.4.1 Foliated metamorphic rocks

In the upper part of the crust the metamorphism is weak and mechanical pressure plays a greater role than temperature. Foliation planes develop as a result of the pressure and claystones in this zone are altered to **slate** (Fig. 7.7). The sheen on slate surfaces indicates that clay particles in the original rock have been changed to tiny flat mica crystals. The crystals, too small to be seen by the

FIGURE 6.6 Limestone layers (light coloured) with intermediate clayey layers (dark coloured) shattered by folding, in the Purbeck Beds (upper Jurassic), Stair Hole near Lulworth, Dorset, England.

naked eye, are arranged in parallel layers, the foliation planes, and the combined effect of their flat crystal surfaces heightens the reflection of light on the foliation planes. Because of this sheen which can be seen on houses roofed in slate, the tiny mica crystals are called sericite from the Latin word for silk. Stronger metamorphism with higher temperatures creates a thin layered **phyllite** out of the slate (Fig. 6.7). This has a much larger mica content and partial recrystallization of other minerals. Still more intensive metamorphism at greater depth in the crust produces **mica schist** in which the individual mica crystals are visible on the foliation planes. Mica schist often also has quartz and other minerals in the parallel structure of the foliation.

Extreme high pressure and high temperatures deep in the earth's crust produce **gneiss** in which there has been a more or less complete melting followed by recrystallization. Gneisses closely resemble plutonites in their mineral composition and size of crystals. Because of the high pressure, however, the crystals in gneiss lie parallel to one another in a foliated pattern (Fig. 6.8).

Gneiss formed from igneous rocks is termed **ortho-gneiss**, gneiss from sedimentary rocks **para-gneiss**. Geomorphologically there is little difference. Granite is changed directly into ortho-gneiss, while basic rocks, such as gabbro, have an intermediate metamorphic phase of green slates.

Slate, phyllite, mica schist and gneiss together form the group of **schistose rocks**. They split more easily along their foliation planes than in any other direction. This cleavage provides surfaces that can be attacked by weathering.

6.4.2 Unfoliated metamorphic rocks

Quartzite and **marble** are the two main types of unfoliated metamorphic rocks. Quartzite results from the metamorphism of sandstone whereby the quartz in the sandstone is recrystallized and, in general, also fills out the pores. Any clay particles that were present become mica crystals. The fracture surfaces of quartzites are smoother than sandstones because they go through the

FIGURE 6.7 Phyllite in Tanzania.

quartz crystals. In sandstones the fracture surface goes around the grains and feels rougher because the grains are relatively harder and less easily fractured than the cementing agent.

Sedimentary quartzite is a non-metamorphic rock formed from sandstone by the later filling in of pore spaces with silica (SiO_2). Tertiary quartzites in western Europe belong to this type, forming large blocks in the Forêt de Fontainebleau near Paris, for example, also the sarsen stones of the Salisbury Plain in southern England and residual blocks of Tertiary age on the Cretaceous plateaus on the border of Germany and the Netherlands. Metamorphic quartzites are light grey to white in colour and sedimentary quartzites often yellowish brown to brown. In some cases they can be distinguished only in thin sections under a microscope.

Some **marbles**, such as the white marble of Carrara in Italy, resemble quartzite quite closely. Marble is formed by the metamorphism of limestone. The calcium carbonate in the limestone is recrystallized as **calcite** and has a hardness of 3 on the Mohs hardness scale which ranges from 1 (talc) to 10 (diamond). It is, therefore, a relatively

soft mineral, easily scratched with a knife, compared to quartz with hardness 7. Marble also fizzes as strongly as limestone in a test with diluted hydrochloric acid (section 6.3.3.3).

6.4.3 The effects of contact metamorphosis

The metamorphic rocks described in the previous sections are the result of regional metamorphosis and often extend over large areas. The effect of contact metamorphism is, by contrast, limited to the vicinity of solidifying molten magmas. The magma heats the neighbouring rock. Gases, hydrothermal solutions and part of the molten magma penetrate and alter the neighbouring rocks. Pressure plays only a small role. The metamorphosed rocks surround the magmatic mass in a **contact zone**. The contact zones are often rich in ores which are precipitated out into fissures in the rocks as **dykes**. Rocks rich in quartz, such as sandstones, are hardly changed

FIGURE 6.8 Gneiss in the Lötschental, Switzerland.

by contact metamorphism, except for the precipitation of new materials in the form of dykes. Shale changes to hornfels in the inner part of the contact zone. As in regional metamorphism, limestones become marble.

The rocks directly below lava sheets as well as above and below sills can also be altered by contact metamorphism. Clayey rocks are especially affected by the heat, acquiring a brick-like character and reddish colour.

7

THE WEATHERING SYSTEM

7.1 THE TERM WEATHERING

In geomorphology, the term weathering has three functional aspects:

1. the effect of atmospheric processes on rocks and minerals;
2. the adaptation of the rocks to the environmental conditions at the earth's surface.
3. the preparation of rock material as a prerequisite for removal by processes of denudation or erosion.

Ollier (1969) and Yatsu (1988) have written comprehensive studies of weathering, its processes and its geomorphological significance. Brunsden (1979) has made a detailed review.

7.1.1 Weathering as the effect of atmospheric processes on rocks and minerals

Weathering includes all processes involving direct or indirect changes in inorganic and some dead organic (shells, coal) substances that are influenced by the weather. Important are temperature fluctuations, the formation of ice, humidity and its fluctuations and the chemical effects of substances dissolved in rainwater, soil water and groundwater. The changes brought about that are of significance in geomorphology are changes in the rocks and the soil material. There are two types of weathering: **mechanical weathering** and **chemical weathering**.

Mechanical weathering includes the processes that change the **state** of the rock, for example, its grain size or its cohesion. The composition of the

material remains unaltered. Chemical weathering processes bring about changes in the **composition** of the rock material by decomposition or corrosion of the substances and the development of new compounds. Part of the decomposed material is removed in solution.

In nature, mechanical and chemical weathering often affect the same rock simultaneously and complement one another. Mechanical weathering produces and enlarges fissures, for example, in which decomposition by chemical weathering can then take place. Chemical changes in rocks also reduce cohesion which contributes to the physical disintegration of the rock.

7.1.2 Weathering as the adaptation of the rocks to the environmental conditions at the surface of the earth

Most rocks have been formed under temperature, humidity and pressure conditions that differ greatly from those prevailing at the earth's surface. Granite, for example, is solidified magma which cooled very slowly at great depth. Its chemical and mineral composition reflects the conditions at depth and, as long as the granite remains there, it does not change. But if, following the denudation of the rocks above it is exposed at the land surface, it is subjected to variable temperature, humidity and pressure conditions to which it is not yet adapted. Mechanical weathering processes soon attack the newly exposed surface, accompanied by chemical decomposition. A layer of unconsolidated material, **regolith**, is produced that covers and protects the underlying unweathered rock from further weathering. The weathered material of the regolith is more stable and, therefore, better adapted to the atmospheric conditions than the original unweathered rock.

All weathering is an adaptation of rock types to new environmental conditions. Only the degree of the adaptation varies. Volcanic lava solidifies at the earth's surface rapidly and without the influence of the high pressure that is present with plutonic rocks but it, too, is decomposed chemically by the effects of rainwater and broken up mechanically by the effects of temperature fluctuations, especially frost.

The conditions under which sedimentary rocks have formed also differ from the environment at the land surface in which their weathering takes place. Marine limestones, such as the Jurassic limestones in Germany or the chalk in southern England, were laid down on the sea bottom under conditions of constant humidity and almost constant temperatures. At the land surface, they are attacked by mechanical weathering processes, particularly frost weathering, and decomposed by carbon dioxide (CO_2) that is present in solution in the rain, soil water and groundwater.

Sandstones were formed under conditions that more closely resembled those at the land surface than limestones. Dune sands and river sands form sandstones when their grains are cemented by a secondary cementing substance. The sand grains themselves are the product of earlier rock weathering and less vulnerable to change than the cement between the grains, which is attacked mechanically and chemically so that the grains become unconsolidated sand again. A sand grain lying today in a river bed, on a beach or on a dune may have been a component of rocks many times in earth history and may have been weathered, removed and redeposited as sand from which, either by diagenesis or by metamorphosis, new rocks were formed.

Siltstone and clays are also composed largely of the stable remnant particles of earlier rock weathering and are not as vulnerable to as many different weathering processes as igneous rocks or many crystalline metamorphic rocks. In general, a rock is more easily weathered and can be attacked by a greater variety of processes, the more complex its chemical and mineral composition and the more the conditions under which it was formed differ from those at the land surface.

Weathering also affects the stones in buildings. Such changes are more conspicuous than on natural rock faces because they are seen more often. The stone surface of a carving is well known and any alteration soon observed. A wide variety of stones have been used to build the cathedral in Cologne in Germany and they are weathering at different rates (Fig. 7.1).

7.1.3 Weathering as preparation of the rock for removal

Processes that wear down the rocks rarely affect fresh bedrock in its unaltered state. Exceptions are landslides, abrasion of pebbles on the stream bed and on rocky wave cut platforms, wind erosion by sand particles and glacial abrasion.

More commonly, denudation or erosion processes remove loose material, either still unconsolidated sediment, or regolith produced from bedrock by weathering. The energy required for removal of this material is much less than that required for downwearing of the fresh bedrock. The geomorphologically most important function of weathering lies in this transformation of the rock from a coherent mass to a regolith and its preparation of the rock for possible removal.

7.2 WEATHERING AS A PROCESS RESPONSE SYSTEM

The intensity of weathering of a particular type of rock at a given location depends on:

1. the specific resistance of the rock, that is, its physical hardness and its chemical and mineral composition,
2. the characteristics of the morphoclimate at that location;
3. the degree to which the relevant atmospheric influences have access to the rock.

Rock resistance is discussed in connection with the individual weathering processes. The relevant morphoclimatic factors and the degree

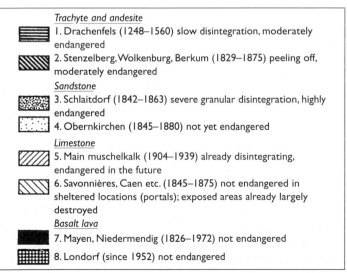

Trachyte and andesite
1. Drachenfels (1248–1560) slow disintegration, moderately endangered
2. Stenzelberg, Wolkenburg, Berkum (1829–1875) peeling off, moderately endangered
Sandstone
3. Schlaitdorf (1842–1863) severe granular disintegration, highly endangered
4. Obernkirchen (1845–1880) not yet endangered
Limestone
5. Main muschelkalk (1904–1939) already disintegrating, endangered in the future
6. Savonnières, Caen etc. (1845–1875) not endangered in sheltered locations (portals); exposed areas already largely destroyed
Basalt lava
7. Mayen, Niedermendig (1826–1972) not endangered
8. Londorf (since 1952) not endangered

FIGURE 7.1 The most important types of building stone in Cologne Cathedral, Germany, and their vulnerability to weathering (Wolff, 1976).

of accessibility are discussed as components of a general weathering process response system.

7.2.1 Morphoclimatic factors and their effect on mechanical weathering

Dominant for mechanical weathering is the effect of **fluctuations in temperature**, especially freeze and thaw which generates volume changes in the rock near the surface This is known as **thermal weathering**. When water freezes, its volume in cracks and pores in the rock also changes and **frost weathering** takes place. Both these kinds of volume change lead to the disintegration of the bedrock.

7.2.1.1 Thermal effect: expansion and contraction of the rock

Alternate warming and cooling of the rock surface expands and contracts the volume of the rock by a small amount. The range in temperature of a rock heated by the sun during the day that cools at night under a clear sky is greater than the range in temperature in the air above the rock. A temperature fluctuation of 40°C in the rock is not uncommon. In deserts the ranges are even greater. Various sources give values of 85°C to almost 100°C for the highest ground temperatures on dark, and therefore highly heat-absorbent, rocks in the Sahara (Barth, *et al.*, 1939) and night temperatures of 20°C or less. Thunderstorms also cool the rock down to similar values, only much more rapidly.

The **linear expansion coefficient** of most rocks lies between approximately 5×10^{-6} and 25×10^{-6}. This means that a block of rock 2 m long heated by about 50°C expands, depending on the rock type, by 0.5 mm to 2.5 mm. If it cools by 50°C, the rock contracts by the same amount.

Within the rock, but close to its surface, the range of temperature fluctuation is considerably smaller, declining exponentially rather than linearly with depth. The same is also valid for the decline in the daily temperature range in regolith and for the number of freeze–thaw changes. Figure 7.2 shows this functional relationship graphically. Curve 1 shows the daily fluctuation of temperature in regolith in south Germany,

$$\Delta T_c = 19.58e^{-0.064C} \tag{7.1}$$

curve 2, the daily temperature fluctuation of temperature in bare rock in the Mojave Desert in the USA,

$$\Delta T_c = 22.69e^{-0.051C} \tag{7.2}$$

and curve 3, the number of freeze–thaws in the regolith in the period January – March 1973 near Clermont-Ferrand, France,

$$F_c = 35.54e^{-0.041C} \tag{7.3}$$

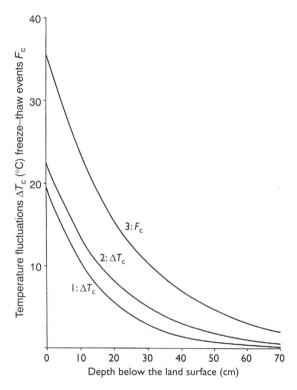

Figure 7.2 The decline in the daily temperature fluctuations and the freeze-thaw frequency with increasing depth below the surface. *Curve 1*: July soil temperature fluctuations in the regolith on a south slope in south Germany, based on data from Goetz *et al.* (1971). *Curve 2*: temperature fluctuations on a bare rock in the Mojave Desert, California, based on data from Roth (1965). *Curve 3*: Frequency of freeze-thaw, January–March 1973, as a function of depth below the surface, based on data from the weather station at Aulnat Airport, near Clermont-Ferrand, France.

where C is the depth (cm) below the surface, ΔT_c is the daily fluctuation of temperature (°C) and F_c is the absolute number of freeze–thaw cycles.

Because of the exponential decrease in the temperature fluctuations with depth below the surface of the rock, the expansion or contraction of the rock a few centimetres below the surface is much smaller than at the surface itself. Tension is created close to the surface, and with each change in temperature the cementing material in the rock is weakened and the rock eventually disintegrates.

In climates in which ground frost occurs, there is a direct relationship between the frequency of the freeze–thaw and the size of the temperature fluctuations. The greater the latter, the more probable it is that the temperature change reaches across the freezing point and causes a freeze–thaw event. The frequency of freeze–thaw in the soil declines, therefore, exponentially with depth in the same way as the temperature fluctuates. About half a metre below the surface the daily temperature fluctuations have disappeared almost completely. Annual fluctuations of temperature do still occur to depths of 5–10 m.

Temperature fluctuations and freeze–thaws influence the intensity of many mechanical weathering processes. A bare rock face is most affected by fluctuations and is weathered mechanically much more intensively than the same rock under a layer of debris or soil. The thicker the overlying layer, the more the underlying rock is protected from mechanical weathering processes. The weathering effect of temperature fluctuations on particles of rock debris (boulders, gravel, pebbles) declines with a decrease in the size of the rock fragments, even at the surface. There are two main reasons for this.

1. Small rock fragments have a relatively large surface area in relation to their mass so that the temperature of the fragments and of the air between them is equalized more rapidly. Their internal temperature differences and consequently the internal shear stresses are, therefore, smaller than in larger fragments.
2. Because the fragments are unconsolidated, they expand and contract individually into the surrounding hollow spaces (pores) without greatly affecting the bulk volume of the mass.

7.2.12 Frost shattering, hydrofracturing and frost heaving

Temperature fluctuations are particularly effective if they cross the freezing point and humidity is also present. Water increases its volume by 10 per cent when it becomes ice. The pressure created in pores and fissures near the surface that contain water causes **frost shattering**. Single grains and entire boulders are broken off by this process and large boulders are fragmented into smaller pieces. Depending on their hardness and porosity, rocks react differently to the pressure produced by the formation of ice. Laboratory experiments at the Centre de Gèomorphologie, Caen, France, have confirmed this (Lautridou, 1971; Lautridou and Ozouf, 1982).

Temperature changes in the soil or rock are brought about by temperature changes at the land surface so that water freezes first near the surface, then progressively at greater depth. When the fissures and pores are filled with water that freezes rapidly, the expanding ice exercises an increasing hydrostatic pressure, or **cryostatic pressure**, on the still liquid water beneath. The pressure in the unfrozen water is transferred with equal intensity, similar to a hydraulic press, through all the interconnected hollow spaces. The shattering effect of frost can, therefore, reach far deeper down into the rock than the freezing itself. This process is termed **hydrofracturing** (Selby, 1982, pp. 16–18) to distinguish it from the pure frost shattering.

If the soil is not saturated and once the water present in the soil is frozen, the water vapour circulating in the pores that are still open and in direct contact with the ice condenses and freezes. This takes place because the saturation water pressure of water above ice is lower, and condensation occurs, therefore, at lower humidity, than above water or wet soil. Ice lenses develop which push up the overlying layers of soil, a phenomenon known as **frost heaving**. It can often be observed on asphalt roads built on a foundation of gravel in which there are a large number of pores. Frost heaving in natural soils occurs most frequently in polar and subpolar climates (section 8.5.6). Ice lenses can also develop in joints that lie parallel to the surface and produce mechanical pressure effects.

7.2.1.3 Salt shattering

The water in pores and fissures contains dissolved salts from chemical weathering processes. If it evaporates, the salts precipitate out in crystalline form. The pressure produced by crystallization is increased when the growing crystals become hydrated (section 7.4.2). Repeated soaking and drying out further increases the pressure which widens cracks and loosens mineral grains. In its mechanical effect, salt shattering resembles frost shattering. It is mainly significant in frost-free climates with pronounced dry seasons.

7.2.1.4 Swelling, shrinking and slaking

Water supplied to material containing clay causes **swelling**; when the material dries out **shrinking** follows. The swelling occurs when water is deposited in and around the clay particles; how much depends on the water content, the clay content and the type of clay mineral present. Three-layer clay minerals, such as smectite (montmorillonite) and vermiculite are able to swell considerably more than the two-layer kaolinite. The shrinkage that follows drying out produces cracks which allow rainfall to infiltrate more deeply into the soil.

The swelling of the clay at the surface as a result of moisture in rocks containing clay (claystone, shale and sandstones with clayey cement) sometimes leads to a type of platy disintegration known as **slaking** (Yatsu, 1988, pp. 110–118).

7.2.1.5 Mechanical weathering by organisms

Organisms stimulate mechanical weathering in a number of ways. Tree roots penetrate fissures in the rock and as they grow and thicken enough pressure is exerted to move large boulders from their original location. Dead lichen leaves a dark colouring on the previously lighter surface of the rock. The darker area absorbs a greater amount of radiation and thermal weathering is encouraged. By contrast, the light-coloured crusts of excrement which often develop on rock walls below bird's nests reflect the sun's rays better than the neighbouring rocks, thus reducing local heating. Digging by animals in the soil increases the perviousness of the soil for air and water allowing the effects of changes in temperature and humidity to penetrate more deeply.

The most important influence in many areas during the past few thousands of years has, of course, been man. Formerly soil-covered bedrock in quarries, mines and road cuts has been exposed. Disruptions have been caused by explosions. The land surface is sealed by buildings and streets so that water flows off and does not seep in. Porosity and thereby infiltration are increased in areas of cropland. Irrigation, drainage and well construction change the flow of water and the water budget in the soil and rock and influence the weathering processes.

7.2.2 Morphoclimatic factors in chemical weathering

Chemical weathering changes the substance of the minerals in the rock. Atmospheric water has a threefold role in this process:

1. It is itself a reagent in that it reacts chemically with the mineral compounds.
2. It transports other reagents, such as carbon dioxide, into the pores and fissures in the rock and promotes weathering.
3. It serves as a means of transport for the removal of the products of chemical weathering, usually in solution.

The presence of water in a liquid state is an essential prerequisite for chemical weathering processes of a geomorphologically significant intensity. They predominate, therefore, in humid climates with temperatures above the freezing point. The water temperature itself also influences the solubility of both the reagents and the weathering products.

In contrast to mechanical weathering, which is confined to a shallow zone near the surface, the movement of water in the soil and rock means that chemical weathering can take place at much greater depth. The regolith thickness also has a different significance for chemical weathering than for mechanical weathering. Bare rock dries rapidly after rain so that chemical weathering processes are often interrupted, even in a humid climate, with the result that the rate of chemical weathering on bare rock is very low compared to that of mechanical weathering. However, a regolith cover absorbs the rainfall and stores it so that it remains in contact with the underlying rock for some time and allows weathering to continue during dry periods. The **critical regolith thickness** C_c stores just enough soil water to keep the underlying bedrock moist continuously. In a climate that remains humid throughout the year, chemical weathering remains uninterrupted under a thinner layer of regolith than in a climate with long dry periods or a desert climate with only rare and sporadic precipitation.

Chemical weathering intensity reaches its maximum under a regolith of optimal thickness. If it is thinner, there is periodic drying out of the regolith and a pause in the chemical weathering process. If the regolith is thicker than the optimum, a large part of the content of dissolved reagents that cause the chemical weathering is used up to weather components in the regolith and is not available for the bedrock.

A curve that shows the chemical weathering rate W of the bedrock as a function of the regolith thickness C in a given climate has, therefore, a low W value for $C = 0$, reaches the maximum W for $C = C_c$ and falls gradually for all $C > C_c$ with increasing C.

There is empirical evidence for this relationship. Rock levees developed under an earlier regolith cover on the flanks of erosion channels because there the regolith was thinner, and chemical weathering less, than in other parts of the land surface (Twidale, 1993).

7.2.3 Mechanical and chemical weathering in a theoretical process response model

The weathering process response system is composed of material components: **rock** and **regolith cover**, and the process component: the **weathering rate**. These components in turn form subsystems which contain properties relevant for weathering. For rocks, the most important are the mineral and chemical composition, the physical hardness and porosity, and for regolith, the composition, thermal conductivity, porosity and especially its thickness. The weathering rate depends on the properties of the rock but is, at the same time, a function of climatic factors, whereby the thermal characteristics are decisive for mechanical weathering and, primarily, the hygric characteristics for chemical weathering (section 7.2.1.1 and section 7.2.2). The extent to which climatic factors affect the weathering rate is determined largely by the thickness of the regolith.

Figure 7.2 and equations (7.1), (7.2) and (7.3) show that the climatic factors most important for mechanical weathering, temperature fluctuations and freeze–thaw frequency, decline exponentially with increasing depth. Based on this empirical experience, the rate of mechanical weathering of the bedrock is formulated in the theoretical process response model as a negative exponential function of the thickness C of the regolith layer (Ahnert, 1973b, 1987b):

$$W_m = W_{mo}\, e^{-k_1 C} \qquad (7.4)$$

where W_m is the weathering rate per time unit, expressed as the thickness of the rock that has been converted from bedrock to regolith, W_{mo} is the mechanical weathering rate per time unit on bare bedrock, C is the regolith thickness and k_1 is a coefficient.

The lowermost curve in Fig. 7.3 shows this functional relationship graphically. W_{mo} can also be understood as a rock-specific **weathering coefficient**, that is a measure of the petrographically controlled susceptibility of the bedrock to weathering, under a given climatic condition. Equation (7.4) is the mathematical expression of the **negative feedback** between W_m and C in mechanical weathering: the larger C becomes, the smaller is W_m. When no regolith is removed by denudation processes, the weathering rate finally approaches zero as a result of the increasing regolith thickness caused by the weathering process itself.

Only when denudation causes the regolith layer to remain thin enough can weathering continue. In this way, denudation has a steering effect on weathering. If the denudation rate is higher than the rate of weathering, the regolith cover gets thinner and the weathering rate increases, but if the regolith thickness increases,

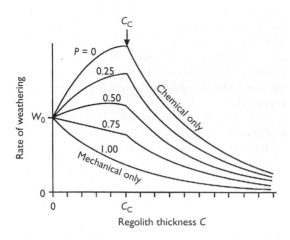

FIGURE 7.3 Mechanical weathering rates, chemical weathering rates and combinations of both rates as a function of the regolith thickness C, from the theoretical model SLOP3D (after Ahnert, 1987b).

the rate of weathering decreases. Because of this association with the regolith, the weathering rate has a tendency to adjust to the denudation rate in the sense of a **dynamic equilibrium**. The material expression of the equilibrium is a constant regolith thickness.

In the humid and subhumid tropics, weathering has created regolith several metres thick in many places. These are areas in which weathering has been more rapid than denudation for a long period. The opposite relationship is indicated by bare bedrock surfaces which occupy only a small part of the land surface. Here the potential for denudation is much greater than the capability of the weathering processes to produce regolith material. Outside the humid and subhumid tropics, soils of moderate depth dominate everywhere because weathering and denudation have been in a state of equilibrium, or at least close to it, for a long time. If this were not the case, very deep soils and areas of bare rock would be more widespread than they are.

In contrast to mechanical weathering, chemical weathering of the bedrock reaches its maximum (uppermost curve in Fig. 7.3) under a critical regolith thickness C_c. Its quantitative formulation in a theoretical model is, therefore, more complex. The bared rock surface weathers chemically with the rate W_{co} (analogous to W_{mo})

a rock-specific coefficient for chemical weathering. From regolith thickness $C = 0$ to $C = C_c$ the weathering rate W_c increases to a maximum, the height of which is determined by another presumably rock and climate specific coefficient ($k_2 \geq 2.0$). The value of C_c is primarily a function of the hygric climate and secondly a function of the specific capacity of the regolith material to store moisture. In wet climates C_c is small and in dry climates, large. Between $C = 0$ and $C = C_c$,

$$W_c = W_{co} (1.0 + k_2 C/C_c - C^2/C_c^2) \quad (7.5)$$

is valid, where W_c is the chemical weathering per time unit (similar to W_m in (7.4), W_{co} is the chemical weathering rate per time unit on the bare bedrock, k_2 is a coefficient (usually > 2.0), C is the thickness of the regolith and C_c is the critical regolith thickness. At the critical regolith thickness $C = C_c$, $W_c = k_2 W_{co}$. For $C > C_c$ the equation

$$W_c = W_{co} k_2 e^{-k_3(C - C_c)} \quad (7.6)$$

is valid, where k_3 is an additional coefficient. Equation (7.6) describes the descending arm of the uppermost curve in Fig. 7.3. It is similar to equation (7.4) in so far as, for regolith thickness $C > C_c$ (section 7.2.2), the chemical weathering is also expressed as a negative exponential function of the regolith thickness C.

Within the validity range of equation (7.6), a negative feedback exists therefore between W_c and C. However, in the validity range of equation (7.5) the feedback for $C < C_c$ is positive, as the ascending arm of the uppermost curve in Fig. 7.3 shows; each increase of C leads to an increase of W_c, every decrease of C to a decrease of W_c. An increase in C occurs when the weathering rate W_c, which is the producer of C, is larger than the local denudation rate. Due to this positive feedback, such a surplus of weathering over denudation leads inevitably to an even larger surplus and therefore to a greater value for C, until $C = C_c$. Conversely, every denudation surplus over weathering causes a further reduction of the latter and leads inevitably to where $C = 0$.

Consequently, the state of the process response system for chemical weathering is very unstable in the value range $0.0 < C < C_c$. Because it does not contain a negative feedback, a trend to a dynamic equilibrium is not possible. Even if an equilibrium did exist for a very short period, on one occasion more or less by chance, then the smallest discrepancy between weathering and denudation would terminate it immediately and cause a development either towards $C = 0$ or $C = C_c$. The land surface is, therefore, made up of regolith covered areas on which the weathering rate has been greater than the denudation rate and of areas of bare bedrock.

The greater C_c is, that is, the wider is the rising arm of the curve in Fig. 7.3, the greater is the probability that the system attains this unstable validity range $C < C_c$. The critical regolith thickness C_c is greater in dry climates than humid climates. This is probably the reason for the more frequent occurrence of bare rock surfaces in regions of dry climates than of humid ones.

It has been assumed for simplicity that $W_{mo} = W_{co}$, so that the curves start at the same point for $C = 0$. This is not necessarily so. The curves between the uppermost and lowermost curves in Figure 7.3 represent the kinds of combinations of chemical and mechanical weathering that occur in many parts of the world on a wide variety of rocks. The proportionality factor $0.0 < P < 1.0$ determines the relative share of both types of weathering in the theoretical model.

7.3 MECHANICAL WEATHERING PROCESSES AND THEIR PRODUCTS

Mechanical weathering leads to the disintegration of rocks without altering their chemical and mineral composition. The geomorphological effect of mechanical weathering is expressed in the size and form of the resulting rock particles. For this reason mechanical weathering processes are divided on the basis of the type of particle they produce rather than the physical mechanisms described in section 7.2.1.

7.3.1 Granular disintegration

The disintegration of rocks into separate grains can take place only if the grains are already components of the unweathered rock. Granular disintegration occurs in sedimentary rocks such as sandstone and in phaneritic crystalline rocks such as granite. In all cases, it is the weakening of the connection between the grains or the bond between crystals in the rock that causes disintegration.

Crystalline rocks

The black biotite crystals, beige feldspars and white quartz crystals on a bare granite surface absorb different amounts of direct solar radiation and expand differently as a result. On cooling, they contract. The thermal change in volume is very small but occurs in different directions for different crystals because the axes of the crystals are not aligned. Frequent repetition of the process weakens the bond between the crystals. The texture is loosened and fine cracks develop which provide a path for water moisture. Frost shattering or salt shattering can then contribute to the disintegration of the rock. Chemical processes, particularly the hydrolysis of the feldspar and oxidation of the biotite, often contribute to the disintegration. The relative importance of mechanical and chemical weathering in granular disintegration of crystalline rocks has not yet been determined fully (Barth *et al.*, 1939, pp. 118–121; Ollier, 1965, pp. 14ff., 81).

Sandstone

Sandstones are composed almost entirely of one mineral, quartz. Differential expansion and contraction is therefore of little importance for disintegration, in contrast to which water is of great significance. The pore volume of sandstone is generally high and the pores relatively large so that water can penetrate the rock easily. The pores in sandstone above the water table are also able to hold water for long periods after precipitation.

On bare sandstone, water causes granular disintegration mainly in the pores immediately below the surface by means of frost shattering or salt fracturing. The expansion pressure of the frozen water or the crystallization pressure of the salt loosens the quartz grains which are transported away by wind or water. There are often small accumulations of sand at the foot of rock walls in early spring, an indication of the importance of frost shattering in granular disintegration. Grains can also be rubbed off easily from the rock surface at this time of year because the cement between the grains has been weakened by frost.

Both mechanical and chemical processes are significant in the granular disintegration of sandstone. The disintegrating mechanism depends, in part, on the kind of cementing material. If it is also composed of quartz, frost shattering and salt fracturing dominate, but if the cement is soluble, such as calcium carbonate ($CaCO_3$) in calcareous sandstone, the cement can be dissolved by rainwater or groundwater containing CO_2 in solution. Clayey cement is removed by dispersion and transport of the clay particles as suspended load in rainwater runoff. A cement material of iron oxide (Fe_2O_3) can be weakened by hydration.

Sometimes organic factors are involved. If a moss cover is removed from a rock surface, for example, sand grains often adhere to the underside of the moss, which means they are more firmly attached to the rhizoids of the moss than to the rock.

7.3.1.1 Differential granular disintegration of sandstones

The grains of most sandstones were originally deposited by flowing water or wind, both of which are subject to rapid changes in their transport capacity. As a result, sands of different grain sizes are deposited in successive layers with different pore sizes which, in turn, influence the amount of cement between the grains and their eventual cohesiveness. Layers that are more firmly bonded are more resistant to disintegration and protrude as narrow ribs aligned with the stratification of the rock. **Differential weathering** reflects the relative resistance of the layers of sandstone (Fig. 6.3).

In conglomerates containing hard pebbles in a sandstone matrix, the pebbles project from the surrounding finer-grained mass which wears back more rapidly than the pebbles; they eventually lose their hold and fall out from the rock face, leaving a hole. Continued back weathering of the rock removes these hollows too.

Honeycomb or **hole weathering** of sandstone is another type of granular disintegration, not to be confused with the holes from pebbles. When they lie densely, rows of more or less hemispherical holes up to a few tens of centimeters in width form a honeycomb pattern. There are numerous rock faces of this type on the Bunter Sandstone in the Dahner Felsenland in the southern Palatinate in Germany. Some of the best developed honeycombs are on the Dahn Castle rocks (Fig. 7.4; Ahnert, 1955, pp. 93–95). The holes are mainly on south-facing walls below rock overhangs where, largely protected from rainwater runoff, they remain dry and free of lichen in a location that is essentially an arid microclimate. Because of its southerly exposure, the rock wall is warmed during the day in winter

Figure 7.4 Hole weathering in the Bunter Sandstone on a south wall of the Dahn Castle rock in the southern Palatinate, Germany.

by direct solar radiation to temperatures above freezing, even if the air temperature remains below freezing, so that there are a greater number of freeze–thaw cycles than on rocks facing in other directions. Freezing of seepage water loosens individual sandstone grains. There are apparently preferred seepage paths in the rock and where they end at the rock surface, granular disintegration is particularly active and the holes form there (Häberle, 1911a). Because the seepage water paths depend on the pore sizes and these, in turn, on the stratification of the various grain sizes in the sandstone, the holes lie in rows along the strata. Salt crusts caused by evaporation of seepage water have developed on some of the sandstone faces, indicating that salt shattering is also taking place (section 7.2.1.3).

Once such a hole has developed, the seepage water converges towards the hole and granular disintegration is increased. The edge of the hole is, however, dryer than before so that in this area disintegration decreases, while within the hole, its depth and often also its width increases. If the interior side walls of neighbouring holes weather through, small galleries develop with the remnants of the side walls forming small sandstone pillars in front.

The depths of the holes decrease when the areal back weathering of the outer rock walls becomes more rapid than the deepening of the holes. This may follow the removal of the protective rock overhang by rock fall which changes the microclimate so that rainwater reaches the holes, and the undifferentiating effect of mosses, lichens and other low plants on the weathering of the rock surface increases.

Table rocks are produced by the differential development of a free-standing rock tower with a platform at the top. The base is a relatively weakly cemented sandstone with a harder rock above that is particularly resistant to granular disintegration. The lower rock weathers by granular disintegration on all sides and forms the leg of the table and the less weathered upper layer, the top. In Germany, several table rocks have been formed in the middle main Bunter Sandstone formation in the southern Palatinate (Fig. 7.5).

7.3.2 Block disintegration

Well-jointed massive rocks weather mechanically by **block disintegration**, The joint limited blocks separate after the joints have been widened. Thermal expansion and contraction, lateral pressure release as a result of denudation of adjacent rocks, ice formation in the joints and root pressure all contribute to the widening of joints (Fig. 7.6).

Tectonic joint systems often run at right angles to one another and the blocks that disintegrate are quadratic in form and limited by flat joint surfaces. When the rock is jointed in more than two directions the forms have acute angles. On steep rock walls the blocks separated from the bedrock by disintegration form a talus at the foot of the cliff if not transported away. Depending on the hardness of the rocks, the fallen blocks either remain whole or fragment into smaller pieces.

7.3.3 Relative intensity of granular and block disintegration

Block and granular disintegration often take place on the same rock face and together determine its form. Their intensity is absolutely and relatively very varied. On a human time scale, granular disintegration is almost continuous although it is not an intensive process because the grains separate singly from the rock. The separation of an entire block although, relatively, very rare removes millions of sand grains in an instant. Little data is available for the absolute rates of either process. Single block falls have often been described (Rapp, 1960; Schumm and Chorley, 1964) but not successive block falls from the same place on a rock wall, which would be necessary to determine a rate.

More information is available on the granular disintegration of sandstone. Häberle (1911b) estimated the weathering back of rock walls carved into bedrock of Bunter Sandstone on the castle rocks in the Dahner Felsenland in Germany to be approximately 10 cm per 220 years. The bedrock walls had been worked by stonemasons in the Middle Ages, but after the castles were destroyed in 1689, the walls were exposed to the weather for 220 years until the estimate was made. At an estimated grain size, including pores, of about 0.5 mm, the 10 cm of sandstone removed is equal to a depth of about 200 grains or a mean annual removal of a layer one grain thick. On 1 cm² of wall surface, this would mean a removal of $20 \times 20 = 400$ grains per year, on a surface 1 m², 4 million grains annually, or about 11 000 per day. During the cold season the daily production is greater and in the warm season smaller than this average so that the retreat of

the rock walls during freeze-thaw periods is more rapid.

The rate of back weathering by granular disintegration is not the same on all parts of a rock wall. Häberle's measurements were made where there were visible signs of active weathering. In other locations, the surfaces have hardly changed over hundreds of years. Nevertheless, as an order of magnitude, the measurements are

FIGURE 7.5 The Devil's Table near Hinterweidenthal in the southern Palatinate, Germany.

FIGURE 7.6 Block disintegration in granite near San Antonio in southern Baja California, Mexico.

useful for a quantitative comparison with rock fall. If it is assumed that joints lie 0.5 m behind and parallel to a rock face so that the wall retreats 0.5 m when a rock fall occurs, then a given location on the wall would need to have a block fall only about every 1100 years to equal the same long-term rate of retreat by means of granular disintegration.

The relative significance of both weathering types can be observed qualitatively on sandstone rocks because the block disintegration, which is bound by the joints in the rock, exposes new joint surfaces that appear as smooth walls. Also, joint surfaces that intersect with one another at various angles create very sharp edges.

Granular disintegration is differential and produces a complex small-scale relief composed of ledges, grooves and also honeycombs on the previously smooth joint surface. Where joint surfaces intersect, the edges are rounded by granular disintegration. On a rock wall, it is clear whether a particular section has been weathered predominantly by block disintegration or by granular disintegration during the last few hundred years.

7.3.4 Slaty disintegration

Slaty disintegration is a variation of block disintegration in shaly or slaty rocks. The discontinuity surfaces in these rocks are not joints but the bedding plane contacts and foliation planes that run parallel to one another only a few millimetres to a few centimetres apart.

The bedding plane contacts in thin bedded shaly sedimentary rocks are the result of cyclical changes of transportation and deposition conditions during the formation of the rock. Fine sands, silts and clays were laid down in thin layers. Pressure and recrystallization subsequently produces the foliation that is characteristic of many metamorphic rocks, including argillaceous slates, phyllite, mica schist and gneiss (section 6.4). In both cases, the hardness of the rock within each sediment layer, or foliation layer, is greater than the cohesion between the layers. Mechanical stress caused by frost shattering, salt shattering or root pressure break up rocks with this type of structure into platy or slaty fragments. Chemical weathering can also be involved, especially in slates composed of several minerals (Fig. 7.7).

7.3.5. Flaking (thermal exfoliation)

In contrast to the disintegration of shales or slates, thermal exfoliation separates fragments about a centimetre thick and a few centimetres across parallel to the surface of the rock (Fig. 7.8). This type of disintegration is independent of the rock structure and, in effect, tends only to occur in rocks that have little or no structure in this order of magnitude.

Flaking is especially frequent where the rock surfaces are heated by intensive solar radiation

FIGURE 7.7 Slaty disintegration in Lower Devonian slate in the Eifel, Germany.

but the rock remains cool not far below the surface because of poor thermal conductivity. As a result the rock expands at the surface during the day and contracts at night (section 7.2.1.1). Shear stresses are created and a very fine initial separation crack develops near the surface. It seems certain, however, that not only thermal expansion and contraction cause these cracks (Rögner, 1987) but that frost shattering, salt fracturing and chemical weathering also participate and separate the exfoliated layer completely from the rock underneath. Since this usually occurs on steep rock walls, the separated piece falls off, exposing a new surface on which renewed exfoliation can take place.

Figure 7.8 Thermal exfoliation (flaking) on a rock wall in the Permian DeChelly Sandstone, Monument Valley, Arizona, USA. Note the rock fragments on the ground.

7.3.6 Pressure release exfoliation

If large masses of rock are removed, the pressure on the underlying rocks that are now at the surface is reduced. Also, when a valley is incised or a quarry is worked, the valley sides and quarry walls are freed from the lateral pressure that the removed rock mass previously exerted. This release of pressure causes rocks to expand in the direction from which the original pressure came. The expansion is directed upwards if overlying rock was removed and laterally towards the centre of a valley that has been incised. In a quarry, too, the expansion is lateral. The expansion takes place with the development of **pressure release joints** that are more or less at right angles to the direction of the pressure release and approximately parallel to the land surface. The joints occur in sets and give the rock an onion-like layer structure. The layers are generally between a few decimetres and a few metres thick, in most cases less than 2 m. Additional longitudinal and transversal tension in the rock produces cracks which break down the layers into blocks. **Pressure release exfoliation** depends on the previous removal of rock and not on any environmental conditions in the atmosphere and is, therefore, only indirectly a form of weathering. Pressure release joints can only develop in massive rocks with few joints. Rocks with tectonic joints or foliated structures parallel to the surface react to the pressure release by widening existing joints rather than by forming new ones.

The massive granites in Yosemite National Park in the Californian Sierra Nevada have very few tectonic joints, and pressure release exfoliation is well developed (Fig. 7.9). On many mountains in the region exfoliated fragments have been transported away by denudation, leaving the summits as large rounded rock domes whose surfaces are formed of pressure release surfaces. Half Dome is a well known example, half of which has been removed by the erosion of Pleistocene glaciers.

The rounded form of exfoliation domes is an expression of a feedback between the form-generating denudation and the structure-generating pressure release stresses in the rock. The denudation leads to the pressure release,

FIGURE 7.9 Pressure release exfoliation in granite in the Sierra Nevada, California, USA.

and the joints produced by the release determine the form of the boundary surface for the next phase of denudation (Yatsu, 1988, pp. 145–150).

7.4 CHEMICAL WEATHERING

7.4.1 General factors

Decisive for determining the intensity of chemical weathering are the morphoclimatic conditions (section 7.2.2), the movement of water in the soil and the rock, the water quality and the mineral and chemical composition of the rocks and soils.

7.4.1.1 Water movement

The movement of the water that infiltrates the soil and percolates down to the groundwater is dependent on the **hydraulic potential difference** or **pressure gradient** and on the **hydraulic conductivity** of the material through which it flows (Darcy equation, section 10.3.1). The more often fresh percolating water moistens the rock, the more intensive, under otherwise equal conditions, is the chemical weathering. Drying out stops the process.

If the rock is permanently saturated, that is, within the groundwater zone, the weathering intensity depends on the speed of water movement. In stagnating water the available reagents are used up and not replaced by the addition of fresh water. The more rapid the movement of water, the greater the possibility that the water is

renewed; this occurs especially where fluctuations in the water table indicate variable flows into and from the groundwater.

7.4.1.2 Water quality

Water quality in relation to chemical weathering refers to the reagent content of the water and to the weathering products already in solution. Fresh rainwater contains carbon dioxide and other chemical compounds. As water percolates through the upper soil horizons, other effective reagents, such as organic acids, may be taken up, which make the water chemically more agresssive. Reagents are used up by the weathering process itself to decompose the rock and to keep the soluble products of weathering in solution and thereby transportable. For this reason the weathering effectiveness of water decreases unless fresh water is added.

7.4.1.3 Susceptibility of the material to weathering

The susceptibility to chemical weathering of the different minerals that make up the rocks varies greatly not only in the specific way they can be attacked but also the degree to which they resist chemical weathering generally. Limestone is an example of selective vulnerability. It is easily attacked by carbonate weathering but practically immune to oxidation or hydrolysis. Minerals with a high iron content are, however, very susceptible to oxidation.

Mineral compounds of calcium, sodium, magnesium and potassium are more easily soluble than those of silicon, aluminium and iron. But solubility is not the only measure and solubility of particular minerals varies for different processes. Table 7.1 ranks frequently occurring minerals according to their resistance to chemical weathering. Iron oxides, aluminium oxides and clay minerals are themselves usually weathering products and therefore particularly resistant.

7.4.2 Chemical weathering reactions

7.4.2.1 Solution and solubility

The solution of a mineral salt in water is a physical, not a chemical process. It involves the dissociation of the molecules into their anions

TABLE 7.1 Minerals ranked according to their resistance to chemical weathering

Most resistant	Iron oxide
·	Aluminium oxide
·	Quartz
·	Clay minerals
·	Muscovite (white mica)
·	Orthoclase (potassium feldspar)
·	Biotite (black mica)
·	Albite (sodium feldspar)
·	Amphibole
·	Pyroxene
·	Plagioclase (calcium-sodium feldspar)
·	Anorthite (calcium feldspar)
Least resistant	Olivine

Sources: after Brunsden (1979, p. 93); Chorley *et al.* (1984, p. 208); Goldich (1938).

and cations, whereby each ion is surrounded by water molecules. Because of the bipolarity of the water molecules and the high dielectric constant of water, approximately 80 times higher than that of air, water is a particularly suitable solvent. In contrast to chemical changes, solution is easily reversible, in part, when the saturation point of the solution is exceeded and a proportion of the dissolved material is precipitated out, and completely when the water evaporates and the dissolved material remains as a solid residue. If the material in solution is transported away in the water, then solution weathering contributes directly to the denudation of the land surface.

Solution has traditionally been regarded as a chemical weathering process largely because the process usually occurs in conjunction with, or as the result of, other chemical weathering processes.

Among the natural minerals, the chlorides of the alkaline metals NaCl (rock salt or halite) and KCl (potash salt or sylvin) are very easily dissolved and found, therefore, only in extremely arid climates. At 10°C the solubility of rock salt is 263 g per litre of water. The solubility of gypsum ($CaSO_4 \cdot 2H_2O$) is only 1.9 g per litre but this is still a large enough amount to have a considerable geomorphological effect. Limestone is also included among soluble rocks, although it can only be dissolved to any great extent when the water contains carbon dioxide (section 7.4.2.4).

The solubility of many minerals depends on the properties of the water as a solvent, especially on the content of free hydrogen ions, expressed as the **pH value**. The pH value is the negative decadic logarithm of the hydrogen ion concentration. In neutral water the pH value = 7 which means a hydrogen ion content of 10^{-7} mol per litre of water. Acid soil water has a low pH value, that is, it contains more hydrogen ions. Basic soil water has a high pH value. At a value of pH > 8, the solubility of silica (SiO_2) increases rapidly. The solubility of aluminium oxide increases at pH > 9 and pH < 5. Good solubility conditions for iron hydroxide lie between pH < 3 and pH > 8. Neither aluminium oxide or iron hydroxide are easily soluble in water with pH values of between about 5 and 8. The solution of mineral substances depends, therefore, to a great extent on the chemical composition of the soil water. The pH value is also influenced by climate so that some minerals dissolve more easily under some climatic conditions, less easily under others; this in turn influences the type of soil that develops. The distribution of the earth's soil zones is therefore related to the climatic zones (section 7.5).

7.4.2.2 Hydration

Like solution weathering, hydration belongs to the transition area between physical and chemical changes. Water molecules are added into the crystal lattice of minerals without otherwise changing the chemical composition of the original material. An example is the hydration of **anhydrite** to **gypsum**:

$$CaSO_4 + 2H_2O \rightarrow CaSO_4 \cdot 2H_2O$$
$$\text{anhydrite} + \text{water} \rightarrow \text{gypsum}$$

Gypsum is therefore also described as a calcium sulphate dihydrate. The storage of water in the crystal lattice causes an enlargement in volume which can lead to hydration folding in gypsum layers lying between other strata. Only the gypsum is folded, not the layers above and below.

The frequent brownish to yellowish soil colouring in humid climates in the middle latitudes is the result of hydration of the reddish iron oxide **haematite** to the rust-coloured **goethite**:

$$Fe_2O_3 + H_2O \rightarrow Fe_2O_3 \cdot H_2O$$
haematite + water → goethite
(iron oxide hydrate)

The bipolarity of the water molecule also plays a significant role in hydration for the bonding of the water on to the mineral substance. Water molecules taken up by clay particles, also a form of hydration, cause clays to swell when wet (section 7.2.1.4).

7.4.2.3 Oxidation and reduction

In geomorphology, the term **oxidation** is limited to the combining of a substance with oxygen. Oxidation weathering mainly affects minerals containing iron. Iron is one of the most important elements in rocks (Table 6.1); the red, brown and yellow colours of iron oxides dominate in soils and also on the chemically weathered surfaces of rocks. Even very small amounts of iron oxide have a strong colouring effect.

The oxidation of iron compounds takes place for the most part when the oxygen is dissolved in water and comes into contact with the mineral. Oxidation weathering on rock walls indicates that occasional moistening by rainwater is sufficient for it to occur.

If the soil or rock is saturated with stagnating (therefore, oxygen-poor) water, **reduction** follows with the help of anaerobic bacteria. This is the opposite of oxidation. The changes that take place in the soil under these conditions are known as **hydromorphization** or gley development. The soil colour is typically grey. Hydromorphic soils develop mainly in valley bottoms but also in areas where a large supply of water, impermeable conditions, low gradient and insufficient drainage allow water to stagnate in the soil.

7.4.2.4 Carbonation

Carbonation is the development of carbonates, the salts of carbonic acid H_2CO_3 which are present through the solution of carbon dioxide (CO_2) in natural water. Carbonation is an independent and dominating chemical weathering process particularly in limestones and dolomites. These rocks are carbonates. Calcite, the mineral of limestones, is composed of calcium carbonate ($CaCO_3$). When it reacts with carbonic acids a double carbonate is produced, calcium hydrogen carbonate:

$$CaCO_3 + H_2CO_3 \rightarrow Ca(HCO_3)_2$$
calcite + carbonic acid → calcium
hydrogen carbonate

In contrast to calcite, calcium hydrogen carbonate is easily dissolved in water. This change is the decisive prerequisite for the solubility of limestone and for the development of solution land forms.

Carbonation also constitutes a component step in the complex weathering of many other minerals, for example, the hydrolysis of feldspar.

7.4.2.5 Hydrolysis

Hydrolysis is the most important chemical reaction in the weathering of silicates. The water, decomposed into cations, H^+, and anions, OH^- (hydroxyl), reacts with the silicate minerals through the exchange of the H^+ ion against a cation (metal) of the minerals, which, in turn, combines with the OH^-. The hydrolytic weathering of feldspar is an example:

$$2KAlSi_3.O_8 + 2H^+ + 2OH^- \rightarrow 2HAlSi_3O_8 + 2KOH$$
orthoclase + hydrogen + hydroxyl →
alumosilicic acid + potassium hydroxide
(two of each because of the following
reactions).

The hydrolysis is only the first step in a more complex weathering process that includes further reactions. The aluminosilicic acid and potassium hydroxide are both unstable and react further. Potassium hydroxide is carbonated as follows (section 7.4.2.4):

$$2KOH + H_2CO_3 \rightarrow K_2CO_3 + 2H_2O$$
potassium hydroxide + carbonic
acid → potassium carbonate + water

The newly formed potassium carbonate is soluble and therefore transported away in solution. The alumosilicic acid reacts further with water:

$$2HAlSi_3O_8 + 9H_2O \rightarrow Al_2Si_2O_5(OH)_4 + 4H_4SiO_4$$
alumosilicic acid + water →
kaolinite + silicic acid

The aluminosilicic acid loses two thirds of its silicon in this reaction to the newly developed soluble siliceous acid H_4SiO_4 which is then transported away in water. This phase of the process is also termed **desilification**. The residue is the clay mineral **kaolinite**.

If the solution equilibrium of the siliceous acid changes, silicon dioxide, or **silica**, is precipitated out of the solution:

$$H_4SiO_4 \rightarrow 2H_2O + SiO_2$$
siliceous acid $\rightarrow$ water + silica

Quartz is the most common crystalline form of silica, and **flint** its most common cryptocrystalline form. They are present in many sedimentary rocks either as cementing material or as concretions.

Table 7.2 shows that after changing from orthoclase to kaolinite only 258/556, or 46 per cent, of the original mass is present. The other 54 per cent has been removed in solution. Other feldspars, such as **albite** ($NaAlSi_3O_8$) or **anorthite** ($CaAl_2Si_3O_8$) weather similarly to orthoclase, except that sodium or calcium takes the place of potassium in the reaction.

Kaolinite generally develops in the humid tropics because silica is more easily dissolved in warmer rather than cooler climates. Weathering of the same feldspars in cool climates more often leads to the development of the clay mineral **illite** which has a much higher silica content (Curtis 1976, 50–53):

$$6KAlSi_3O_8 + 4H_2O + 4CO_2 \rightarrow$$
$$K_2Al_4(Si_6Al_2O_{20})(OH)_4 + 12SiO_2 + 4KHCO_3$$

orthoclase + water + carbon dioxide $\rightarrow$ illite + silica + potassium hydrogen carbonate (soluble)

Local water supply and drainage conditions also influence the type of clay mineral that develops (Curtis 1976, 50–53; Trudgill 1976, 66–67; section 7.5). **Montmorillonite**, important geomorphologically because of its capacity to swell, is formed in poorly drained areas where the alkaline and earth alkaline metals and silica are less strongly washed out and so retain a larger share in this clay mineral. If, on the other hand, in a tropical humid climate, where normally kaolinite is the typical clay mineral, the flow through of water is large, these metals are dissolved and removed. Only the aluminium remains, which together with the hydroxyl OH^- forms the clay mineral **gibbsite**, the basic substance of **bauxite** and raw material for aluminium.

Not all clay minerals are a final end product of weathering. The more complex three-layer clay minerals such as montmorillonite can disintegrate further to become a two-layer mineral such as kaolinite or halloysite. If it loses all minerals except aluminium, it could also become gibbsite. In general, two-layer clay minerals represent a greater degree of weathering than three-layer clay minerals.

Although the weathering processes are similar for the various feldspars, in the initial phases of clay mineral development the composition of the original material does also play a role. For

TABLE 7.2 Loss of mass by hydrolytic weathering of two orthoclase molecules to a kaolinite molecule

		Orthoclase		Kaolinite	
Element	**(a) Atomic weight**	**(b) Number of atoms**	**(a)×(b)**	**(c) number of atoms**	**(a)×(c)**
K	39	2	78	0	0
Al	27	2	54	2	54
Si	28	6	168	2	56
O	16	16	256	9	144
H	1	0	0	4	4
Total			556		258

example, the hydrolysis of mica crystals leads usually to the development of illite.

Millot (1970) describes the geochemical and mineral characteristics, the development and behaviour of these and other clay minerals in detail.

7.4.2.6 Chelation

Chelation is the separation of metal ions from solids. The process binds them to an organic acid to form soluble metallic organic complexes. In soil science, the process is also known as **complexing**. The chelatizing acids are themselves partly decomposition products of plant remains and partly secretions from living roots.

The process allows plants to receive nutrients from otherwise insoluble metal compounds. Geomorphologically, chelation promotes chemical weathering and the transfer of metals in the soil profile both vertically and also horizontally with the lateral movement of water in the soil. The metals most commonly affected by chelation are aluminium, iron and manganese.

7.4.3 Rates and degree of chemical weathering

Like mechanical weathering, chemical weathering progresses very slowly and direct measurements of the rate are very difficult to obtain. Only a few reliable values are available most of which are based on artificially created surfaces such as the pyramids in Egypt or gravestones whose date and original form is known and on which changes can be measured. On natural surfaces, chemical weathering is usually measured on bare rock surfaces which are not representative for most chemical weathering processes, as these take place largely in the soil or on rock covered by soil.

A summary by Brunsden (1979, p.102), shows that the majority of estimates of chemical weathering rates have been made for limestones. Their range is very large and, apart from some exceptional cases, lies between 0.01 mm and 0.4 mm annually (Burger 1992). The great variety of limestone types and of local weathering conditions at the measurement locations is reflected in this wide range. Values determined indirectly for a long period are usually lower than those measured directly for a short period. Perhaps this is because direct measurements are made at particularly advantageous locations whereas finding a suitable location for long term indirect estimates is largely a matter of chance. The lower rates are therefore probably more generally valid than the higher.

There is also a considerable difference between weathering on bare rock and weathering of the same rock under a soil and plant cover (Fig. 7.10). Bauer (1962) estimated the long term mean rate on bare rock in the Austrian Limestone Alps to be 0.01 mm per year and in the same rock under soil and plant cover at 0.03 mm per year. The rates differ because the bare rock dries out frequently, interrupting the chemical weathering process more often than on rocks under a moisture retaining soil (section 7.2.3).

In addition to absolute values for chemical weathering rates, relative weathering rates have also been estimated using a nominal scale for classification (Brunsden 1979, 100–101). Such a scale usually has from four to six stages; fewer would give too coarse a subdivision and more would make the drawing of qualitative boundaries between categories difficult. Table 7.3 is an attempt to provide a useful basis for judging the degree of weathering in the field. It is based on scales by Ollier (1965) and Melton (1965).

7.5 SOILS AS PRODUCTS OF WEATHERING

7.5.1 Saprolite, regolith and soil horizons

As long as weathered rock material has not been transported away either in solution or mechanically, it covers the bedrock. Where mechanical weathering has dominated, this **waste cover** is made up of rock fragments of varying size. If chemical weathering has dominated, there is a vertical sequence of horizons with different degrees of weathering between the unweathered rock below and the land surface.

Above the unweathered **bedrock** is an area of weathered rock in which the structural details of the rock such as joints, quartz veins or individual large crystals can still be seen clearly in their original location in the rock. The material in this zone is known as **saprolite** (Greek *sapros* = decayed, *lithos* = rock). It is so decayed that it crumbles in the hand or can be scraped out. Saprolite is always **autochthonous**, that is, developed in the place where it lies.

The lack of firmness in saprolites developed on crystalline rocks such as granite or gneiss is mainly due to the hydrolytic weathering of the feldspar to clay and the oxidation of minerals containing iron (section 7.4) which cause the rock to lose its cohesion. Quartz crystals and other harder, less weathered mineral components, remain as grains in an already soft mass which disintegrates easily into coarse crumbs or **grus**.

Above the saprolite is the **regolith**, an unconsolidated cover in which the original rock structure is no longer identifiable. The regolith

FIGURE 7.10 Deep chemical weathering along joints in a granite near Ensenada, northern Baja California, Mexico. The unweathered cores between the joints are rounded to form woolsacks. On the horizon there are a number of blocks of this type exposed by denudation and therefore now more resistant.

TABLE 7.3 Estimated stages in the degree of weathering of rock

Stage	Characteristics
1	Fresh unweathered rock, hammer blow has a ringing sound and a hammer springs back, no sign of colouring by oxidation
2	Surface coloured with thin weathering crust, unweathered interior
3	Easily fragmented with hammer, dull sound, interior partially coloured by oxidation
4	Can be broken up using hands, interior decayed, but rock does not disintegrate in water
5	Disintegrates after minimal pressure with hands, falling to the ground or immersion in water

Sources: After Ollier (1965) and Melton (1965).

material can have developed following further weathering of the saprolite or it can be **allochthonous**, that is, transported into the area from elsewhere. A waste cover produced by mechanical weathering is also a regolith.

The **soil** is developed on the upper part of the regolith. It is characterized by its differentiation into several **soil horizons**. The uppermost, the **A horizon**, contains humus, although this may be partially washed out into the deeper parts of the horizon. Below the A horizon, in most soils, is the **B horizon** in which the material (iron oxide, humus, clay minerals) washed down from the A horizon is redeposited. The B horizon can also be present as a result of *in situ* weathering, when the brown iron oxide hydrate goethite (**soil oxidation**) and new clays (**loam formation**) develop. The **C horizon** is the deepest horizon. It is the parent material of the soil and is either bedrock, saprolite or deep regolith. These three main horizons are subdivided into a large number of subtypes; in special cases, such as salt soils and gley soils, additional horizon designations are used.

7.5.2 Grain size classes of soils

An important property of soils is their grain size composition. Table 7.4 shows the main grain size classes and their subdivisions. In addition to the main grain size classes clay, silt and sand, in which one of the three clearly dominates, there are intermediate classes based on varying proportions of each type of grain size (Fig. 7.11). Loam, for example, is a mix of 15–45 per cent clay, 15–70 per cent silt and the rest sand. The names of the intermediate classes are based on their mix of grains: silty sand soil, loamy clay soil, sandy silty loam soil and sandy silt soil are some. The term skeletal refers to the presence of particles greater than 2.0 mm.

7.5.3 Soil types

In the narrow sense, soil formation is the physically, chemically and also biologically determined change of the parent material caused by the action and interaction of seven factor complexes: climate, rock type, landform, vegetation, soil animals, human influences and time. There

TABLE 7.4 Grain size classes

Type		Grain size (mm)
Clay		< 0.002
Fine		< 0.0002
Medium		0.0002–0.0006
Coarse		0.0006–0.002
Silt		0.002–0.063
Fine		0.002–0.0063
Medium		0.0063–0.02
Coarse		0.02–0.063
Sand		0.063–2.0
Fine		0.063–0.2
Medium		0.2–0.63
Coarse		0.63–2.0
Skeletal soil		> 2.0
Angular	Rounded	
Grus	Fine gravel	2–6
Small fragments	Medium gravel	6–20
Medium fragments	Coarse gravel	20–63
Large stones	Pebbles	63–200
Blocks	Boulders	over 200

Source: Arbeitsgemeinschaft Bodenkunde 1971, pp. 34–5.

is also a close relationship between the development of soils and the geomorphological process response systems. The factor complexes bring about the differentiation of the parent material that influences the soil horizons, the characteristics of which form the basis of the classification of the soils into **soil types**.

Podsol is a major soil type. It is widespread in the old moraine landscape of the north German plain and at higher altitudes in the central uplands in Europe, in the evergreen forests of Scandinavia, northern Russia and Canada. Podsol has a raw humus top layer formed of decomposed plant remnants, an A_h **horizon** (h = humus) which contains both organic and mineral components, the result of water movement, soil organism activity and the presence of roots.

The A_h horizon of podsols is usually only 10–20 cm thick, Below is the light grey, often thicker A_e **horizon** (e = eluvial). Water moving

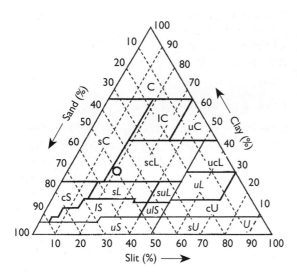

FIGURE 7.11 Triangular diagram of soil types according to their grain size relationships, S: sand; s: sandy; U: silt; u: silty; L: loam; l: loamy; C: clay; c: clayey. The point o, for example, is a sandy clayey loam with 50% sand, 20% silt and 30% clay (from Scheffer and Schachtschabel, 1976, p. 21).

through this horizon washes out the **sesqui-oxides** (two metal ions combined with three oxygen ions) of iron and aluminium together with the humus residue and clay particles and transports them downwards. The Russian word pod (= ash, the colour of this horizon), has given the soil its name.

The B horizon of podsol is characterized by intensive illuviation and is seldom more than 20 cm thick. The upper B_h **Horizon** has been enriched by humus particles and is dark brown to black in colour. Below is the B_s **horizon** which has been enriched with the sesquioxides. If the iron content is high in this horizon, the pores may be closed and a **hard pan** of sesquioxides develops which cannot be penetrated by roots. In iron-poor material, the B_s horizon may be absent. Depending on whether only a B_h horizon, or a B_s horizon, or both, are present, the soils are described as **humus podsols**, **iron podsols** or **iron–humus podsols**. The C horizon is formed of the parent material.

Podsols develop in cool humid climates of the higher middle latitudes where there is a large

supply of water, low evaporation and temperatures that still allow a predominance of chemical reactions. Moisture and decomposed vegetation together provide tools for hydrolysis and chelation (section 7.4). The water supply must also be combined with high perviousness so that eluviation and illuviation can occur. The parent material is therefore more likely to be sandy than clayey, and formed of acid rather than basic rocks if podsols are to develop. They require a relatively long formation period. Based on estimates of podzol development and the age of glacial and glaciofluvial deposits in north Germany, several tens of thousands of years appear to be needed on an impervious parent material such as loam but only several thousand years on sands.

In central Europe, **parabrown earth** soils are the most closely related to podsols. The washing out of substances from the lower part of the A horizon in parabrown soils, in this case termed the A_l horizon (l = lessivated), is less pronounced than in podsols. The material delivered to the B horizon is mostly clay which thinly coats the soil particles and soil aggregates in that horizon.

The B horizon of the **brown earths** is not the result of clay deposition but only of iron weathering and the development of clay, that is, soil oxidation and loam formation (Semmel 1977, p.24) and is named B_v **horizon** ('v' for the German word for weathering).

In some soils the B horizon is not present. These **AC soils** occur mainly in areas with pronounced dry periods. During the dry period water moves upwards in the capillaries of the soil, which means that transport of material is also mostly upwards. With sufficient precipitation they are very rich in nutrients and have a thick, humus and dark-coloured A horizon. Typical are the **black earths** (Russian = *chernozem*) of the Ukraine.

In the AC soils in deserts and semi-deserts, upward movement of water in capillaries and its evaporation on the surface dominate. Humus development is reduced or absent. Evaporation residues, such as rock salt, gypsum and calcium carbonate, concentrate on the surface and may form crusts. The **solonetzes** belong to this group as well as the **solonchaks**, which have covers of salt crusts.

Other AC soils develop when a humus layer forms on a substrate that is not easily weathered, such as gravel, or on a rock whose weathering products are being removed largely in solution so that no mineral remnants are present to differentiate other soil horizons. Soils of this type that have developed on hard rock are termed **rankers**, and on regoliths, **regosols**. Soils similar to rankers but formed on limestones or gypsum are **rendzinas**.

Local conditions produce other types of horizons. **Gley soils** form where the water table is high. In these soils reduction dominates in the part of the profile continuously in the groundwater and causes the grey colour of the soil. Above the water table, there is the usual brown oxidation horizon. **Pseudogley** occurs when long periods of blocked drainage, generally because of an impervious layer in the soil, alternate with occasional periods of drying out. The soil is grey in colour with brown iron oxide patches.

In the humid tropics, mainly on, but not limited to, basic rocks, large areas of soil have developed that are known variously as **red earths**, **red soils**, **laterite** or **latosols**, also as **oxisols**, **ferrallitic soils** or **ferralsols**. They are often several meters thick and reddish brown to red in colour in the upper part, which is rich in sesquioxides and poor in silica. For this reason the clay component is largely kaolinite (section 7.4.2.5). At greater depth, the soil becomes spotty and grades into a lighter-coloured bleached horizon. How these soils form is not completely clear. Their development does, however, take place over a very long period and involves hydrolysis, the washing out of alkaline and alkaline earth minerals as well as intensive desilification with the formation of kaolinite that together with the iron oxide haematite, the source of the red colour, the brownish goethite and the aluminium oxide compose the substance of the soil.

The terminology of soil types was not unified internationally for many years. In 1961, with the publication of the World Soil Map by the FAO, a world-wide new terminology was available which has found increasing acceptance. The most important terms are shown in Table 7.5.

The Soil Taxonomy of the U.S Soil Conservation Service (U.S Department of Agriculture, 1975) is also used frequently (Bronger, 1980; Fitzpatrick, 1980; p.126–145).

7.5.4 Soil catenas

A **catena** in soil science is the regular and, therefore, representative spatial sequence of soil types that develops in a landscape segment as a result of the spatial differentiation of hydrological conditions, weathering and the transport of material, all of which are mainly functions of landform development. Differences of rock type can also play a role but do not belong to the original definition of the concept of catena (Milne, 1935, cited by Ollier, 1976). More narrowly defined it refers to the sequence of soils from the crest to the foot of a slope. At the crest the soil is usually thin because denudation processes transport soil material downslope and there is no supply other than by local weathering. The transport catchment area increases downslope and the local removal of material is more readily compensated by supply from upslope. The amount of water available for weathering from both surface runoff and interflow also increases downslope (section 8.6.5).

If differences in rock type, fossil waste covers and spatial and historical differences in soil use by man are added as factors, the soil sequence along the slope can become very complex. Semmel (1977, p.46) describes a catena from the summit downwards in the Rhenish Slate Mountains in central Germany as follows:

1. ranker on slate fragments;
2. brown earth developed on late Pleistocene debris cover over slate fragments or older Pleistocene basal debris;
3. parabrown earth on cover debris above loess;
4. parabrown earth on loess.

If the slope crest is formed from remnants of an older land surface, a podsol can also be present, and if poor drainage occurs at the slope foot, the high groundwater level can lead to gley soils or, with blocked soil drainage, to pseudogley.

TABLE 7.5 FAO classification of the most important soil types

Type	Characteristics
Fluvisol	Meadow soil with minimal profile differentiation
Regosol	Raw soil of clayey and loamy unconsolidated sediments
Arenosol	Weakly humic sand soil
Gleysol	Hydromorphic soil (gley)
Rendzina	AC soil on carbonate-rich rock (e.g. limestone)
Ranker	AC soil on rock with little or no carbonates
Andosol	Humus-rich soil on volcanic ash
Vertisol	Soil rich in montmorillonite, self-mulching soil
Yermosol	Desert soil, very poor in humus
Xerosol	Semi-desert soil, poor in humus
Solonchak	Salt soil
Solonetz	Alkaline soil with clay enrichment
Planosol	Soil with blocked drainage (pseudogley) with bleached grey A horizon and brownish patches of oxidation
Castanozem	AC steppe soil with chestnut coloured A horizon
Chernozem	AC steppe soil with thick dark A horizon
Phaeozem	Degraded steppe soil
Cambisol	Soil with loam formation and oxidation (e.g. brown earth)
Luvisol	Lessivated soil with high base saturation (e.g. parabrown soil)
Podsoluvisol	Lessivated soil with strongly washed out A horizon that has tongue-shaped protrusions into the B horizon
Podsol	Podsolized soil
Acrisol	Strongly weathered, lessivated kaolinitic soil
Nitisol	Lessivated soil with limited exchange capacity of the clay fraction
Ferralsol	Ferrallitized soil (latosol)
Histosol	Organic soil (e.g. peat soil)
Lithosol	Raw soil of bedrock

Sources: After Scheffer and Schachtschabel (1976, pp. 320–321) and Schultz (1995).

The soil catena is of importance in geomorphology because soil genesis is linked to landform development. Ollier (1976) has identified and compared the catenas of different climatic zones.

7.5.5 Crusts

Metal compounds, in particular soil horizons, harden so much in some cases during the course of soil development that they form a **crust**, a rock-like hardness and thickness which makes the horizon more or less impervious. Depending upon the topographical location, the parent rock and the processes involved, the crust is either an autochthonous residual concentration of weathering products or formed of substances that were transported in solution by the soil water and precipitated out at particular locations (Chorley *et al.* 1984, p.480f.).

Crusts are distinguished according to the substances that produce them. Calcium carbonate crusts are known as **calcrete** or, in Mexico and in the southwestern USA, **caliche**. **Silcrete** crusts are formed of silica and **ferricrete** of iron oxide. Bauxite, which is a concentration of silicate-free aluminium oxides, is also a crust, although it is not as hard as the others.

The type of crust that develops depends not only on the parent rock but also on the climate. Calcrete is a secondary precipitate of $CaCO_3$ in either soil or sediments (Blümel, 1982). Sometimes it is produced already during the sedimentation process (Wenzens, 1975). It occurs mostly in dry climates that have a short rainy season. During the long dry season, the water rises by capillary action. The calcium hydrogen carbonate in solution is precipitated out as $CaCO_3$ when the rising water is warmed near the land surface and part of the dissolved CO_2 escapes into the atmosphere, thus disturbing the ion equilibrium of the solution (section 7.4.2.4). This occurs either at the surface or a few centimetres or tens of centimetres below it.

Silcrete is formed by the precipitation of amorphous or microcrystalline SiO_2, usually from siliceous acid $H_4SiO_4 \rightarrow SiO_2 + 2H_2O$. Silica is precipitated out between the soil grains and binds them to a precipitate quartzite. The siliceous acid is derived from the hydrolysis and desilification of silicate minerals, processes that are particularly effective at high pH values (section 7.4.2.5). Silcrete crusts are common in the dry areas of Australia and Africa. Where they are underlain by kaolinite, they were very probably produced by capillary action. However, there is also clear evidence that in tropical humid regions some silcrete crusts were formed as a result of horizontal movement of siliceous acid. In southeastern Brazil, for example, silica precipitates out primarily on exposed slope locations at the upper edge of valley slopes on the edge of a plateau (Römer, 1993). These crusts increase the resistance of the plateau edge and slow down its denudation.

Ferricrete occurs in the middle latitudes mainly as goethitic hardpans formed during podsolization (section 7.5.3). In the tropics, it is a late phase in the development of red earths or latosols primarily in iron-rich basic igneous and metamorphic rocks, in humid or at least subhumid climates with relatively high annual precipitation, generally 1000 mm or more. The concentration of iron oxides, especially haematite, that remains after hydrolysis and desilification, hardens in some latosols when in contact with air. This material has been used for hundreds of years in India as building material. In 1807, it was named **laterite** by Buchanan (Latin *later* = brick). The name laterite has long been applied generally to tropical red soils. This has led to confusion because not all tropical red earths harden in contact with the air. Hardened red earths are, therefore, now termed **plinthites**. In East Africa they are known as **murram** and also serve as building material.

The thickness of ferricrete crusts, often several metres, and their distribution in relation to the landforms, indicate that horizontal transport played a significant role in the concentration of the iron oxides.

Bauxite develops as a residual concentrate from basic rocks, such as syenite, that have a low iron and quartz content, in the humid tropics and subtropics under especially favourable weathering conditions. It is composed of several clay minerals, especially **gibbsite**, a pure aluminium hydroxide which is the raw material for aluminium. Bauxites are relatively rare because of the long period of stable environmental conditions required for all the iron and silicon compounds to be removed by hydrolysis, desilification, carbonation and chelation.

Calcrete, silcrete and ferricrete are not mutually exclusive in their formation and can occur in combination. Goudie (1973) and Thomas (1974, pp.42–84; 1994, pp.88–122) have described crust types, their possible developmental processes and geomorphological importance.

Not to be confused with crusts in the soil are weathering crusts of precipitated mineral substances that are at most a few millimeters thick and occur on rock surfaces. Iron oxides and manganese oxides are most common. They develop when water seeps out on the surface of the rock, evaporates and deposits its solution load. Compared to the underlying rock, which usually has been chemically weathered and is partially disintegrated, these crusts are relatively hard. If they have been destroyed in places by, for example, thermal exfoliation, denudation by water and wind removes the underlying disintegrated rock, leaving weathered out holes, or **tafoni**.

Weathering crusts are often visible on crystalline rock and debris in arid regions. Their iron and manganese content gives them a dark, slightly shiny colouring known as **desert varnish**. In humid climates they often occur on joint surfaces.

7.5.6 Stone lines

A **stone line** is defined as a horizon in the soil that has a significantly greater concentration of stones than the soil above and below. In profile they appear as a line of stones. Some consist of a thin, in some cases discontinuous, scattering of stones which lie next to, but not on top of, one another. Others are several tens of centimetres thick.

Stone lines are found in many parts of the tropics and have also been observed in the USA (Parizek and Woodruff, 1957; Ruhe, 1959). The stones in tropical stone lines are usually quartz fragments which originate in resistant quartz veins and have been left during chemical weathering of the bedrock. More seldom they are rounded quartz concretions, the result of secondary precipitation from mobile silicic acid (Stocking, 1978).

Stone lines show a range of forms indicating that they can develop in various ways (Vogt, 1966; Thomas, 1974; Stocking, 1978; Ahnert, 1983). One cause is **bioturbatio**n, the redisposition of material in the soil through the activity of organisms. Termites are of particular importance. Their mounds can reach a density of 100 per hectare in the savannas of the subhumid tropics (M.A.J. Williams, 1968). The termites bring large quantities of fine material out of the soil to the surface and effect a vertical sorting of grain sizes. In a soil in which quartz pebbles were initially evenly distributed, they are gradually moved downwards and are redeposited, collecting at the lower boundary of the termite activity as a stone line. Its thickness depends on the skeletal share of the stones in the original soil and the total depth of the sorting process.

A second possible cause of stone line development is the accumulation of stones as residual debris of a previously existing soil whose fine components have been washed away areally or removed by the wind. The residual material has later been covered by other fine material transported in by water, wind or mass movements. Instead of an autochthonous skeletal soil residue, one developed in place, the stone lines can also be allochthonous in origin, that is, transported in from elsewhere and then covered by fine material.

The third cause of stone lines are episodic and repeated local mass movements, such as soil flowage or small mudflows whose flow lobes of fine material and stones lie on top of one another. If the water content of the flowing mass is high

FIGURE 7.12 Allochthonous stone line near Cajati in Sao Paulo State in Brazil.

and its viscosity therefore low, stone lines can develop when the coarse material sinks during the flow and collects at the bottom of the mass. If the process is repeated in the same location, several stone lines can develop one above the other, each separated by fine material.

The fine material in an unsorted flow lobe deposit can also be washed out, leaving a residual layer of debris, if the time period between flow events in one place is very long. This type of allochthonous stone line is then covered by the next mass movement (Figure 7.12).

7.6 THE RELATIVE SHARE OF MECHANICAL AND CHEMICAL WEATHERING IN DIFFERENT MORPHOCLIMATES

Mechanical weathering predominates on rocks and in environmental conditions dissimilar to those on which chemical weathering predominates. The lithological, structural and textural characteristics that are particularly advantageous for mechanical weathering are clastic sedimentary rocks with many joints, relatively large pores and weak cementing between the mineral grains. Chemical weathering is particularly effective on crystalline rocks and carbonate sedimentary rocks that are partially or wholly formed of easily decomposed minerals.

Favourable environmental conditions for mechanical weathering include large temperature fluctuations with frequent freeze–thaw events. Its effect is greatest on bare rock. Even a thin regolith cover reduces the amplitude of the fluctuations considerably. Chemical weathering requires, above all, moisture in contact with the bedrock. In a permanently humid climate, this condition is fulfilled already by the storage of water in only a thin regolith cover. In the savanna climate of the subhumid tropics or the winter rain climate of the Mediterranean area, the regolith must be thicker if it is not to dry out in the dry seasons. Rainwater seldom penetrates deeply into the soil in deserts and semi-deserts, and much of it rises again by capillary movement and evaporates on the surface.

In general, mechanical weathering decreases in importance where a regolith cover is present and there is insufficient denudation to remove it. Under such a cover mechanical weathering is reduced and initially, at least, chemical weathering increased.

In the extreme cold climates of the polar regions or in the desert and semi-deserts of the tropics and subtropics chemical weathering, and with it soil formation, tends to dominate in low-lying areas where there is little or no denudation, even though it is very weak in absolute terms because in polar regions the water flows only during the summer and in the deserts it is rarely available at all. An indication of chemical weathering in these locations is the presence of sparse vegetation which obtains its dissolved nutrients from the chemically weathered soil. On level land in valley bottoms in northern Greenland at 82–83° N. plants grow in sufficient quantities to feed small herds of musk ox for the entire year. Vegetation is not completely absent in deserts where salt crusts indicate chemical processes in the soil. However, on the slopes in these extreme climates, denudation is usually intensive and the bedrock often exposed. Water drains off quickly and the debris is moved downslope so that there is little chance for chemical weathering to take place or for vegetation to grow. In both polar and desert climates, therefore, mechanical weathering predominates on the slopes and chemical weathering on the level areas, in the valley bottoms and at the foot of the slopes. This contrast which is expressed by the spatial distribution of vegetation and soil characterizes the landscape of both climatic regions.

In the humid and subhumid tropics, the temperate climates of the middle latitudes and the subpolar forest and tundra, the bedrock is largely covered by soil and vegetation. Outcrops of bedrock on which intensive denudation can take place are few and limited to areas such as coastal cliffs, eroded river banks, steep slopes with ledges and spurs and the summits of mountains. Chemical weathering dominates almost everywhere, though more strongly in the warm humid climates than in the cooler regions.

The high mountain areas of the world are a special case. Above the treeline in the alpine subnival and nival altitudinal zones, climatic conditions are similar to those in the polar regions with the difference that at low latitudes, the daily, and at high latitudes, the annual, temperature fluctuations are significant. In high mountains in the tropics, the freeze–thaw frequency is higher but penetration below the surface is less than in polar regions, with the result that mechanical weathering rates are different in the two areas. In addition, in high mountains, the level areas on which chemical, rather than mechanical, weathering may dominate, together with soil formation and vegetation growth, are generally fewer and smaller than in the less mountainous polar regions.

PROCESS RESPONSE SYSTEMS OF DENUDATION

8.1.1 Denudation and erosion

Denudation includes all processes of **areal downwearing**. Denudation processes remove the regolith and eventually leave the bedrock bare if the material that has been removed is not simultaneously replaced by material supplied from upslope or by new regolith production through local weathering. **Erosion** is downwearing along approximately linear erosion paths such as streams (fluvial erosion) or valley glaciers (glacial erosion).

Denudation is the link between the areally distributed process of debris production by weathering and the linear erosion paths that transport the material they receive and eventually deposit or bring it as dissolved load to the sea.

The term denudation is often applied to downwearing over large areas and erosion to the downwearing of individual slopes and parts of slopes. The removal of soil from agricultural land, for example, is known as soil erosion. There is no clear boundary in every case but it is useful to make a distinction between the areal and linear aspects of downwearing.

8.1.2 Types of denudation processes

All denudation processes involve the transport of rock material and are, therefore, subject to the force of gravity. They can be subdivided by the type of material transported.

1. Gravity determined mass movements of rock and debris: block fall, rock fall, landslides, blockslides, debris slides in coarse material;
2. Mass movement of the regolith, usually in combination with pore water, ice or snow: mudflows, avalanche denudation, earthflows, creep denudation;
3. Movement of the regolith with a substantial freezing effect, usually with a permanently frozen subsoil: cryoturbation, gelifluction (solifluction), rock glaciers, block streams;
4. Removal of dissolved substances in the soil and groundwater;
5. Downwearing and transport of material resulting from the impact of raindrops and the unconcentrated runoff of rainwater: splash, wash denudation;
6. Denudation and transport by the wind (deflation or aeolian denudation);
7. Denudation and transport by glacier ice (glacial denudation).

The removal of dissolved substances in the soil and groundwater are discussed in Chapter 21 on karst forms and the erosion and transport by glacier ice in Chapter 22 on the glacial system. The other process types are discussed in this chapter.

8.2 PHYSICAL BASIS OF DENUDATIVE MASS MOVEMENTS

8.2.1 Slope angle and the effect of the force of gravity

All substances on the earth are subject to the pull of the **earth's gravity**, K, which is directed downwards to the centre of the earth. The general physical definition of mechanical force is mass times acceleration. In this case, therefore, the product of the **mass**, m, of a particular substance and the earth's **gravitational acceleration** $g = 981$ m/s² is

$$K = mg \qquad (8.1)$$

Almost the entire surface of the earth is composed of slopes that border on one another. The effect of gravity on the movement of material on these slopes is a function of the **slope angle**. Because the movement of the material usually takes place parallel to the slope but, at the same time, the material rests on the slope's surface, it is useful to separate the vertically downward directed gravitational acceleration g into two vectors: one vector (τ) parallel to the downslope surface and the other (σ) which is also directed downwards but at right angles to the surface. The vectors σ and τ are at right angles to one another as the sides of a vector parallelogram whose resultant is g (Fig. 8.1). If the vector τ is large enough there is a shearing off and movement of the material downslope. The size of the vector is

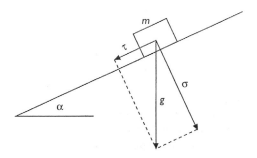

FIGURE 8.1 Vector parallelogram of gravity acceleration on a slope.

$$\tau = g \sin \alpha \qquad (8.2)$$

where α is the slope angle in degrees. Related to a unit of surface area, for example 1 cm², and a unit of mass, for example 1 gram, τ is the **shear stress**. Using equation (8.1) on a mass m such as a block on the slope, the **shear force** acting parallel to the slope is

$$K_s = m\tau = mg \sin\alpha \qquad (8.3)$$

The other vector is the component of the gravity acceleration which brings about the pressure of the mass m on a slope

$$\sigma = g \cos\alpha \qquad (8.4)$$

This vector, which is expressed per unit of surface area and of mass, is termed **normal stress** because it is normal, i.e. at right angles, to the slope surface. The related pressure force or **normal force** K_n of the mass m is therefore

$$K_n\ m\sigma = mg \cos \alpha \qquad (8.5)$$

On a perpendicular wall ($\sin\alpha = 1.0$, $\cos\alpha = 0.0$) the normal stress is equal to zero and the shear stress equal to the gravitational acceleration g; on a horizontal surface ($\sin \alpha = 0.0$, $\cos \alpha = 1.0$), the reverse is true. Slope surfaces are expressed physically as inclined planes on which both τ and σ are greater than zero.

8.2.2 Plastic flow and Coulomb's Law

Most denudational debris movements on slopes involve **plastic flow**. This is a mass movement of unconsolidated material that begins when the shear stress τ exceeds a certain threshold. The threshold is the **critical shearing resistance** σ. In a material that is not cohesive, such as dry sand, s depends only on the gravitational acceleration g (constant) and the internal friction of the material which is a function of the shape of the grains and the way in which they lie in the deposit. A loose mass of rounded grains has a lower internal friction than one of angular grains because the latter become more closely wedged together. If angular grains have been deposited only loosely, the internal friction is less than in a densely packed layer in which the edges of the grains fit more closely into the neighbouring pore spaces.

TABLE 8.1 Representative values of the angle of internal friction

Material	φ (degrees)	
	Loose	Dense
Silty sand	27–33	30–34
Round-grained sand with uniform grain size	27.5	34
Angular sand	33	45
Sandy gravel	35	50

Source: Terzaghi and Peck (1967, p. 107).

The **angle of internal friction** is the angle at which a cohesionless mass of loose material is set in motion, that is, at which the shear stress becomes greater than the internal friction of the material. It is usually designated with φ (phi) and expresses the natural angle of repose which, as a general rule, is greater the more angular the grains of the material, the more densely packed the grains and the greater the variation in the grain size of the material (Table 8.1).

The critical shear stress or shearing resistance s of cohesionless loose material per surface unit and unit of mass is therefore (see equation (8.2))

$$s = g \sin \phi \qquad (8.6)$$

The corresponding critical normal stress is (equation (8.4))

$$\sigma = g \cos \phi \qquad (8.7)$$

Because $s/\sigma = (\sin \phi / \cos \phi) = \tan \phi$, equation (8.6) and

$$s = \sigma \tan \phi \qquad (8.8)$$

are equivalent. In cohesive material, (cohesion c) the various ways in which the grains of the material are bound together, form an additional impediment to mobility, so that

$$s = \sigma \tan \phi + c \qquad (8.9)$$

is valid. This equation for the shearing resistance, named **Coulomb's Law** after its originator C.A.Coulomb (1776), is of fundamental importance in the investigation of soil mechanics in mass movements (Terzaghi, 1943). The critical shear stress for mass movement with a constant angle of friction and constant cohesion c (both characteristics of the material), is a linear function of the normal stress, which in turn, is a function of cos α. The steeper, therefore, the slope, the less is the normal stress and the less the critical shear stress necessary for movement.

The material constants φ and c are usually determined experimentally with a shear box in which samples of the material are pressed down by a vertical load (Fig. 8.2a). The load exerts the normal force K_n on the sample (equation (8.5)). A shear force K_s is applied at right angles, horizontally, (equation (8.3)) and increased gradually until the sample shears off. The shear stress at the moment of shearing off is then equal to the shearing resistance s. If the tests are repeated with other loads K_n, a straight line $s = f(\sigma)$ can be constructed from the additional measurement values, the gradient of which is equal to tan φ and the ordinate segment equal to the cohesion c (Fig. 8.2b).

The **cohesion** is defined as the critical shear stress of the material in the absence of any

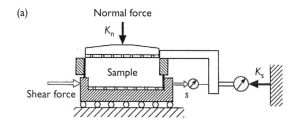

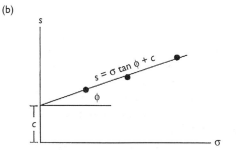

FIGURE 8.2 (a) Shear box; (b) critical tangential stress s as a function of the normal stress σ (after Gudehus, 1981, pp. 102–3).

normal stress ($\sigma = 0$, Fig. 8.2b), for example, on a vertical wall ($\cos 90° = 0.0$, therefore also $\sigma = 0$) from which part of the rock overhangs. The overhang is prevented from falling only as long as its cohesion with the rock wall is greater than the shear stress exerted on it, which, in this case, is equal to the gravitational acceleration g.

8.2.3 Variability of cohesion and critical tangential stress

Cohesion and critical tangential stress are subject to environmental changes. In bedrock, pressure release joints (section 7.3.6) may develop parallel to the slope and weaken the internal cohesion of the rock. Widening of tectonic joints by frost weathering has a similar effect. In both cases, the cohesion between the neighbouring sections of the rock is weakened. By contrast, chemical weathering of the cementing material between the grains of sedimentary rocks reduces cohesion within the rock itself. Cohesion can also be increased gradually by the precipitation of additional cementing material (iron oxide, calcium carbonate, silica) into the pores of the rocks.

In regolith, the presence of soil water changes the cohesion and the critical shear stress. Seasonal freezing and thawing loosens the soil structure; the pressure or suction effect of the water in the pores in the regolith also influences the cohesion.

Negative pore water pressure is present when the pores in unconsolidated material are only partially wetted and still have some air in them. The water in the pores between the particles forms numerous small surfaces which, because of their surface tension, bind the particles together more strongly than if the material were totally dry. A type of suction, or negative pressure, develops in the pores which increases the critical shear stress and the stability. In an unconsolidated material, such as pure sand, this effect is known as apparent cohesion (Terzaghi and Peck, 1967, p. 108) and is evident in the steep slopes and small tunnels which can be dug in moist sand, something impossible in dry sand.

If the pores are completely filled with water, there are no internal surfaces in the pore water and the surface tension with its bonding effect is no longer present. Instead, because the pores are filled with water, there is the addition of a buoyant force that further reduces the coherence of the grains. A **positive pore water pressure** exists which reduces the critical shear stress and may lead to viscous flow of the soil.

In determining the critical shear stress in material containing water, the modifying effect of the pore water pressure (u) must be taken into account. This is done by replacing the normal stress in the Coulomb law (equation (8.9)) by the effective normal stress ($\sigma - u$) so that

$$s = (\sigma - u) \tan \phi + c \qquad (8.10)$$

With negative pore water pressure u, the critical shear stress s and thus the stability of the material become greater (and with positive pore water pressure, smaller) than in a completely dry state.

If completely saturated sand is compressed, by stamping on wet sand on a beach, for example, the sand grains are shifted so that the pores between them become smaller and the pore water pressure increases. If it becomes as great as the normal stress so that $(\sigma - u) = 0$, then the sand loses its shear resistance completely and becomes **quicksand**, which behaves like a fluid.

8.2.4 Viscous flow

The most important difference between viscous flow and plastic flow is that plastic flow must exceed a critical shear stress before the material flows. No such threshold exists for viscous flow; movement takes place as long as the shear stress $\tau > 0$. The material must also have an internal friction that is close to zero and be practically without cohesion. This shear stress is expressed in equation (8.2).

Viscous flow is, in effect, a movement of fluids. It occurs as a slow mass movement in clayey regolith and in clay sediments (Brunsden, 1979, p. 177). In contrast to the discontinuous, often only seasonally active, plastic flow, viscous flow takes place more or less continually and is known as **continuous creep**. Creep processes are described in section 8.4.

8.2.5 Critical height of slopes

The stability or instability of a slope depends on:

1. the properties of the material, which are the **cohesion** c, the **density** γ and the **angle of internal friction** ϕ;
2. the geometric characteristics, the **slope angle** α and **the relative height** H of the slope, which is the height difference between the slope crest and the slope foot.

The density γ is, in this case, the **bulk density** of the material and incorporates all components including the pores. Its dimensions are mass per unit volume, in this case tonnes per cubic metre.

The **critical height** H_c of a slope is the threshold value of the relative height which when exceeded, under given conditions of cohesion, bulk density, internal friction and slope angle, would lead to instability, for example, to landslides. This is expressed quantitatively as

$$H_c = 4c \sin \alpha \cos \phi / \gamma(1.0 - \cos (\alpha - \phi)) \quad (8.11)$$

(Carson, 1971, p. 101). A slope is stable when its actual height H is lower than its critical height H_c; it is unstable when $H_c < H$. Equation (8.11) is valid for all values of $\alpha > \phi$. When $\alpha = \phi$, then the denominator in equation (8.11) has the value zero, that is, the critical height H_c becomes infinity. In other words, a slope whose inclination is equal to, or smaller than, the angle of internal friction of the material remains stable whatever its relative height.

When $\alpha > \phi$, the slope stability depends on the parameter values in equation (8.11). The greater the cohesion c or the angle of internal friction ϕ, the greater is H_c and, therefore the slope stability. The steeper the slope α or the greater the bulk density of the material, the lower is H_c. The variability of the cohesion c was discussed in section 8.2.3. The bulk density γ of the material can also change the absorption of water into the pores.

Slopes of loess material in the Middle West of the USA can be used as an example. The loess has the following parameters (data from Lohnes and Handy, 1968).

cohesion $c = 0.091$ kg cm^{-2} = 0.91 t/m^2
angle of internal friction $\phi = 25°$.
        bulk density = 1.208 t/m^3

TABLE 8.2 Critical height of slope in loess at different slope angles

Slope angle $\alpha(°)$	Critical height H_c(m)
90	4.7
80	6.3
70	8.8
60	13.0
50	22.3
40	51.5
30	358.8
25	Infinite

Source: Based on data from Lohnes and Handy (1968)..

Insertion of these values in equation (8.11) gives the critical height values H_c for the slope angles shown in Table 8.2. They indicate that vertical walls in loess up to more than 4 m in height or 70° slopes up to more than 8 m high can be stable. This agrees, in general, with observations of the steepness of loess slopes in sunken roads in the Kaiserstuhl region and other loess areas in Germany. Higher loess slopes on which the critical height at slope angles of less than 70° could be tested in field comparisons are rare but might be found in north China where the loess deposits are much thicker than elsewhere.

Critical height is of major importance for any analysis of slope stability, especially in the investigation of landslides.

Nash has discussed a large number of methods of slope stability in a volume on the problems of slope stability edited by Anderson and Richards (1987, pp. 11–75,). Brunsden, Prior and co-authors (Brunsden and Prior, 1984) have discussed all types of mass movements on slopes.

8.3 ROCK FALL DENUDATION AND LANDSLIDES

8.3.1 Block fall

A block falls from a rock wall when the gravity determined shear stress τ on the block exceeds the shearing resistance σ (equation (8.9)). For this to occur the wall must be steep although not necessarily perpendicular. The detachment fissure separating the block from the outcrop is usually already present as a zone of weakness in the rock, a joint, for example. Root pressure, repeated freeze–thaw events or chemical weathering further weakens the connection to the outcrop along the joint (section 7.3.2).

The actual fall of the block can be caused by a seemingly unimportant event such as a mild night frost followed by a morning thaw. Investigations by Rapp (1960) in Kärkevagge in northern Sweden have shown that there are clear frequency maxima of block fall during periods of thaw. During the intervening days of frost without thaw, there was almost no block fall at all.

If block falls are sufficiently frequent, a **talus** collects below the rock wall as an intermediate storage area where the blocks are disintegrated further by weathering and from where they are transported away by other denudation processes. The larger blocks generally lie on the lower part of the talus. With their greater mass they roll or spring farther down the slope than smaller blocks.

Where the paths of the blocks are confined to a narrow zone, a **talus cone** forms with its apex adjoining the lower part of the rock wall. The gradients of taluses and talus cones are more than 20° and vary depending on the angle of internal friction of the material.

8.3.2 Rock fall

Rock falls differ from block falls in that an entire rock wall, or a large part of it, falls at one time rather than individual blocks. The detachment surface or **shear plane**, also usually follows an existing joint. A scar remains on the rock. The freshly exposed, smooth back wall is generally a joint surface and the upper limit of the scar a vaulted overhang (Fig. 8.3). The vaulted form is an expression of the distribution of shear stresses and the shearing resistance of the rock at the time of the rock fall. It is a stability form in that the tension exerted from above is diverted to the sides, similar to the vaulted ceiling of a Gothic cathedral or in the arch of a stone bridge.

Many rock falls are caused by undercutting and steepening of the rock wall because of lateral stream erosion or accelerated back weathering of the wall foot due to seepage or small springs. A joint running parallel to a rock wall can also be widened by pressure release or other weathering processes and become the shear plane of a rock fall.

FIGURE 8.3 Rockfall on a cuesta scarp spur in the Jurassic Entrada Sandstone near Fort Wingat, New Mexico, USA.

The triggering of a rock fall after an extended period of preparation is also often a minor event such as a small earthquake, lightning or heavy rainfall. In contrast to the talus of block falls which is of various ages and at different stages of weathering, **rock fall talus** is of the same age. This has consequences for the further weathering and subsequent morphological development of this type of talus.

8.3.3 Landslides and landslips

There is no very sharp boundary between block falls and rock falls or between rock falls and **landslides**, except that rock falls only occur on bare rock walls and landslides also take place on less steep, soil-covered slopes. To be defined as a landslide, four criteria must be fulfilled:

1. the movement must be rapid, lasting seconds or at most a few minutes;
2. the sliding surface, the shear plane, must go through the bedrock so that bedrock composes the largest part of the landslide;
3. the sliding mass must disintegrate during the movement;

4. the slope area affected by the movement and the volume of the rock mass in motion must be large enough for it to be defined as a landslide by the people living in the area. A relative definition of the size of the mass seems more appropiate than one expressed by an arbitrary number of square metres, tonnes or cubic metres.

Once the landslide has begun to move, an air cushion may develop on its subsurface that can further reduce friction enormously and make speeds of several hundreds of kilometres per hour possible (Shreve, 1968).

Landslips move much more slowly than landslides (Fig. 8.4) and the sliding mass can, in part, maintain its coherence. Both are set in motion when the critical height H_c becomes smaller than the actual height of the mountain slope. The causes for the reduction of H_c are very varied (section 8.2.5, especially equation (8.11)) and depend on local conditions. They may include:

1. Increase in the actual relative height of the mountain slope above the critical height by

Figure 8.4 Landslide in deeply weathered saprolite in the Sierra da Mantiqueira, Sao Paulo State, Brazil.

incision of a stream flowing along the slope foot.

2. Slope steepening by undercutting of the slope as a result of (a) lateral erosion against the slope by a stream at the slope foot, (b) undercutting by springs or seepage water at the slope foot, or (c) erosion by surf at the foot of a coastal slope.

3. Reduction of cohesion on a large scale in the rock caused by progressive weathering along joints running parallel to the slope, widening of the joints as a result of lateral pressure release, loosening of the rock strata by an earthquake, or increased water circulation on the joints or in bedding planes that might act as potential sliding surfaces. In the latter case, sliding is particularly likely to occur when the strata dip downslope at an angle of dip that is smaller than the angle of the slope itself. The dipping bedding planes, or interbedded strata with a lower shear resistance act, in this case, as a zone of weakness that intersects with the steeper slope surface along the strike of the rock.

4. Increase in the bulk density of the rock, and thereby its mass, because of an increase in water content after a period of high precipitation.

The additional load on the land surface of buildings or other structures increases the mass and has the same effect as an increase in density in the rock. In California, hill slopes that were previously stable have slid after houses were built on them, often following heavy winter rainfalls which penetrated the soil and increased its density.

The characteristic landforms associated with landslides and landslips have three components: the **landslide scar**, the **slide track** or **sliding plane**, and the **landslide (or landslip) mass** or **landslide debris**.

If the scar does not reach to the summit or crest of the mountain but begins lower down, the landslide scar is a concave hollow in the bedrock. Downslope the scar merges into the slide track, the bottom of which is covered with the landslide debris. The landslide track is also concave but flatter than the landslide scar. The concavity is determined by the distribution of stresses in

the rock before movement was initiated. The rocks in the lower part are under pressure from the rocks in the upper part of the slope and yielding to this pressure is only possible by a movement away from the slope foot in the direction of the valley floor. The direction of movement and therefore the sliding surface along which the moving mass shears off, is flatter and more outwards directed in the lower slope than in the upper slope.

Landslide debris usually lies as an irregular humpy mass at the foot of the slide track. If a landslide takes place in a narrow valley, the force of the slide event may carry debris high up on the opposite valley side. Slow-moving landslips tend to form a debris tongue with convex sides and bulges in which the debris is piled up locally because of the varying rates of movement in the tongue. The crests of the bulges are transversal to the direction of movement.

In a narrow mountain valley, landslide debris can block drainage and cause the development of a landslide dammed lake, which may be filled later by stream deposits. Often the pressure of the water behind the dam is too great and the sudden emptying of the lake causes a flood disaster downstream. Major floods of this type occurred in the Brenno valley above Biasca (Ticino, Switzerland) in the 16th century, in the Himalayas in 1893, in the Pamir in 1911 and in Wyoming, USA in 1925 (Zaruba and Mencl, 1969, pp. 12–13).

Some of the largest slides in this century include the Hebgen landslide in Wyoming in 1959, the Hope landslide in 1965 in British Columbia, Canada (Fig. 8.5), the Vaiont landslide, 1963, Friaul, France and at Sondrio in 1987 in the Veltlin in Italy.

The Vaiont landslide in a side valley of the Piave River near Longarone, had a volume of 2.6×10^8 m^3. Most of this mass fell into a reservoir lake. The concrete dam held, but the water was displaced by the landslide and poured over the dam in a 100 m high shock wave. The wave destroyed several villages and almost 2000 people died.

Numerous landslides occurred in the Alps at the end of the last ice age about 10 000 years ago. Glacial erosion had transformed the valleys into

Figure 8.5 Hope landslide in southern British Columbia, Canada. The landslide occurred on 9 January 1965 and was probably triggered by a minor earthquake. The landslide mass was more than 10^8t and the volume more than 4×10^7 m^3 (photograph taken in 1972).

steep-sided troughs which the glacier ice stabilized and protected from major temperature fluctuations. Once the ice melted, many of the slopes were oversteepened and became unstable. The largest post-glacial landslide, with a slide debris mass of over 12 km^3, occurred at Flims, near Chur in Switzerland in the Vorderrhein Valley. The river has since cut a narrow V-shaped valley into it.

8.3.4 Slumps

Soft rocks, such as marls, clays, claystones and slates, that can be deformed plastically and have a relatively low stability, may slide along a single almost cylindrical shear surface with a backwards rotation when their critical height H_c is exceeded (Fig. 8.6).

The shear surface need not have formed on a discontinuity in the rock but may result directly from the existing distribution of stresses. In large slides of this type, the lowest point of the shear surface lies lower than the original slope foot so that part of the slope foreland is included in the movement. The rotation during the sliding and the position of the shear surface have the effect that, while the upper part of the mass is sliding and tipping backwards, the rock at its base moves forward and is pushed upwards in the slope foreland. The upper part of the slide maintains its internal structure. This type of slide is termed a **slump**.

Slumps are the typical slides on cuesta scarps where nearly horizontal layers of hard sandstone or limestone overlie slump-prone rocks containing clays. The hard rocks above the clays produce an additional burden which increases the instability of the clay strata. In the south German cuesta scarpland, these conditions are present in the Keuper formations (Blume and Remmele, 1989, and Fig. 8.6a).

The slumps at Folkestone Warren in England on the Channel coast are also of this type (Ward, 1945; Hutchinson, 1969; Fig.8.6b). An approximately 50 m thick layer of clay and marl lies below 120 m of chalk at sea level. The surf has eroded the clay and undercut and steepened the slope, so that the critical height decreased. In addition, the bulk density and therefore the burden of the chalk is increased by groundwater

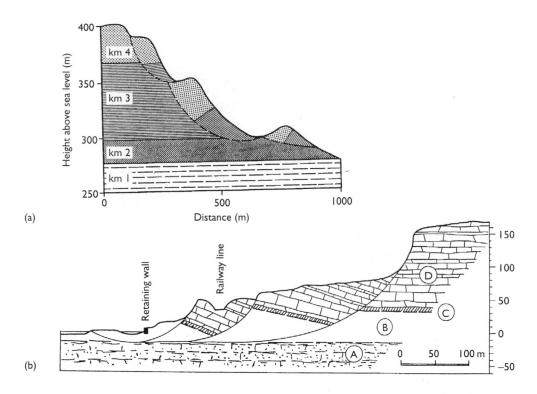

FIGURE 8.6 (a) Slumps in the Keuper upland of the Kraichgau, Germany (schematic, after Blume and Remmele, 1989); (b) Slumps at Folkestone Warren (after Ward, 1945, from Zaruba and Mencl, 1967, p. 63). A: glauconitic sandstone (upper Greensand); B: claystone (Gault clays); C: glauconitic marl; D: marly chalk (lower chalk).

stored in the pores of the chalk above the impermeable clays. Below the clays, and below sea level, is a sandstone that limits the possibility of sliding below.

Figure 8.6b shows several slumps one behind the other. After slumping, the base of the older slump block lying further seaward was attacked by the surf, and its size reduced. It lost its function as a sufficiently heavy obstacle to further slumping, and a second slump block slid behind the first which was moved forward again. The railway line from Dover to London lies along the slump and an attempt has been made to stop movement by building a heavy retaining wall to prevent wave erosion and provide a counterweight to further rotational movement on the slump block. Despite this, the railway lines shift now and then and their position has to be corrected. Movements still take place but their intensity has decreased.

The forward rotation of blocks is termed **cambering**. When the less resistant rock in a cuesta scarp is deformed by the burden of the overlying more resistant rock, without developing a shear surface, the front edge of the resistant rock tilts forward in a downslope direction.

8.3.5 Debris slides and avalanche transport

Talus is built up by blocks falling from the rock walls above. Accretion in the upper part of the talus accumulation means that the talus slope becomes steeper until the angle of internal friction of the fragments in the talus is exceeded and they become unstable. The blocks arriving at the upper end are then transported downslope by **debris slides**. The talus foot advances as a result and the angle of inclination of the talus surface is again below the critical value. Debris slides

Figure 8.7 Melting avalanche containing debris in the Stubai Valley, Austria (photograph taken in April 1975).

belong to the normal process response system of talus development.

In high mountains, debris transport on the talus and also the form of the talus itself can be considerably altered by **avalanches**, particularly in spring when the avalanches of wet snow transport soil and rock debris with the snow. The avalanche tends to transport the load over the greater part of the talus and deposit it only at the lower end of the talus area (Fig. 8.7). Frequent occurrence of this type of avalanche decreases the slope angle of the talus and its profile becomes concave (Caine, 1969; Embleton, 1979). In the upper part of the slope, avalanches generally follow pre-existing hollows such as dells or small valleys which are further eroded by the avalanches into **avalanche trenches**. Because of the damage caused by avalanches, they are an important research area in applied geomorphology. When large quantities of meltwater are mixed with avalanches of wet snow, **slushflows** develop whose flow is similar to that of mudflows (Barsch *et al.* 1993).

Talus formation and the transport of debris influence the development of forms on bedrock slopes because the talus covers and protects the bedrock from further denudation, while the debris-free upper slope continues to be worn back by weathering and denudation.

8.3.6 Mudflows

Precipitation or meltwater flows rapidly off a talus composed of large-pored coarse debris. If the talus contains a large component of fine material, the pores may periodically become filled with water. Positive pore water pressure then develops (section 8.2.3) and the talus debris moves downslope as a **mudflow**. Water flows out from the sides of the mudflow so that friction increases in this part of the flow and the rate of flow is lower than at the centre. Because of this damming of the debris on the sides of the mudflow, **mudflow levees** or debris ridges form, while along the axis of the mudflow, where movement is more rapid, a channel like hollow develops.

The prerequisites for a mudflow to occur are:

1. an extensive accumulation of debris with a sufficiently large proportion of fine material;
2. a relatively infrequent supply of heavy rainfall or meltwater in quantities large enough to fill the pore spaces in the accumulated debris and create positive pore water pressure;
3. a steep gradient.

Mudflow tracks lie mostly in hollows or channels on the slope which become further deepened and have their form accentuated by abrasion during repeated mudflow events. At the foot of the slope, the mudflow slows and widens into a tongue until movement ceases. If a number of mudflows occur in the same track, the tongues overlie one another and form a **mudflow cone**, a half cone at the slope foot, the upper pointed end of which indicates the boundary between the transport track above and the accumulated material below.

Mudflow cones have a gradient of 8–12°, less steep than the 20° angle of a talus cone formed of fallen blocks and steeper than alluvial fans of stream sediment (Fischer, 1965). The three types of cone also differ in their composition and the sorting of their material. Talus is composed of mostly unsorted coarse material. In mudflow cones the individual tongues are an unsorted mixture of coarse and fine material but there is some layering of the tongues within the cone. Alluvial fans have well sorted layers of sand and coarser material. The long-term frequency of mudflow events has been investigated by Strunk (1988, 1991) in various parts of the eastern Alps using annual growth rings of trees whose growth has been disturbed by mudflows.

8.3.7 Earthflows

Earthflows, first defined by Sharpe in 1938, are flow movements of regolith on a slope, similar to mudflows but generally slower, shallower and covering a shorter distance. They are triggered by positive pore water pressure. In temperate climates they occur in late winter and early spring when snow meltwater has saturated the soil and freeze–thaw has also loosened the soil structure (Fig. 8.8).

FIGURE 8.8 Shallow earthflow in a road cut in Washington DC, USA, the result of melting groundfrost and positive pore water pressure in late winter.

Figure 8.9 Slide and earthflow in a road embankment in Aachen, Germany. Notice the block slide zone (above), the flow lobe (middle) and the alluvial fan (below). An older slide scar is visible on the right of the photograph.

On a slope, the saturation of the soil increases downslope so that the pore water pressure becomes positive some distance below the crest of the slope. Flow movement begins at the point at which it becomes positive and soil material is removed, causing slope steepening at this location. Instability is then created in the adjoining area above, which also begins to flow. The field of movement extends upslope, as a form of headward denudation (Ahnert, 1954). Upslope the pore water pressure is not positive so that the coherence of the soil is greater and the earthflow movement takes place as movement of soil blocks whose internal structure remains intact. The upper edge of the field of movement is a concave scar from which the uppermost soil blocks have slid away with some backward rotation, similar to a slump (section 8.3.4) but much smaller in size.

The moving soil mass flows downslope as a tongue from the area on the slope where the earthflow began. The movement presses the pores together and part of the pore water exudes on to the surface and flows down eroding fine material, especially sand and silt, which is deposited as a fan at the slope foot. The three form and material components of earthflows are therefore (Fig. 8.9):

1. the concave niche-like scar with the uppermost, often backwards-rotated soil blocks;
2. the actual flow tongue in which positive pore water pressure dominates and in which the original soil structure has been destroyed by the movement;
3. the alluvial fan at the bottom of the slope formed by water flowing on the surface of the earthflow.

Earthflows are particularly common on artificial road cuts in unconsolidated material.

8.4 CREEP DENUDATION

8.4.1 Definition and types

The term **creep** in geomorphology covers all very slow downslope movements of unconsolidated material or of soft rock that is easily deformed under pressure. The annual rate of movement of creep is rarely more than 1–2 cm, usually much less. Three types are defined:

1. continuous creep;
2. creep due to expansion and contraction of the material;
3. splash creep.

8.4.2 Continuous creep

Continuous creep is a type of viscous flow (section 8.2.4) and occurs in material containing clay. It is the only creep that takes place both in unconsolidated material and in shales and claystones. The load of the scarp-forming rocks overlying clays and claystones may cause creep that leads to cambering or slumping (section 8.3.4). An important prerequisite for continuous creep is the extent to which the material can be deformed. This, in turn, depends on:

1. the clay content;
2. the water content;
3. the mass (the thickness and the bulk density of the regolith or rock layer);
4. the existing shear stress (the slope angle, equation (8.2)).

8.4.3 Creep due to freeze–thaw in the soil

Creep caused by expansion and contraction is very widespread and usually limited to the regolith. In climates in which there is freeze–thaw and also sufficient moisture in the soil, the soil water expands upon freezing and increases the soil volume. However, because there is material below and to the sides which prevents expansion downwards and laterally, the entire volume increase raises the soil surface at right angles to the slope. On thawing the regolith material sinks back, not at right angles to the slope but in the direction of gravity, perpendicularly downwards (Fig. 8.10). Depending on the slope angle α a net movement L takes place parallel to the slope surface

$$L = h \tan \alpha \qquad (8.12)$$

The expansion h is equal to the expansion of the frozen water in the pores. As a rough estimate, it can be assumed that in a saturated soil with 30 per cent pore volume, a frost event freezes the soil to a depth of 20 cm. The water volume is equal, therefore, to a layer of water 0.3×20 cm = 6 cm which expands about 10 per cent on freezing and brings about an elevation of the surface of $h = 6$ mm. On a slope with an inclination of $\alpha = 20°$, the net downslope movement of a particle at the soil surface after thawing would be

$$L = 6 \text{ mm} \times \tan 20° = 6 \times 0.36 = 2.1 \text{ mm}$$

In reality, the net movement with such a freezing depth is usually less because:

1. the soil is not completely saturated;
2. the cohesion present in the soil when it thaws causes some contraction in the direction of the original position of the particles, so that there is also an upslope component, usually very small and known as retrograde movement (Washburn, 1979, p. 200), which reduces the net downslope movement.

The distance particles move is greatest at the soil surface and decreases with increasing depth,

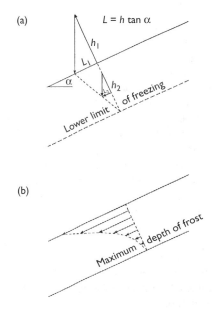

FIGURE 8.10 Movement of soil particles as a result of expansion and contraction due to freeze–thaw: (a) actual and net movement of regolith particles during one occurrence of freezing and thawing; (b) cumulative net movement of the regolith after several freeze–thaw events which penetrated to various depths.

reaching zero at the lower boundary of the soil layer that had been frozen. For a single freeze–thaw event, the decrease in the distance with depth is linear (Fig. 8.10a) but, because the soil freezes and thaws more often near the surface and the freeze–thaw frequency declines with increasing depth, particles near the surface, affected by this type of movement, move more often than those lying deeper in the soil. The cumulative distance a particle moves during a period of several years is, therefore, a non linear negative function of the depth and becomes zero at the greatest depth to which freeze–thaw has reached during the period (Fig.8.10b).

8.4.4 Creep as a result of swelling and shrinking

Expansion and contraction creep can also take place in a soil containing three-layer clay minerals capable of swelling, such as montmorillonite, which are alternately soaked and dried out. The net downslope movement is smaller than with freeze–thaw, mainly because the shrinking of clay also produces a movement component that is directed backwards, that is, upslope.

8.4.5 The effect of needle ice

The effect of needle ice on creep is limited to soil particles near the surface When the soil temperature sinks below freezing on clear nights, ice crystals form on unvegetated ground in the uppermost pores of the soil following sublimation of the moisture in the cooled air. The needle-shaped crystals grow upward and lift soil particles at right angles to the surface. The ice needles lie close to one another, similar to the teeth of a comb. Thawing the following day moves the particles vertically downward, or downslope if the surface is inclined or the ice needles themselves tip downslope. Needle ice can only develop in the uppermost layer of topsoil and its direct denudational effect is very small. More importantly, needle ice loosens the soil significantly at the surface and increases the effectiveness of other denudation processes such as wash denudation and deflation.

8.4.6 Splash creep and splash

Splash creep is dissimilar to other creep processes. It is the movement that results from the off-centre impact of heavy raindrops on small and medium-sized gravel particles, from 2 mm to 2 cm. The particles move a very short distance in the direction opposite to the side on which the raindrop impacted; on horizontal surfaces the particles are moved with equal probability in any direction and on a slope, with a net downslope directed movement (Moeyersons, 1975, 1983).

Splash affects finer grain sizes. As the raindrop strikes the soil, particles of sand, silt and clay, as well as small soil aggregates, are thrown in all directions in parabolic flight paths. The flight paths are longer downslope than upslope which results in a net material transport downslope approximately proportional to the sine of the angle of the slope (De Ploey and Savat, 1968; De Ploey, 1972; Moeyerson and De Ploey, 1976).

8.4.7 Qualitative evidence of creep processes in the field

A considerable effort is required to measure creep quantitatively because its rate of movement is so low. Qualitative signs of its presence are particularly important. The downslope-directed convex knees in the lower section of tree trunks are an indication. During their early years of growth, the young trees, trying to grow vertically, are pushed downslope by creep. The oldest, lowest section of the trunk is bent most strongly. Later the tree is rooted more firmly and the trunk thicker so that it resists the pressure from the downslope movement. The knee remains as a sign of the initial pressure and existence of creep during the tree's lifetime (Fig. 8.11).

Because creep declines with increasing depth (Fig.8.10b), rock fragments separated from the bedrock by weathering are transported further downslope the nearer they are to the surface. Particles near the surface have also been separated earlier and moving longer than particles at greater depth in the regolith. If the bedrock is composed of recognizable layers that outcrop at

FIGURE 8.11 Knees in tree trunks near Spiez, Switzerland.

an angle to the surface of the slope, the fragments in the soil have a pattern similar to that shown in Fig. 8.10b, displaced downslope in a curve. This pattern is termed **outcrop curvature** and can be observed in quarries and road cuts. Its presence does not, however, indicate whether creep is continuing at the present time. Outcrop curvature has been reproduced in the laboratory by Coutard, *et al.*, (1988).

Other apparent signs of creep such as the displacement of gravestones or garden walls and leaning power poles may not be evidence of natural creep but be due to the additional load of the objects themselves.

8.4.8 Creep on the moon and the planet Mars

In contrast to all other denudation processes, creep processes caused by expansion and contraction depend only on the direction of gravity and not on its strength. In principle, therefore, equation (8.12) is valid on the moon or Mars as well as on the earth. Instead of the freezing and thawing of water, the large temperature fluctuations at the surface give rise to alternating expansion and contraction of the surface material. Temperatures on the moon range from $+130°C$ to $-150°C$ and on Mars from $+15°C$ to $-85°C$. Splash and splash creep also take place, the result of the frequent impacts of small meteorites.

8.5 PERIGLACIAL DENUDATION PROCESSES

8.5.1 The term periglacial

Periglacial is a general term used to describe the natural conditions in regions of cold climates that are unglaciated but where the subsoil remains frozen for the entire year.

The morphoclimatic characteristics of periglacial regions are:

1. mean annual temperature below $0°C$;
2. sufficient summer heat to melt the entire snowfall so that no glaciers can develop.

The periglacial regions are primarily in the high latitudes of the Arctic regions of Europe, North America and Asia and on the relatively small unglaciated margins and islands of the Antarctic. Periglacial altitudinal zones occur in mountain ranges outside of the Polar regions. In the tropics they lie at 4000 m above sea level, and in the high mountains of the middle latitudes at more than 2000 m. Polewards the altitude of the zones declines further and eventually merges with the periglacial regions in the high latitudes (Fig. 8.12).

Because of the continuous summer sunshine in the high Arctic the region is more strongly heated than around the Polar Circle where the sun remains above the horizon for 24 hours on only one day of the year. The summer air temperatures in the low-lying areas of the high Arctic often remain above freezing for several weeks at a time with freeze–thaw occurring mainly at the beginning and end of the warm season, in contrast to the Polar Circle where it takes place almost daily throughout the summer

The mean annual temperature in the high Arctic remains below $0°C$, with the result that the deeper soil layers are permanently frozen, a condition known as **permafrost**. Nearer the surface, a layer of soil thaws in summer, known as the **active layer** because it is mobile in summer and geomorphologically active. In northern Greenland at 75–80°N the summer depth of thawing is about 40–50 cm, in Canada at 60°N more than 1.5 m. At this latitude in Canada it is present in only some areas and is termed **discontinuous permafrost**.

In the periglacial zones of mountains in the tropics, the annual fluctuation in temperature is small but the diurnal range is large. Instead of the seasonal thaw, a daily thaw prevails which lasts only a few hours and does not reach to any great depth. On the other hand, the frequency of freeze–thaw cycles is much greater in the tropics

than in polar regions, which causes differences in the periglacial processes.

Middle latitude mountain ranges, such as the Alps, show widespread evidence of periglacial activity at altitudes that are well below the permafrost limit and which result in **seasonal permafrost**. In these areas the subsoil thaws completely during the middle and late summer, but in the spring and early summer it is frozen and periglacial conditions predominate.

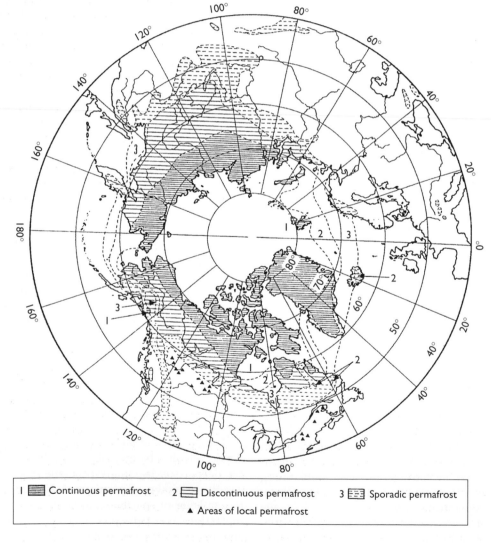

| 1 Continuous permafrost | 2 Discontinuous permafrost | 3 Sporadic permafrost |
| ▲ Areas of local permafrost |

Figure 8.12 Distribution of permafrost in the northern hemisphere; after Ives (1974) from Karte (1979).

The geomorphologically most important environmental conditions in periglacial regions are

1. the intensity and frequency of freeze–thaw events;
2. the imperviousness of the permafrost, which causes a retention of the precipitation and meltwater in the active layer and an increase in its mobility (section 8.2.3). Freezing and thawing of the soil water also produces pressure and movement of material.

Typical of phenomena in periglacial areas are gelifluction and the development of patterned ground, including stone nets, stone stripes and ice wedge patterns. Periglacial systems have been described by French (1976) and Washburn (1979).

8.5.2 Gelifluction (periglacial solifluction)

At the beginning of the century, Andersson (1906) used the term **solifluction** to describe downslope movements of water-saturated soil material. Today the term is generally limited to movement of this type above permafrost in periglacial areas. The permanently frozen subsoil is the prerequisite for the saturation of the thawed upper soil layer that makes this active layer mobile. To emphasize the importance of the frozen subsoil and at the same time to retain the more general meaning of the term solifluction, Baulig (1957, cited by Washburn, 1979) used **gelifluction** to describe the periglacial variation of soil flow.

In contrast to the expansion and contraction movement of creep (section 8.4), gelifluction is not only dependent on the direction of gravity but also its magnitude because the flow is similar to a viscous flow as long as the active layer is water saturated. For it to become saturated, however, the share of fine-grained material (fine sand, silt and also clay) in the active layer must be large enough for the pores to be small. Gelifluction cannot occur in coarse rock debris that has no matrix of finer material.

The annual movement by gelifluction depends on the slope angle, the grain and pore size and the pore volume of the regolith and also the amount of available water and the length of time that water is available. Gelifluction can take place on slopes with very low angles, 2–3°, if conditions are favourable.

High precipitation areas such as the west coast of northern Norway, Alaska or southern Patagonia receive considerable amounts of snowfall which are supplemented in summer by a large rainfall. In these areas, the active layer is saturated throughout the warmer months and gelifluction takes place continuously during this period. In northern Greenland and the northern Canadian archipelago where precipitation is very low, widespread gelifluction occurs only in the early phases of the summer thaw when there is enough meltwater from snow and soil ice to saturate the soil on the slopes. Later in the summer, the thawed layer often becomes too dry, especially in the upper parts of the slope where less water is available than lower down.

In Arctic periglacial regions, gelifluction moves the soil material at the surface downslope to a maximum of about 10 cm per year during the summer months. Below the surface the movement in the active layer decreases with increasing depth, although not as rapidly as creep. Mean rates of movement of 5 cm per year have been measured (Washburn, 1979, p. 213). Measurements of gelifluction in periglacial areas, including mountains outside the polar regions, always include some frost creep by expansion and contraction movements. In the Arctic, gelifluction often moves a regolith layer about 0.5 m thick. With a mean rate of 5 cm per year, the volume transport rate is 250 cm^3 per centimetre width of slope per year.

Local differences in the speed of flow cause **gelifluction tongues** or lobes to develop which move with a convex front downslope (Fig. 8.13). The front edge is usually steep and has a larger concentration of stones than the interior of the tongue (Fig. 8.14). Within a tongue, the long axis of the stones is generally oriented downslope.

Dells are a widespread form on slopes affected by gelifluction (Schmitthenner, 1926). They are elongated hollows with a shallow trough-shaped cross-section up to several hundred metres in length (Fig. 8.15). Downslope they adjoin a valley or valley head. The dells in the uplands and hilly areas of central Europe developed under the periglacial conditions of the last ice age when gelifluction dominated, although wash denudation by seasonal melt-water flow probably also affected their formation.

In the periglacial altitudinal zones of middle- and lower-latitude mountains, frequent diurnal freeze–thaw interrupts viscous gelifluction movements so that they occur in much smaller steps than in the polar regions. Because the diurnal thaw depth at higher altitudes is less than the seasonal thaw depth, the thickness of the regolith affected by gelifluction in mountains is also less. Steepness of the slopes does, however, increase the rate of movement of the shallow gelifluction layer unless slowed down by a dense grass vegetation cover and its root system which tends in general to impede gelifluction.

FIGURE 8.13 Periglacial gelifluction slope near Thule, north Greenland.

FIGURE 8.14 Late Pleistocene gelifluction lobes in the Cairngorm Mountains, Scotland.

FIGURE 8.15 Dell near Lindelbrunn in the Dahner Felsenland, southern Palatinate, Germany.

Periglacial wash denudation was probably a major factor in the formation of dells in central Europe. At the present time, meltwater flowing in early summer on the initially shallow active layer can remove considerable amounts of fine material in the Arctic. Because the depth of the thawed layer is limited, sheet wash dominates, with perhaps the formation of some small rills.

8.5.3 Nivation hollows and cyroplanation terraces

Nivation is a variation of gelifluction occurring in the immediate area of snow drifts that last well into the summer. The meltwater saturates the active layer on the downslope margin of the snow patch and causes increased gelifluction at this location, which, because of the repeated soaking, continues longer into the warmer months than on the snow-free slope area (Fig. 8.16). As the snow melts, the margin of the snow patch retreats upslope, and with it the area of intensive nivation. The process is repeated annually and a **nivation hollow** develops in the area of the snow drift, caused by the locally increased removal of material. With the enlargement of the hollow the amount of snow collecting annually in the hollow increases over the years, which in turn increases the intensity and seasonal duration of the nivation process. The back of the nivation hollow is steeper than the neighbouring slopes but the floor is relatively flat and forms a **nivation terrace**. Large nivation terraces are produced by the growing together of several nivation hollows lying at more or less the same height. They are **cryoplanation terraces** (Demek, 1968, 1969; Priesnitz, 1988) if intensive seasonal gelifluction continues after the snow patch has melted in late summer.

8.5.4 Stone nets and stone stripes

Stone nets, **stone polygons** or **stone rings** are stones sorted in patterns like nets with finer material within the mesh, which occur on approximately level ground over wide areas in periglacial regions (Fig. 8.17). They develop in a regolith cover that has a wide range of grain sizes from large stones to fine sand and silt. The proportion of coarse material must be large enough for closed nets to develop. The diameters of periglacial stone polygons range from 10 cm to 20 m.

The source material is often the unsorted ground moraine of a glacier. The coarser fragments are gradually brought to the surface by repeated frost heaving and accumulate there because, when thawing occurs, they sink back into the soil hollows less easily than the finer material. This heaving process is common in temperate zones where there is freeze–thaw and is a reason for the renewed presence each year of

FIGURE 8.16 Nivation near Moraine Lake in the Canadian Rockies.

FIGURE 8.17 Periglacial stone nets near Thule, north Greenland.

stones in ploughed fields, despite annual removal.

On the surface, the stones are moved horizontally with every freeze–thaw, usually away from the areas of fine material, which, because of the water stored in its pores, expands on freezing and arches upwards. The stones move, therefore, from the centres radially until they meet the stones moving from the neighbouring centres. With the continuation of the centrifugal movement of the stones, the areas of fine material become larger and gradually the stones that have moved to their periphery form a connected and more clearly defined polygonal stone net. Flattish stones often stand on their edges. In many cases the concentration of stones limiting the polygons extends below the surface almost to the base of the thawed layer. Various causes have been suggested for this, from convection type movement of the soil (cryoturbation) to pre-existing frost wedges or cracks caused by drying that are filled with coarse material. The latter seems, at least in some cases, very probable (Washburn, 1979, pp. 141–6 and 167–73). Differentiated cyroturbation within the soil leads to the disturbances of the soil structure in which former soil horizons are compressed and interrupted.

Interconnected nets require a high enough density of stones at the surface. When too few stones are present, incomplete nets, short stone strands or stone nests develop (Stingl, 1969, 1974; Schunke, 1975).

Small mesh stone nets with correspondingly small stones can be found outside the periglacial zone in areas with little or no vegetation. In this case the movement of the stones is usually random both in their direction and frequency of movement. The causes of movement are generally accidental kicks by moving animals or humans, and sometimes also irregular movements of vehicles, for example on unmarked parking lots covered with gravel (Fig. 8.18). The polygon pattern develops when the friction, either between the neighbouring stones or by already existing fissure patterns in the soil, prevents further movement (Ahnert, 1981a, 1994c).

On inclined surfaces in periglacial areas, the general downslope movement is superimposed on the centrifugal movement. Stone polygons on

FIGURE 8.18 Small sorted stone nets on a parking lot in Aachen, Germany. Total length of ruler is about 16 cm.

slopes are, therefore, elliptical with the long axis oriented in the gradient direction and elongate into **stone stripes** on steeper slopes (Washburn, 1979, pp. 153–6) forming downslope-directed narrow bands of stones separated by bands of fine material. The fine material moves two or three times faster than the stones in the stripes (Pissart, 1977, p. 152). There is also a lateral movement from each fine material stripe outwards towards the neighbouring stone stripes, evident from the upright position of many of the flatter stones. The lateral distances between stone stripes are generally similar to the diameter of stone polygons on flat ground.

8.5.5 Ice wedge nets

Water increases its volume by about one tenth when frozen. With further cooling below the freezing point, the ice, similar to all solids, contracts. In the cold winters of the higher latitudes, cooling of the frozen ground leads to this type of contraction, especially when there is a high content of ground ice and cooling is rapid (Washburn, 1979, p. 140). Contraction fissures, known as **frost cracks**, develop a polygon network over the surface (Fig. 8.19). During the summer thaw, they are filled by both meltwater and soil particles which fall into the fissures from their edges and walls.

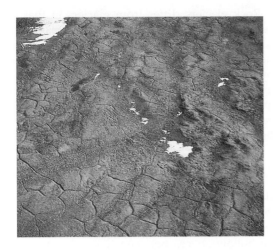

Figure 8.19 Ice wedge net and pingo-like upwarping near Thule in north Greenland. Photograph taken from low-level flight. Mesh width of the net is 10–20 m.

In the following winter the water in the fissures freezes with an approximate increase of 10 per cent in volume and forms an **ice wedge** which widens the fissure and at the same time pushes its edges upwards so that a furrow appears on the surface. Further cooling at low temperatures causes more contraction and fissure development, this time along the already existing wedge. During the subsequent thaw period more material falls into the fissure, which now becomes a permanent discontinuity in the soil that is widened each year. Ice wedge nets have a mesh width of up to several tens of metres. Stone polygons also develop within the ice wedges. The ice wedges remain as a fossil form long after the periglacial environmental conditions have ceased and are often visible in sand and gravel pits.

8.5.6 Pingos and palsas

A **pingo** is a mound that upwarps above an ice core. Pingos develop under fully periglacial conditions and are from 30 m to several hundred metres in diameter and can be several tens of metres high. Their development is a direct consequence of the formation and growth of ice cores.

Ice cores form in two ways: cryostatic and artesian. When the development is **cryostatic**, the pores in part of the soil lying above the permafrost are filled initially with water. The sides of this soil water body are enclosed by frozen soil or ground ice. At the beginning of the cold season, the land surface above the soilwater body also freezes and the freezing progresses downward. The volume increase of the water upon freezing exerts a **cyrostatic pressure** on the remaining water below. This water cannot be squeezed or reduced in volume and is, therefore, pressed upwards. The upper soil layers are pushed up and the water within the upwarped soil freezes as an ice lens. If the upper soil tears before the process is finished, part of the water may flow out and freeze on the surface. With the repetition of this process from year to year, the ice core within the pingo and the pingo itself increases in size.

Artesian formation of ice cores is attributed to the pressure produced by groundwater rising towards the surface (section 10.3.2). The water is injected into the upper soil zone and freezes there as **injection ice**, forming a body of ice that is more or less free of soil substances. This type of ice core also increases in size as long as the artesian water supply is maintained and the soil continues to be upwarped at the surface. Detailed accounts on pingos and their development have been published by Mackay (for example, Mackay, 1978).

Palsas are formed mainly on the margins of periglacial areas where permafrost is discontinuous (Schunke, 1975; Seppälä, 1988). They usually contain an ice lens but it is developed in peat and contains large amounts of organic material.

Small ice lamellae and ice lenses form in the unsaturated pores of periglacial soils following sublimation of the moisture that is in the air circulating within the soil. The ice forms on the ice crystals already present in the pores and displaces the soil material so that a more or less pure body of ice develops. Compared to pingos, these ice masses are very small.

8.5.7 Rock glaciers

Rock glaciers are masses of rock debris resembling glacier tongues. The pore spaces within the debris mass are filled with ice and this makes the glacier-like movement of the debris possible. Rock glaciers develop out of particularly thick debris accumulations, such as those in cirques or on large talus slopes in the periglacial zones of high mountains, where low soil temperatures cause the snow meltwater seeping into the pores in the debris to freeze.

The rock glaciers in the Val Muragl near Pontresina in the upper Engadine in Switzerland are more than 600 m long and 200 m wide. In the area of their long axes they are more than 10 m thick. The front edges are steep, lying at the angle of internal friction of the debris which is unfrozen here and, therefore, not solidified by ice (Fig. 8.20). Rock glaciers in the Engadine move up to 30 cm per year, faster than most gelifluction tongues. Compared to a gelifluction tongue, the moved mass of a rock glacier over a similar area is considerably greater. However, the transport of debris by rock glaciers is confined to very small parts of the periglacial zone and affects much smaller areas than gelifluction (Barsch, 1969, 1996; Barsch and Hell, 1975).

Rock glaciers differ from glaciers in that the ice they contain formed in the pores of the existing debris and filled them. Glaciers are made of ice that has developed from snow over a long period and has had an interim stage as firn. A glacier completely covered with debris can appear similar to a rock glacier but has a core of clear ice under the debris. In some cases a precise distinction may not be possible.

8.5.8 Block streams

Block streams are not as thick as rock glaciers and contain hardly any fine material which, if present at all, lies at the base of the block mass in the hollow spaces between the rock fragments. The block stream is elongated downslope and its lower end is often convex both in ground plan and cross-section. The front edge is steep, particularly in block streams that are moving.

Active block streams are not limited to the periglacial zone. In the Central Massif in France, for example, large numbers of block streams lie on the southern and southwestern slopes of the Puy d'Angle near Le Mont Dore on volcanic

Figure 8.20 Rock glacier in the Val Muragl near Pontresina in the Grisons, Switzerland, photographed in 1977.

rocks at a height of about 1700 m above sea level (Fig. 8.21). They have perhaps developed during earlier colder climatic phases, although they move at the present time, if not as a whole at least in part. Indications of movement are

1. blocks standing on their edges in a number of areas with their long axes oriented toward the edge of the block stream and in a downslope direction;
2. the absence of lichen growth on the blocks in these areas;
3. the continuation of the A horizon in the soil beneath the forward edge of the blocks;
4. bushes in the process of being run over by the block movement (Fig. 8.22).

The block streams of the Puy d'Angle begin some distance below the summit on a grass-covered slope on uniform volcanic rock. There is no bare rock surface at their upper edge the weathering of which could have supplied the rocks. The causes and mechanism of their formation and the further development of the block streams in this region of the Central Massif is still to be researched in detail.

Block streams also occur at lower levels in the central uplands of the temperate latitudes. There they are generally inactive, the products of periglacial processes during the last ice age. Rock faces of resistant rock that have served as sources

FIGURE 8.22 Bush overrun by a block stream on the Puy d'Angle, photographed May 1978.

of the blocks, often rise above them. Examples are the block fields beneath the steep faces on the Acker-Bruchberg quartzite ridge in the upper Harz in Germany and the block fields on the slopes of quartzite ridges in the folded Appalachians in the USA.

Similar in appearance but of quite different origin are the block fields on the Felsberg in the Odenwald in Germany. They are elongated downslope like block streams. The bedrock is a granodiorite. The blocks are several metres in diameter, well rounded and result from chemical weathering along the joints in rock that has been reduced to grus. This deep weathering took place in the Tertiary. Wash denudation removed the weathered materal from the joint areas leaving the unweathered rock between as a collection of residual blocks (Braun, 1969; Zienert, 1989). There is no significant movement of individual blocks. Wilhelmy (1958, p. 29) has described similar block fields in the warm humid tropics.

8.5.9 The significance of periglacial processes for denudation

Not all the periglacial processes described in the previous section are true denudation processes, that is, processes of denudation and transport of rock material. Gelifluction, in terms of the volumes moved and its widespread effectiveness on the surface, is the most important periglacial process. Nivation and stone stripes are only

FIGURE 8.21 Block streams on the west flanks of the Puy d'Angle (1738 m) in the Mont Dore region, Auvergne, France. The convex lower section of the slope appears to be an old landslide mass (*see* Fig 8.4).

locally effective variations of gelifluction and are often combined with part areal and part linear wash denudation by flowing meltwater.

Rock glaciers contribute to denudation in periglacial altitudinal zones wherever there is enough debris for their development and the ground temperature remains low enough. The denudation capacity of a rock glacier 200 m wide and 10 m thick with a mean rate of movement of 0.1 m/year can be estimated approximately. Through a cross-section of the rock glacier, 200 m³ of debris is transported each year. Assuming the debris supply area above the cross-section is 400 m × 500 m = 200 000 m², a size that corresponds to the sizes of block glaciers in the Engadine in Switzerland, the mean denudation rate of the debris supply area would then be

$$d' = \text{debris movement/supply area}$$
$$= 200(\text{m}^3/\text{year})/200\,000\ \text{m}^2$$
$$= 0.001\ \text{m/year} \qquad (8.13)$$

or 1000 mm per 1000 years. This rate is two or three times higher than the mean denudation rates found in the inner Alpine stream drainage basin areas such as the Kander, the Alpine Rhône, the Isère, and the Alpine Rhine (Table 3.1). The effectiveness of the denudation is, however, limited to the small area of the rock glacier.

Based on a debris movement rate of 200 m³/year, it is possible to estimate the approximate existence duration of the rock glacier. If the length of the entire rock glacier is 800 m and its total volume 800 × 200 × 10 = 1.6 million m³, it would require $1.6 \times 10^6/200 = 8000$ years to produce a rock glacier with this rate of movement. The development of rock glaciers, including the production of the debris by weathering, is likely to have begun after the last ice age and is apparently continuing at the present time.

Rough estimates of the transport and denudation by gelifluction produce much smaller values, although its effect is much more widespread and it has a much greater impact in periglacial areas. For example, a 20 m wide and 1 m thick gelifluction lobe, with a debris catchment area of 4000 m² and a mean rate of movement of 5 cm/year, which could represent gelifluction in the periglacial zone of mid-latitude mountains, is equal to a mean denudation rate of 0.00025 m/

year or 250 mm/1000 years, only a quarter of the amount estimated for rock glaciers.

Forms produced by other periglacial processes – stone nets, ice wedge nets, pingos and palsas – are largely stationary and with no significant transport of material. Their most important functions are to sort material locally and to change the vulnerability of the surface to denudation processes. When, for example, the pingo surface is broken up, due to the over-expansion of the upwarped soil cover, the summer meltwater from the ice core is released and local wash denudation takes place. Similarly, the development of ice wedge fissures interrupts the continuity of the soil cover. Stone nets, on the other hand, are a sign of long-term stability and absence of movement.

8.6 WASH DENUDATION

8.6.1 Definition and hydrological prerequisites

Wash denudation is the areally effective denudation and transport of regolith material by water flowing over the land surface, away from streams and rivers. The water generally is rainwater and snow meltwater. There are two types of overland flow.

In **Horton overland flow** (Horton 1945) the ground is not saturated and can still absorb more water. However, when the precipitation rate, in millimetres per time unit, is greater than the possible infiltration rate of the water in the uppermost soil layer, the surplus water runs off on the surface.

With **saturation overland flow**, the soil is already saturated at the beginning of the precipitation event and can absorb no more water. Even a very low precipitation intensity then produces overland flow. It occurs more often in humid rather than arid regions and more often on the lower part of a slope where water is supplied from the upper slopes, than at the watershed.

The **infiltration capacity** of a soil is expressed in millimetres per hour (mm/h) and represents the maximum possible infiltration rates during long continuous precipitation. It depends on the pore volume of the soil as well as the size and

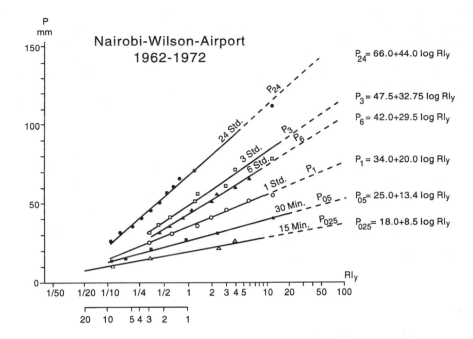

FIGURE 8.23 Magnitude-frequency of heavy rain events of different durations for Nairobi, Kenya, 1962–1972 (after data from Lawes, 1974; Ahnert, 1982, p. 27). 24 hours = total daily rainfall (not continuous), all other durations refer to continuous rainfalls of the durations indicated. RI_y = recurrence interval in years.

interconnectedness of the pores. The pore size is, in turn, a function of the grain size. Loamy sandy soil, a frequent type of soil, has an infiltration capacity of about 25–30 mm/h. For pure sand it is higher. Runoff occurs only rarely in these types of soil. There are exceptions. In sands whose surfaces repel water because of organic substances and products of decomposition originating, for example, from algae, the infiltration rate is reduced (Jungerius and Dekker, 1990; Witter *et al.*, 1991). The infiltration capacity of loam soils is between about 12 and 25 mm/h and for clayey loam about 2–5 mm/h (Morgan, 1969).

Impacting raindrops and flowing water cause some of the silt and clay particles to be redeposited in the previously open pores on the soil surface which then become sealed. This natural sealing process, known as **crusting**, limits infiltration and increases runoff. In the humid middle latitudes, it develops mainly in summer and disappears again during winter because of frost action and needle ice. In the frost-free areas the effect of crusting is counteracted by the formation of cracks during the dry period. Burrowing animals also contribute to the loosening of the soil.

In semi-arid and subtropical regions where wash denudation is particularly effective, a precipitation intensity of 25 mm/h seems, as a general rule, to be sufficient to produce overland flow. It is not necessary for 25 mm of rain to fall in an hour, 12.5 mm in half an hour is as effective because it is the intensity that matters, not the duration. Intensity values cannot be derived from the mean annual or mean monthly data found in most climatic tables. Magnitude–frequency analysis is more useful (Fig. 8.23; section 5.1.1).

8.6.2 Flow velocity and runoff discharge

The **Manning equation** (Manning, 1889) is valid for the velocity of flowing water:

$$V = n'D^{\frac{2}{3}}S^{\frac{1}{2}} \qquad (8.14)$$

where V is the flow velocity (m/s), D is the water depth (m) (in the original Manning formula this is the hydraulic radius), S is the gradient, expressed as the tangent of the slope angle, and n' is the smoothness coefficient; on natural soil surfaces $n' = 20$ or less (in the original Manning equation, instead of n' a roughness coefficient $n = 1/n'$ is used (Chow, 1964, Chapters 7, 25)).

Even older but similar to the Manning and still useful is the **Chezy equation** (Chezy, 1775)

$$V = CD^{\frac{1}{2}}S^{\frac{1}{2}} \qquad (8.15)$$

where C is a coefficient dependent on the relationship between water depth and surface roughness. Figure 8.24 shows that the influence of slope clearly increases with increasing depth of water. The water depth D, the flow velocity V, the tangent of the slope angle S and the overland flow discharge q', measured in m^3/s per metre of slope width (per unit of slope width) together

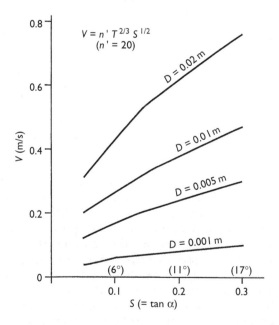

FIGURE 8.24 Velocity V of the surface runoff as a function of the gradient S in accordance with the Manning equation for water depth D from 1 mm to 2 cm with a constant smoothness coefficient.

form a system of functional relationships in which

$$V = f(D,S) \text{ and } D = f(V,q')$$

D and V are, therefore, both dependent and independent variables, that is, they are in a feedback relationship with one another. S and q' appear only as independent variables and are, therefore, determined by outside factors. Of the two, S changes only very slowly with the long-term form development, a change which does not affect the present-day overland flow events. By contrast, the overland flow discharge q' changes over a very short period, even during a single precipitation event. It is a function of the catchment area of the slope point in question, the precipitation intensity, the infiltration rate and the evaporation rate.

The **topographic catchment area** of a surface point reaches from this point upslope to the local watershed. The **hydrological catchment area** of the point is the area from which the point actually receives runoff, generally only part of the topographic catchment area, particularly in the case of Horton overland flow.

The difference is particularly significant in arid and semi-arid tropical regions where short, showery rainfall events with Horton overland flow dominate. On long slopes in these areas the rain falling at the slope crest flows only a short distance downslope and seeps into the slope surface after the end of the rain; it does not contribute to the runoff at the slope foot.

A hydrological catchment area can be estimated using the **effective runoff distance**. This is the length of slope that the water flows down during a rain of a particular duration. It depends on the flow velocity, that is, on the slope angle and water depth, as shown in Fig. 8.24. A water depth of 5 mm and a slope angle of 11° would give an estimated flow velocity of about 0.23 m/s which, with a rain of 15 min duration, would mean an effective runoff distance of 200 m.

Depending on the rainfall duration, therefore, the hydrological catchment of a surface point for overland flow is either equal to, or smaller than, the topographic catchment area. When smaller, it is a **partial contributing area** for the relevant precipitation event.

8.6.3 Drag force, sediment transport and denudation

The drag force of flowing water is its ability to set in motion a soil particle of a particular weight. The product of the water depth and the sine of the slope, which is roughly proportional to the square of the flow velocity (compare equation (8.15)), are of major importance in determining the drag force (Simons, 1969, p. 316).

Empirical field experiments by Zingg (1940) and others have shown that with a constant intensity of precipitation, the total denudation on a slope section with a length L and a gradient S, expressed as the tangent of the slope angle, is proportional to the product $L^{1.5}S^{1.5}$. Instead of L, the length of the limited test field or the runoff distance, the runoff discharge q' can be substituted and instead of S, the sine of the slope angle α, which is physically more correct; at low angles sine and tangent are virtually identical. In order to determine the denudation rate R at a point on the slope, rather than a section of slope, the q' that was substituted above for the runoff distance must be differentiated, from which results

$$R = kq'^{0.5} (\sin \alpha)^{1.5} \qquad (8.16)$$

where k is a coefficient. This equation is also used in theoretical models of the development of wash denudation on slopes (Ahnert, 1987d).

8.6.4 Sheet wash, rills and gullies

High precipitation intensity and low slope angles, which would mean a high supply rate and relatively slow velocity of overland flow, have the effect that the depth of the flowing water is large enough to cover the surface as a **sheet flood** and, if the velocity is sufficient, for areal denudation to take place. If the sheet of water is thin enough to be penetrated by falling raindrops, their splash effect (section 8.4.6) delivers additional soil particles to be removed by the overland flow.

With moderate runoff discharge, the water moves between clumps of grass, stones and other surface unevenness, flowing lineally over short stretches and washing out small **ephemeral rills**. If the rills are formed again with every runoff event, or their position is at least changed,

areal denudation results which can still be defined as sheet wash (Fig. 8.25). This type of rill can also be produced under a sheet flood if uneven ground creates variations in the water depth.

More pronounced and more permanent rill systems develop where **rill wash** predominates over sheet wash because greater surface roughness and steeper slopes concentrate the flowing water in narrow paths. Once rills are established, they serve as lines of runoff during subsequent precipitation events and are deepened further. A distinct linear erosion process accounts for each rill (section 8.1.1) but the collective effect of densely spaced rills is one of areal denudation.

If the rill becomes overloaded with eroded material, any surplus is deposited at various points either in the rill itself or beyond its margins, as small alluvial fans which interrupt the continuity of the erosional form. The rill becomes a **discontinuous rill**.

Gullies are deeper, larger rills eroded into the saprolite, sometimes down to the bedrock. Their steep side slopes generally reflect the maximum possible slope angle of the material.

8.6.5 Interflow and piping

A highly porous unconsolidated upper soil often lies above a denser, less pervious subsoil. On a sloping surface precipitation infiltrates only the upper soil layer and then flows downslope on the boundary plane of the denser subsoil. This type of flow within the soil is termed **interflow**. Observations on a test slope of 11–15° in the Bunter Sandstone area near Heidelberg in Germany showed that almost half the total flow on the slope is interflow (Barsch and Flügel, 1989). Since the boundary surface of the subsoil is not level and the perviousness of the upper soil layer at the boundary is not the same everywhere, interflow is often concentrated more strongly along preferred paths. Finer material may be transported away as suspended load by the interflow and hollow spaces, known as **pipes**, develop if the soil is cohesive enough not to collapse. The development of pipes greatly accelerates the discharge rate in the interflow.

Tunnels formed in the soil by burrowing animals can also have the function of pipes. Gullies

FIGURE 8.25 Forms resulting from areal wash denudation on the north foot of Oljeto Mesa, Monument Valley, Arizona/Utah, USA.

may develop when the pipes collapse. The gullies frequently have an upslope continuation as a pipe.

8.6.6 Badlands and earth pillars

When the settlers in North America reached the northern Great Plains in the 19th century, they found large areas on clayey marls in Dakota that were dissected by systems of gullies several metres deep and so closely spaced that only sharp ridges remained between them. The land could not be used and the area became known as the **badlands** (Fig. 8.26). Today this term is used to describe densely gullied areas, regardless of whether the erosional dissection was of natural origin or the result of poor agricultural practice.

Large stones that are too heavy to transport are bypassed during gully formation. They protect the fine-grained material lying below them from removal while the surrounding slope surface is lowered. Eventually the circumvented stone and its fine-grained substrate remain as an **earth pillar** with a stone cap. With time the fine

material pillar is undercut by gully wash and the cap stone falls. Later, the fine material is also largely removed.

Earth pillars, or earth pyramids, are often well developed on moraines in high mountains. The moraine material is made up of stones embedded in a matrix of loam which is given stability by its clay content. Lowering of the loam by wash denudation exposes new stones which serve as cap stones for a period (Fig. 8.27).

Miniature earth pillars under small stones a centimetre or so in size occur frequently on the sides of sunken paths. The pillars are made of fine material and are usually produced by the splash effect of raindrops, protected by the stones.

8.7 THE AEOLIAN PROCESS RESPONSE SYSTEM

8.7.1 Introduction

The physical and geomorphological knowledge of **aeolian processes** (Greek *Aiolos* = god of the winds) is based in large measure on the funda-mental observations, experiments and theoretical considerations of R.A. Bagnold (1941). Cooke and Warren (1973, pp. 229–327) have summa-rized later research.

Bagnold (1941, p. 37) distinguished between two dominating, but also associated, types of sand movement: saltation and reptation. In **salta-tion**, the sand grains are lifted from the ground and moved by springing in a low curved path in the wind direction. Each impacting grain loses part of its energy at the point of impact. On hard rock the amount of energy lost is relatively small; on a sand surface it is larger and enough to move other sand grains that are being hit some milli-metres forward on the ground. The impact effect of numerous saltating grains leads to a type of creep movement of sand grains, termed **repta-tion** or **surface creep**. Experiments by Bagnold (1941, p. 34) indicated that, over a given distance,

FIGURE 8.26 Badlands in South Dakota, USA. The relief in the photo-graph is about 20 m.

reptation accounted for 20–25 per cent of the sand movement and saltation for about 75–80 per cent.

The height of a saltation jump rarely exceeds 1 m. Walking through dunes in a strong wind on the coast, the pricking of sand is often felt on the legs but seldom on the arms or face.

Sandstorms remain close to the ground. The sandstorms that travellers report darkening the sun and making breathing difficult are in reality dust storms. Dust is not moved by saltation but as a suspended load, held in suspension by turbulence in the air. It is carried to great heights and moved over considerable distances. Only the finest sand grains can be transported in suspension and then only at very high wind velocities (Bagnold, 1941, p. 37).

The higher the wind velocity, the larger the sand grains that can be lifted from the ground and transported. Decisive is the **wind shear**, the

FIGURE 8.27 Earth pillars in Pleistocene moraine material, Val d'Herens, Valais, Switzerland.

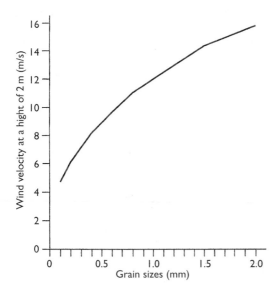

FIGURE 8.28 The functional relationship between wind velocity and the initiation of aeolian transport of different grain sizes (after Bagnold, 1941).

gradient of change in the wind velocity, near the ground. Wind shear depends on the wind velocity at greater height and the roughness of the ground surface which, in the absence of any other obstacles, depends on the size of the grains that are present. The estimating curve in Fig. 8.28 is derived from the relationship described by Bagnold (1941, pp. 85–95) and based on the wind velocity at a height of 2 m because wind measurements made with a hand-held anemometer in the field are generally carried out at this height.

The values shown in Fig. 8.28 are rough estimates for the initiation of wind transport. Once the movement is in progress it continues at somewhat lower wind velocities because the sand grains are thrown up again into the air stream by saltation and driven foward. Winds strong enough to transport sand are anyway usually gusty.

8.7.2 Deflation and wind abrasion

Aeolian denudation takes place by means of **deflation**, the blowing out of loose material, and by **wind abrasion**, the abrasive effect of sand grains on rock surfaces and stones. Deflation attacks the surface with varying intensity

because of spatially varying resistance over short distances determined by local differences in grain size, natural crusting or vegetation growth (Hagedorn, 1968, 1988). Deflation hollows develop on the more easily attacked parts of the surface. If the wind blows mainly from one direction, elongated furrows develop separated by ridges, known as **yardangs**, with streamlined outlines and profiles.

Wind abrasion on rock faces is limited to the zone up to 1–2 m above the ground, the height reached by saltation. Sand grains are blown against the rock and corrade notches in the rock. The effect is greatest when the wind meets the rock at a sharp angle and is directed along its surface. If a wind blows against the rock at right angles the wind force is dissipated by eddying and the potential abrasive effect reduced.

Stones lying on the land surface are faceted by abrasion on their windward side. **Wind-faceted stones** are found in both low-latitude deserts and the cold deserts of the polar regions (Fig. 8.29).

If fine material is blown out from an unsorted regolith of coarse and fine material, the coarse material remains as a cover of stones, or **desert pavement**, on the surface (Cooke and Warren, 1973, pp. 120–9). A desert covered with gravel is known as a **serir** in the eastern Sahara and **reg** in the western Sahara, in contrast to the rock desert

FIGURE 8.29 Wind abrasion on stones at Jörgen Brönlunds Fjord, Peary Land, north Greenland.

or **hamada**. The sand deserts or **erg** in the Sahara are deposition areas of blown out sand.

8.7.3 Aeolian transport and accumulation forms

The land surface formed by transport and deposition of the sand is the boundary surface between two movable media. The velocity of the propelling medium, air, is greater than the propelled medium, sand. This difference causes friction and turbulence between them and local changes in the velocity and direction of the wind immediately at the surface of the sand. Sand movement is accelerated or slowed locally and hollows and elevated forms develop in a wave-like spatial pattern which further increases turbulence near the ground. Initially, wind, sand transport and surface forms create positive feedbacks with one another, a self-reinforcing process response system that builds up the sand waves until an approximate dynamic equilibrium between the building up and the removal of the wave crests has been reached. The development has some similarity to the formation of waves at the boundary surface of water and air or current ripples on the sand under a body of water.

Small waves with a wavelength ranging from a few centimetres to a few decimetres and with more or less parallel running crests are **wind ripples**. Larger more complex forms are **dunes** (section 8.7.3.2).

8.7.3.1 The development of wind ripples, cover sands and loess

Sand with uniform grain sizes develops very shallow wind ripples with a very small ratio of height to wavelength. The ratio becomes larger if the grain sizes vary and the coarser grains collect on the crests of the ripples, and the fine grains in the intervening troughs.

Unlike wind ripples, the grain sizes in dunes always decrease from the base to the crest. Bagnold (1941, p. 152) suggests that differences in the effect of saltation cause this contrast. Small grains lying on the crest of the ripples are more easily knocked into the neighbouring ripple trough than large grains by the impact of saltating grains so that the proportion of larger grains on the crests increases. Also, the wind

velocity is higher on the crests than in the troughs and this encourages the concentration of larger grains on the crests of the ripples. Dune forms, by contrast, are much larger than the distances covered by saltation and many individual movements are required to transport a sand grain to the crest of a dune.

Very high wind velocities move the sand from the ripple crests to the ripple troughs, flattening the sand surface. The finer the sand the lower the wind velocity required. Some of the cover sands that were widespread in the non-glaciated areas during the last ice age and preserved in some areas of central Europe, probably developed in this way. Others are the result of periglacial wash denudation.

Aeolian-transported dust, **loess**, covered the existing landscape without developing any particular accumulation form. In northwestern China the loess cover is over 100 m thick (Liu Tungsheng, 1988) and originated in the dry areas of central Asia. Loesses are pervious and can develop very steep slopes (section 8.2.5). The progressive dissection by gullies and ravines in the loess regions of China interferes with agricultural land use. Loess accumulations in central Europe and North America are only a few metres thick and were deposited during the ice ages (section 22.8.3).

8.7.3.2 Dunes

Dunes are sand hills accumulated by the wind. There are several types of fully developed dunes (Fig. 8.30). Cooke and Warren (1973, pp. 229–327), Mainguet (1976) and Besler (1987, 1992) have descriptions of their shapes and origins.

Common to all active dunes is the contrast between the low angle of the windward side and the steep angle of the lee slope. On the windward side, the sand is driven upwards by saltation, whereby, in general, the grain size decreases towards the crest. Because of the continual impacting of saltating sand grains, the sand surface on the windward side is firmer and more densely packed than on the lee side. Once at the top, the sand falls over the edge of the dune crest on to the lee slope which has an angle of about 30°, the angle of internal friction of loose sand.

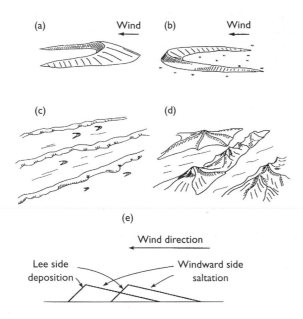

Figure 8.30 Types of dunes: (a) barchans; (b) parobolic dunes; (c) longitudinal dunes; (d) star dunes. (e) Shifting of dunes (schematic). (Partially based on Cooke and Warren, 1973, pp. 288–289).

The contrast in the slope angles of a dune reflects the contrast in the stratification in its interior. The steeply dipping layers of the leeward deposition are overlain by the covering layers dipping less steeply in the opposite direction, to the windward side. Both the stratification and the sand movement on the surface show that the dune shifts in a leeward direction as a result of a slow orbital movement of the sand grains (Fig. 8.30). A sand grain at the windward slope is driven by saltation to the crest of the dune, falls over to the lee side and remains in the newly formed deposit layer. Other grains are deposited on it and the original grain becomes part of the dune interior. The deposition on the lee side and the simultaneous removal of the material from the windward side shifts the dune form as a whole, although the sand grain in the dune remains in the same place until exposed again on the windward side and its next movement cycle begins.

The greater the volume of a sand dune the higher it generally is and the more slowly it moves. Seven years of measurements on 60 dunes in the Peruvian desert near Arequipa indicated a mean annual shift of about 32 m for 3 m high dunes and 22 m for 6 m high dunes (Hastenrath 1967). The dominating wind came from only one direction. For dunes of less than 10 m in this area, an estimating equation can be derived

$$A = 41 - 3.1H \qquad (8.17)$$

where A is the rate of shift (m/year) of the dune and H is the height (m) of the dune crest above the dune foot ($H < 10$ m).

The range of variation in the data is quite large because factors other than height have an effect, including the variation in sand volume of dunes of equal height because of differences in their outline, differences in the grain sizes and also differences in the frequency and strength of the wind blowing the grains.

Barchans are a widely occurring dune form (Figs. 8.30a and 8.31). Necessary for their formation are a sufficiently strong wind from one main direction, a relatively flat land surface compared to the dune height on which the dunes can shift leeward without impediment, and a relatively

FIGURE 8.31 Barchan dunes near Huatabampo on the coast of Sonora, Mexico.

limited supply of sand. The lee slope of the barchan is concave in plan because during the development phase the lower sides or horns of the dune shift more rapidly to leewards than the higher centre. The barchans measured in Peru had a mean width, measured as the distance between the horns, about 8–10 times greater than the crest height above the surroundings; the gradient on the lee side was more or less constant at about 32° and on windward slopes ranged from about 10° for 3 m high dunes to about 17° for 8 m high dunes.

Barchans are often staggered so that the horn of one barchan is aligned more or less with the centre of the barchan on its lee side. In this way, sand is moved from barchan to barchan in a leeward direction.

Parabolic dunes are also curved but in the opposite direction to a barchan (Fig. 8.30b). The lee deposition slope has a convex outline and the horns curve to windward. They have this shape because the sand transport is impeded by the roughness of the surface over which the dune moves, usually by dwarf bushes and other plants which slow the movement of the lower horns more than the centre of the dune.

In the trade wind belt deserts such as the Sahara, **longitudinal dunes** tens of kilometres long have formed. Their crests are aligned in the direction of the dominating wind which is from northeast to southwest in the trade wind belt of the northern hemisphere. In reality, the trades vary between north and east with the net effect that the dune ridges run in a southwesterly direction. In addition, helical turbulences in the northeast–southwest air currents of the trades sweep the sand together transversally to form the ridges. Large longitudinal dunes are also described as **linear dunes** or **draa** (Besler, 1987, p. 425). Small longitudinal dunes, often only a few metres high and usually formed of a series of curved ridges oriented in the wind direction, are known as sif or **seif**. Their development has apparently been considerably different from that of the draa (Besler, 1987, p. 426).

The highest dunes of all are the variously termed star-shaped dunes, pyramid dunes, sand mountains or **ghourds** (Cooke and Warren 1973, pp. 304–6). In the Sahara they rise to more than 100 m, in individual cases even to several hundred metres, above the surrounding area and are made up of a number of ridges running together in a star shape to form a summit similar to a pyramid in the centre. They develop where there is a very large amount of sand available and the wind blows from different directions. Star dunes also occur as a secondary form on linear dunes if there is a plentiful supply of sand.

FIGURE 8.32 Beach with kupsten dunes near Hayle, Cornwall, England.

In humid regions, dunes are confined largely to coastal areas where there is a sand beach as the source of supply. During the last ice age, sea level was lowered eustatically and sand was probably blown on shore from the exposed sea floor. In their natural condition, coastal dunes in humid areas are, at least partially, covered with dune grasses which limit the movement of sand. Blowouts develop where rabbit burrows or footpaths destroy the vegetation (Jungerius and Van Der Meulen, 1989). Coastal dunes usually have an irregular shape and are known as **kupsten dunes** (Fig. 8.32).

8.8 THE DETERMINATION OF DENUDATION RATES

8.8.1 Ways of determining rates

The denudation rate expresses the amount of mineral substances removed per time unit. It may be determined principally in three ways:

1. by directly measuring the lowering of the land surface;
2. by determining the amount of mineral substances transported from an area of the surface of known size, expressed as a mass or volume of material removed;

3. by determining the rate of movement of the material transported away by the denudation processes.

Which of the methods is used depends on the type of denudation and the purpose of the investigation.

8.8.2 Measuring the lowering of the land surface

This procedure determines the effect of denudation, that is the change in the landform, most directly and is particularly useful when the denudation only removes the uppermost layer of the soil material, in wash denudation (Dunne et al., 1978) and deflation, for example, the latter especially in association with dunes.

One of the simplest measurement methods involves the use of erosion pins. (Haigh, 1977, Statham, 1981, p. 176). They are about 60 cm long and at most 5 mm in diameter and must be placed deep enough not to be moved by frost heave. Usually a small holed disc (a washer) is put over the pin to mark the height of the soil surface and the length of the pin above the disc.

The procedure was first used by Schumm (1956) in an area in which wash denudation dominated. Although it has the advantage of being a direct measurement, there are also disadvantages. Denudation is determined at only one point of the surface. In order to include the spatial variability of denudation on a large surface, as many erosion pins as possible have to be used, but there is also the danger that the erosion pins themselves alter the erosion process being measured. Surface runoff is often accelerated around the pin so that denudation is increased immediately at the point of measurement. In addition, swelling of the ground surface as a result of saturation, or shrinking due to drying out, can alter the height of the surface without addition or removal of mineral substance. On dunes, very local turbulence of the air stream can be generated by the pin in its immediate vicinity, which may cause a change in denudation at the measurement point. Erosion pins placed horizontally on river banks can help to determine lateral erosion by the river (section 13.3.1).

8.8.3 Estimating the amount of material removed from a stream catchment

In a stream catchment, the rock material removed from the slopes is transported to the stream and carried out of the area by fluvial transport. The material consists of bed load, suspended load and load in solution. The latter can also reach the stream from the slope by infiltration into groundwater, emerging in springs and seepage.

If the entire bed load, suspended load and solution are measured at a given point on the stream over a long enough period, at least several years, a mean denudation rate for the catchment area above this point can be determined. This rate is a mean value, in both a temporal and spatial sense: the mean value for the entire observation period and the mean value for the entire catchment basin above the measurement point.

The measurement determines only the weight (mass) G of the stream load, expressed in tonnes, directly. The transported volume V in cubic metres is

$$V = G/\gamma \qquad (8.18)$$

where γ is the specific density of the rock material, which is usually assumed to be 2.5. The

transported volume becomes the transport rate $R(\text{m}^3/\text{year})$, when it is divided by the duration of the measurement period T:

$$R = V/T \qquad (8.19)$$

If the transport rate R is now divided by the size of the surface area $A(\text{m}^2)$ of the catchment basin, the mean denudation rate d' results:

$$d' = R/A \qquad (8.20)$$

Expressed in metres per year, d' is too small an amount to be easily imagined and the value is usually converted to

$$d = d' \times 10^6 \qquad (8.21)$$

where d is the mean denudation rate of the catchment basin in millimetres per 1000 years. In large catchment basins d is a function of the mean relief, the vertical height difference between the watershed and the valley bottom (Ahnert, 1970a). In section 3.3 the mean denudation rate of 20 large catchment basins in Europe and North America served as the basis for a simple model of relief development.

The denudation rate determined in equation (8.21) includes, in addition to the denudation on the slopes of the catchment basin, the direct stream erosion resulting from the downcutting of the stream bed and the additional stream load this produces. It is, therefore, a **total denudation rate** and not a denudation rate in the narrow sense of a slope denudation for which the term denudation rate is often used. When necessary it is possible to estimate approximately the proportion of stream erosion in the total denudation rate (Ahnert 1970a, p. 253).

8.8.4 Measurement of the removal of material on slopes by wash with sediment traps

Material transported by wash denudation on slopes can be trapped by trough-shaped sediment traps known as **Gerlach troughs** (after Gerlach, 1967) set into the ground surface transversal to the gradient direction.

When the catchment area of the slope (A) of the Gerlach trough is known, the amount of sediment trapped (R) can be transformed into the mean denudation amount for this area using equations (8.18)–(8.21), although only the solid wash load and not the load in solution would be included, nor any transport that might result from creep and other mass movements of the soil.

8.8.5 Determination of denudation by measuring the velocity of slow mass movements on the slope

Mass movements such as gelifluction and creep are too slow to be caught by a sediment trap and movements of this type are measured using markings in or on the soil. Washburn (1979, p. 213) and others have determined gelifluction rates of up to 5 cm per year with this method.

Similarly, Schumm (1967) measured surface creep movement of small rock fragments on shale slopes in Colorado, USA for seven years. The mean rate of movement varied between 0.1 cm/year and 9.2 cm/year with a clear dependence on slope gradient. The values are relatively high, which is probably due to the fact that many of the measured rock particles were moved individually by splash creep or short wash denudation impulses. In any event, the mass of the moved material remained small because only the uppermost layer of the soil was affected.

In order to measure the total movement of the regolith below the soil surface, Young (1960, 1972) dug 1.5 m by 1.5 m pits down to the bedrock on slopes in the southern Pennines in England. Horizontal marking rods were placed in the side walls of these **Young pits** in regular vertical distances above one another and the pits refilled (Young, 1972, pp. 54–6; Goudie, 1981, p. 171). In one pit on a 25° slope that was opened after 12 years, the rods near the surface had clearly shifted more downslope than those at greater depth. In the uppermost 5 cm of the soil the rate of movement was about 1–2 mm/year. On the basis of this movement, a downslope-directed net transport of 0.61 cm^2 per centimetre of slope width and year was estimated (Young, 1978). In order to convert this transport rate into a denudation rate, it would be necessary to know the slope catchment area for the measurement point (Ahnert, 1980). Block streams and rock glaciers (section 8.5.9) can also be measured using this method.

8.8.6 Measurement of denudation by rapid mass movements

Rapid mass movements such as landslides and mudflows are temporally and spatially discontinuous. They are process events that take place in a short period of time and are limited spatially. Only rarely can the event itself be directly measured. In most cases instrumentation that had been set up would probably be destroyed by the movement itself. Instead, measurements are generally made after the event, particularly of the accumulated volume of debris at the foot of the landslide or mudflow and of the size of its area of origin. From these it is possible to estimate the denudation effect. A direct estimate of a longer-term denudation rate at the same location would require a determination of the local frequency and the magnitudes of earlier events. Usually this is not possible because later events tend to destroy the traces of previous smaller events (Rohdenburg, 1989, p. 135). In a mountain area that is uniform enough geomorphologically, lithologically and structurally this difficulty might be overcome if the spatial frequency and magnitude were examined rather than the temporal frequency of events of different magnitude. With magnitude–frequency analysis, it would then perhaps be possible to derive some information about the temporal magnitude–frequency (section 5.1.1). A prerequisite would be that no significant changes in the exogenic and endogenic conditions had taken place in the relevant time period.

9

DENUDATIONAL SLOPE DEVELOPMENT

9.1 SLOPES AND DESCRIPTIVE SLOPE CLASSIFICATION

The greater part of the natural landscape is made up of slopes inclined at different angles and in various directions. The characteristics, the development and the function of slopes as components of geomorphological process response systems are, therefore a central theme in geomorphology.

Depending on their form, slopes can be classified into **slope types**. The form of the slope is described both by its fall line, its **profile**, from the divide to the slope foot and by its lateral form, expressed on topographic maps by contours. Young (1972, pp. 178–193) has summarized the various methods of classification. In a descriptive slope classification there are three basic types of profile:

- rectilinear profile: R
- convex profile: X
- concave profile: V

Many long profiles are made up of more than one of these types and the entire profile is expressed by combining the letters that describe each section from the divide to the slope foot. The slope is described from the top down because the divide is the natural upper limit of a slope while the slope foot is a more or less arbitrary limit on the valley floor or in a stream bed.

A slope profile with a convex upper segment and a concave lower segment is an XV slope. Six combinations of two segment slopes are possible: XR, XV, RX, RV, VR, VX (Fig. 9.1a) and 12 with three segments (Fig. 9.1b). Of the latter, XRV occurs often in nature. It resembles the letter S and is referred to as a **sigmoid slope**, after the Greek letter sigma. The divide or summit of a slope is always convex and the foot slope nearly always concave, even on an otherwise straight slope; but, if the lengths of the upper and lower segments are insignificant, they do not have to be included in the overall profile description.

The profile segments in Figs 9.1a and 9.1b merge into one another with no sharp boundaries. If the profile segments are separated by a sharp convex or concave slope break, a small x or v is used at this point in the profile notation (Fig. 9.1c).

It is also possible to characterize the lateral shapes of slope surfaces in the same way. A slope spur is lateral convex (X_l), a slope dell is lateral concave (V_l) and a slope crossed by straight contours is lateral straight (R_l). Abrupt lateral changes can be indicated, for example, sharply convex bedrock ribs with x_l or small valleys or gullies with v_l.

The classification allows a rapid description of the form characteristics of slopes using a simple notation. Information about slope angle, length of slope segment, soil, rock material, vegetation

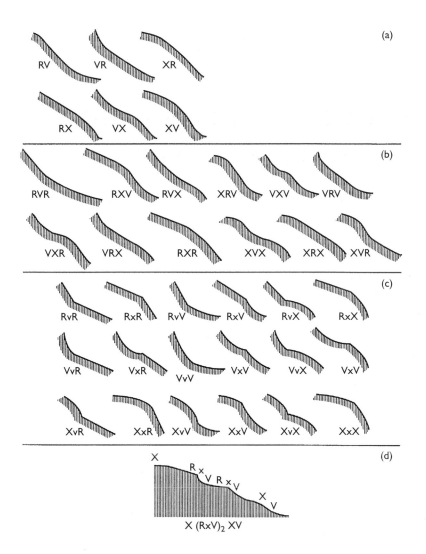

FIGURE 9.1 Slope profile form types (schematic): (a) two-segment profiles; (b) three-segment profiles; (c) two-segment profiles with slope break; (d) complex profile (Ahnert, 1970b).

or land use can be added easily, depending on the detail required or purpose of the description (Ahnert, 1970b).

9.2 THE MASS BALANCE OF SLOPE DEVELOPMENT

The areal denudation, transport and deposition of rock material are the most widespread of all

geomorphological processes. The form of the surface changes when the amount of net denudation or net accumulation at different points on the surface varies. The spatial differentiation in the type and intensity of these processes is decisive for the development of forms.

The denudational mass balance is the quantitative expression of the removal and/or deposition of the material and can be expressed simply for each slope point as follows:

$$C = C' + (W + A - R)\Delta T \qquad (9.1)$$

where

C is the thickness of the regolith at the end of the time interval ΔT;
C' is the regolith thickness at the beginning of the time interval ΔT;
W is the rate of local regolith production by weathering of the bedrock (per time unit);
A is the rate of the material supply: the gross increase in C as a result of denudative transport processes from upslope (per time unit);
R is the rate of material removal: the gross removal from C as a result of denudative transport processes downslope (per time unit);
ΔT is the length of the time period: the number of time units.

The components of equation (9.1) are linked in several ways by functional dependencies and feedbacks (Fig. 9.2). The regolith thickness C is a positive function of W and A but a negative function of R. In most cases W is a negative function of C (equations (7.4) and (7.6)). This interdependence produces the first negative feedback in Fig. 9.2: the higher the weathering rate W, the more rapidly the regolith thickness C increases, but as a result of this greater regolith thickness, the weathering rate W decreases.

The feedbacks shown in Fig. 9.2 also link the regolith thickness to the local slope angle, α_R downslope and α_A upslope, and thereby with the transport rate of the material removal R and the

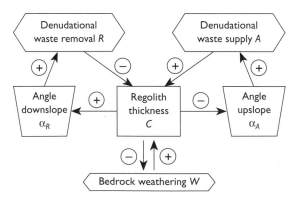

FIGURE 9.2 The functional structure of the mass balance system at a slope point (after Ahnert, 1987b).

material supply A that are largely determined by these slope angles (section 8.2).

An increase in the supply of material A to this point on the slope, without a simultaneous and similarly large increase in the removal R from the point, produces a material surplus which increases the height of the surface. At the same time the angle is increased downslope and decreased upslope. The local net accumulation of the transported material leads to an increase in removal of material and to a reduction in the supply from upslope. Both tendencies counteract further accumulation of the material at this point. The increase in the regolith thickness therefore declines and eventually ceases if there are no further changes due to other causes.

Conversely, an increase in the removal of material R without a simultaneous and equally large increase of the supply A, leads to a reduction in the regolith thickness, that is to a lowering of the surface at the slope point with a corresponding decrease of the slope angle α_R downslope and an increase in the slope angle α_A. The supply is consequently increased and the removal reduced until there is no further change in the regolith thickness.

The material components C and C', the process components R, A and W as well as the form characteristics α_R and α_A together form a process response system (section 2.2.3) that steers itself by means of the negative feedbacks that develop between its components. At each slope point therefore the system moves toward a **dynamic equilibrium** between the supply $A + W$ and removal R, that is, towards a steady state in the mass balance shown in equation (9.1). Linking all process components is the regolith thickness C (Ahnert, 1964). When the dynamic equilibrium between the process components is reached then

$$C' = C \text{ and } R = A + W \qquad (9.2)$$

is valid. The tendency to equilibrium is a basic characteristic of geomorphological process systems. It was already recognized by Gilbert (1877) in his pioneering work on the geology of the Henry Mountains in Utah.

The exogenic process response system on slopes is itself only a subsystem of a larger process response system that includes regional crustal movement and regional denudation. The overall tendency to dynamic equilibrium is shown in models of relief development in sections 3.5–3.7.

An equilibrium state is reached only when there is enough time available for the system to adapt to it before there is a new disturbance of one or several other system components. However, the recognition of the tendency to a particular state of equilibrium is an important diagnostic element in the interpretation of form development in a landscape.

9.3 WEATHERING-LIMITED AND TRANSPORT-LIMITED DENUDATION AND THEIR INFLUENCE ON SLOPE FORM

If the weathering rate remains, in the long term, less than the potential denudation rate, that is, when denudation processes could transport more material away than is delivered by weathering, the denudation is **weathering-limited** (Young, 1972, p. 24; Chorley *et al.*, 1984, pp. 220–3). On slopes of weathering-limited denudation, the bedrock is exposed and the form of the rock slope is determined primarily by the type of weathering that dominates locally: block disintegration on rock walls, hole weathering on honeycomb surfaces and rill lapies where solution weathering predominates. A dynamic equilibrium with the self-steering described in section 9.2 cannot develop on a slope surface on which denudation is weathering limited.

Denudation is **transport-limited** if the local rate of delivery of material by weathering and from upslope is at least as great as the potential rate of denudative removal. The bedrock remains covered, the slope has a tendency to develop a dynamic equilibrium between the processes involved and to develop the **characteristic slope form** for these processes.

Weathering-limited denudation is more often found at the divide and on the upper part of the slope than on the lower parts of the slope because the denudative catchment area and consequently the rate of debris delivery from the upper slopes increases in a downslope direction.

The lower boundary between the zone of weathering-limited denudation and the transport-limited zone is also the boundary between the bedrock slope and the regolith-covered slope below. The boundary is particularly sharp at the foot of a rock wall formed by block disintegration and rock fall because here the steep wall rises above the talus at a well-defined break of slope. If the bedrock surface is inclined but not particularly steep it may merge into a slope of talus with bedrock outcrops and a regolith-covered lower section with not much overall change in slope angle.

Although dynamic equilibrium is not possible on the weathering-limited rock surface of the upper slope, equilibrium can be established for the slope as a whole if the debris production by weathering of the bedrock, in accordance with equation (7.4), decreases with increasing thickness of the regolith cover and, if at the slope foot, there is a transport-limited removal of the debris, that is delivered.

On a bedrock slope with talus at its foot, for example, as long as the rate of debris removal at the boundary between rock slope and talus is lower than the rate of supply, the volume of the talus increases and its upper limit moves upslope. The exposed rock surface supplying the talus is then reduced and the supply decreases. The change continues until the rate of debris supply from the rock surface equals the rate of debris removal from the upper margin of the talus and until the total rate of debris production by weathering on the bare rock and below the talus equals the removal rate at the slope foot. The boundary between rock slope and talus then remains in the same position and the volume of the talus does not change, although debris continues to be weathered, transported downslope and removed at the slope foot. The rock slope and the talus slope are in a steady state, with the possible exception of the divide.

On the divide, only weathering and debris removal take place. The divide is lowered by denudation, if there is no simultaneous uplift by the same amount to compensate. Uplift is a prerequisite if a steady state is to be reached here too, since it directly affects the vertical erosion of streams and is, in its role as supplier of eksystemic energy, the cause of the relief development and, therefore, of slope evolution (section 3.5).

All other conditions being equal, the type and extent of the lateral curvature of the slope surface has a considerable effect on the location of the boundary between the rock slope and the talus. On a laterally convex slope, a spur at the end of a ridge, for example, the movement of debris diverges and material arriving from upslope is spread over an increasingly larger surface area downslope. Each square metre receives less debris than if it were on a straight slope and, to compensate for the deficit, the boundary between the rock slope and the talus shifts downslope.

On a laterally concave slope, the debris movement converges and the material, which arrives from a larger delivery area lying upslope, is concentrated in a narrower transport path. In hollow forms, such as dells, the thickness of the debris increases more rapidly downslope so that the boundary between the rock and the debris slope is located higher up than on a laterally straight slope.

The geometrically determined relationship between lateral slope form and diverging, parallel or converging debris transport explains why on projecting spurs and at the ends of mountain ridges bare rock surfaces extend further down than on lateral straight slopes and why bare rock surfaces are often absent in hollow forms, even if they are present on the intervening spurs.

9.4 PROCESS-SPECIFIC SLOPE FORMS

9.4.1 Introduction

Denudational slopes on valley sides usually develop following dissection by fluvial erosion on an uplifted block. Stream erosion incises the valleys and denudation attacks the valley flanks and causes them to retreat. If the upper slopes of neighbouring valleys intersect at the divide, this is also lowered by denudation. The rate of downcutting E of the stream and the rate d_s of removal

at the divide determine whether the relative height of the slope, the local relief h, increases or decreases. Per time unit,

$$h = h' + (E - d_s) \qquad (9.3)$$

is valid, where h is the local relief at the end of the time unit, h' is the local relief at the beginning of the time unit, E is the net rate, that is the net amount per time unit, of the lowering of the slope foot by stream erosion, and d_s is the net rate of denudation (lowering) of the slope divide or summit (section 3.3–3.6).

If the relief increases, the slope becomes steeper overall; if the relief decreases, the slope angle also decreases. In both cases the mean denudation rate is affected (equation 3.2).

Slope form is determined by the spatially differentiated denudation on the slope itself. The form changes as different parts of the slope are lowered by denudation or increased in height where material is transported in. The exogenic processes that cause the changes are for the most part determined by the morphoclimate (section 5.2). Morphoclimatically caused differences in the shape of the slope exist between, for example, regions in the high and middle latitudes where slow mass movements such as gelifluction and creep predominate and areas in the tropics and subtropics where wash denudation dominates.

9.4.2 Characteristic slope profile with slow mass movement

Slopes on which slow mass movements dominate are generally convex. Figure 9.3 shows the characteristic development of this type of slope form in a theoretical model. In the model a stream incises an initially horizontal land surface at the left side of the model profile with a constant gross rate from time $T = 1$ to $T = 1201$ (the length of the model time unit is arbitrary). The net downcutting rate is the difference between the gross rate and the removal rate by the stream of the debris that has arrived from the slope. After $T = 1201$, the downcutting ceases but the stream in the model continues to transport away all the material brought to the stream from

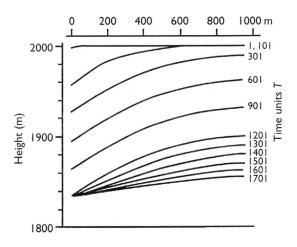

FIGURE 9.3 Characteristic development of the slope profile form by slow mass movement with lowering at the slope foot by erosion until $T = 1201$. Theoretical model (Program SLOP3D) after Ahnert (1987b) (*see* Fig. 8.13).

the slope. The height of the slope foot remains constant from this point onward in time.

In the early phases of development (until $T = 301$), the vertical erosion of the stream is more rapid than lowering in the middle section of the slope and at the slope divide. The slope becomes higher and steeper and delivers increasingly more material to the stream whose net downcutting rate E decreases progressively as a result.

The denudation rate d_s at the slope divide or summit also increases progressively during this period of slope steepening. If a sufficiently long period is available without changes in the eksystemic supply of energy (crustal movement, climatic factors) for this development, it continues until no further slope steepening takes place, that is until the stream downcutting rate E and the summit lowering rate d_s are equal. At that time the rate of lowering at all intermediate slope points is also the same and the entire slope is in a steady state with dynamic equilibrium between debris supply and debris removal at all points on the slope, from the divide to the slope foot.

The equilibrium includes the relationship between the weathering rate W and the net

denudation rate d at each slope point which in this case also corresponds to the summit denudation rate d_s so that

$$d = E = d_s = W \qquad (9.4)$$

If W, the rate of debris production by weathering of the bedrock, is a function of the regolith thickness on the bedrock (equations (7.4) and (7.6)), then the fact that the regolith thickness is the same at all slope points also means that the weathering rate is the same over the entire slope. Because weathering delivers material to all parts of the slope, a uniform regolith thickness is only possible when the transport rate increases downslope. However, where slow mass movement dominates, the transport rate is a function of slope angle (section 8.2) and can only increase downslope if the angle also increases. The result is the characteristic convex slope form of mass movement slopes which is also clearly developed in the theoretical model in Fig. 9.3.

In the model profile the steady state begins at about $T = 301$ and continues until the end of the fluvial incision at $T = 1201$. The profile is lowered parallel to itself along its entire length. After $T = 1201$ there is a basic non-equilibrium in so far as the rate of erosion E is now at zero, while the summit denudation rate d_s is still effective. The relief is therefore lowered and the slope angle decreases together with d_s and the mean denudation rate d. Even though the slope is not in equilibrium, the slope remains convex because it is still necessary to transport downslope increasing quantities of material (*see also* Fig. 8.13).

It is possible to calibrate the initially arbitrarily chosen time scale of the model approximately by comparing the model data in Fig. 9.3 with the empirically estimated summit denudation rate $d_s = 0.2h$ per million years (section 3.3). Between $T = 1201$ and $T = 1401$, during 200 model time iterations, the relief in the model is lowered from about 68m to about 43 m, by about 0.37 of the relief at $T = 1201$. This is roughly similar to the denudation rate for about 1.8 million years, which in the theoretical model would mean that a time unit was equal to about 9000 years.

9.4.3 Characteristic slope profile with wash denudation slopes

In contrast to the convex slope forms of slow mass movement, wash denudation produces concave slopes if surface runoff increases progressively downslope from the divide to the slope foot. This is valid whether or not there is fluvial incision at the slope foot. The only condition is that the downcutting rate at the slope foot cannot be greater than the maximum possible rate of lowering by weathering and denudation. Figure 9.4 shows the development of a slope profile by wash denudation in a theoretical model.

The initial profile is horizontal and a stream incises on the left side of the profile. In the beginning phases, as long as part of the initial profile is retained, the profile has a convex segment, one in which lowering in the lower part is more rapid than in the upper part. Already after $T = 100$ time units, a concave profile is apparent in the lower part of the segment which gradually extends upslope. By $T = 201$ the initial profile has been removed and the entire profile is concave. It is lowered parallel to itself, an indication that the dynamic equilibrium shown in equation (9.4) is present.

Once vertical erosion has ceased by $T = 601$, the summit denudation rate declines gradually, although more slowly than on the mass movement slope in Fig. 9.3 because on a concave slope a relatively steep gradient and the related high denudation rate at the summit are present for a longer period.

The characteristic concave form of wash denudation slopes develops if the surface runoff increases over the length of the slope from summit to slope foot. Because the rate of wash denudation at each slope point depends on both surface runoff and the local slope angle (equation (8.16)) the dynamic equilibrium in equation (9.4) can only occur on the entire slope if it becomes concave, which means that the higher runoff on the lower slope is compensated for by a lower slope angle, and the lower runoff rate on the upper slope by a higher slope angle.

This rule is valid for suspended load wash in which the soil material is transported from the slope without intermediate deposition, and also for bed load wash in which there is wash transport of coarse-grained particles for short distances, with intermediate deposition on the slope, as long as the downslope increase of the bedload that accrues is being removed by an increase in runoff.

The characteristic slope form is different, however, if the runoff does not increase progressively over the entire length of the slope. In dry climates, for example, precipitation events

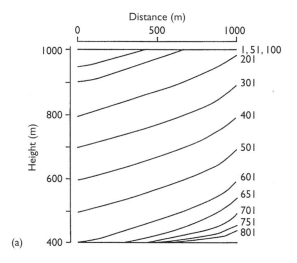

FIGURE 9.4 (a) Development of the slope profile form by wash denudation with lowering at the slope foot by erosion until $T = 601$. Theoretical model (program SLOP 3D, after Ahnert, 1987b). (b) Slope formed by wash denudation near Guaymas, Sonora, Mexico.

are often of short duration and the runoff increases downslope only as long as the rainfall lasts. When the duration of a typical geomorphologically effective precipitation event is shorter than the time required for the runoff to flow from the divide to the slope foot, then the runoff increases progressively from the divide downslope only over the distance it flows during the precipitation event, the **effective runoff distance** (section 8.6.2). Below this stretch the runoff is constant until it reaches the slope foot. The compensation for an increased runoff by a reduced slope angle, necessary for the concave equilibrium profile, is therefore absent. Only the upper slope has the characteristic concave profile because only here does the runoff increase downslope. The length of the concave section depends on the length of the effective runoff distance during typical runoff events (Fig. 9.5). Schumm (1961) has observed wash profiles similar to the convex form derived in the theoretical model in temporarily active channels in dry areas of the western USA.

The actual duration of precipitation events on a given slope varies. In general they are shorter in the dry climates of the tropics and subtropics than in more humid climates. Table 9.1 shows this relationship for weather stations with greatly varying annual rainfalls in Kenya. A regression analysis of the data results in

$$T_p = 0.00255P_y - 0.295$$
$$(r = 0.8) \qquad (9.5)$$

where T_p is the mean number of rain hours per rain day and P_y is the mean annual precipitation (mm).

The effective runoff distance of wash denudation in dry areas is also a function of the intensity of the precipitation event, the infiltration rate, evaporation, gradient, surface roughness and other factors such as the interception of rainfall by plants before it reaches the soil. All these factors can influence slope form. K-H. Schmidt (1987) has shown that convex wash slopes develop if there is a decrease in runoff as a result of a downslope increase in infiltration and piping in bentonitic claystones with a high montmorillonite content (Brushy Basin Formation of

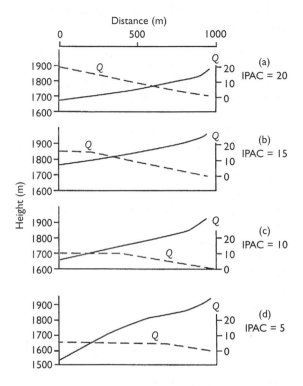

FIGURE 9.5 Theoretical model profiles in steady state, formed as a result of bed load wash denudation with different characteristic rainfall durations (IPAC) and therefore different effective runoff distances. (a) rainfall duration so great that the runoff increases from the divide to the slope foot; (b) runoff increase from the divide downslope over three-quarters of the slope; (c) runoff increase in the upper half of the slope; (d) runoff increase only in the upper quarter of the slope. Q is the local runoff along the profile in relative units (model program SLOP3D, after Ahnert, 1987b).

the Petrified Forest Series) on the Colorado Plateau in the southwestern USA.

9.4.4 Characteristic profile form on slopes with a combination of mass movement and wash denudation

Figure 9.6 shows the development of a model profile affected by simultaneous slow mass movement, wash denudation and vertical incision at the slope foot until $T = 1001$. Initially the influence of stream erosion dominates and the

TABLE 9.1 Mean annual precipitation and mean duration of precipitation events at selected stations in Kenya

Station	Mean annual precipitation (mm)	Mean rain hours per rainy day
Nairobi (Dagoretti Corner)	1079	2.4
Nairobi, Embakasi	785	1.9
Nairobi (Wilson Airport)	909	2.4
Embu, Mwea	907	2.2
Voi	549	1.0
Narok	736	1.9
Nakuru	956	1.5
Naivasha	620	1.0

Source: East African Meteorology Dept., Nairobi (cited by Ahnert, 1982, p. 25).

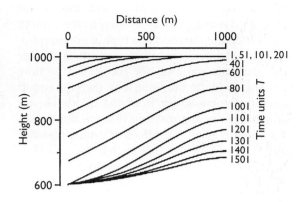

FIGURE 9.6 Characteristic development of the slope profile form as a result of the combination of slow mass movement and wash denudation with lowering at the slope foot until $T = 1001$. Theoretical model (program SLOP3D, after Ahnert, 1987b).

profile is convex. From $T = 201$, before the divide has been reached by the denudation extending up the slope, a weak concavity is apparent in the lower slope. This remains part of the profile and becomes accentuated after the end of the downcutting phase ($T = 1001$). Between $T = 601$ and $T = 1001$, the slope is being lowered almost parallel to itself and is close to dynamic equilibrium.

In contrast to the completely convex mass movement slope (Fig. 9.3) and the concave wash denudation slope (Fig. 9.4), the characteristic slope form affected by the combination of both processes is convex in its upper part, similar to the mass movement convex profile, and concave in the lower section, similar to the wash denudation slope. Mass movement dominates in the upper part of the slope because there is very

little runoff available for wash denudation. Downslope wash denudation increases and dominates on the lower slope. How much it dominates depends on the amount of runoff that is typical for a particular slope and the wash denudation produced, compared to the mass movement transport rate. In Fig. 9.6, the relative intensity of the wash transport is not very high so that the concavity on the lower slope can develop fully only after the slope steepening effect of fluvial downcutting has ceased. This sigmoid profile form (XRV, section 9.1) occurs frequently in the middle latitudes.

A sigmoid slope form also develops if regolith material from the upper slope is deposited as colluvium at the slope foot. The colluvium slope is gentle, especially if the sediments are not removed by a stream.

THE HYDROLOGICAL AND HYDRAULIC BASIS OF THE FLUVIAL SYSTEM

Fluvial geomorphology investigates the forms and processes that relate to the shaping of the land surface by streams. The morphological effect of a stream at any location depends in large measure on its gradient and its discharge. **Discharge** is the volume of water that flows past a given point per time unit; it is usually expressed in cubic metres per second (m^3/s) and is a function of the size and the water balance or water budget of the stream's catchment area.

10.1 GLOBAL WATER BALANCE AND WATER BUDGET

The terms **water budget** and **water balance** are often used interchangeably. It is, nevertheless, useful to define water balance as the amount of water in a region that is stored in seas, glaciers, lakes etc. and the water budget as the transfers or movements of water between these stores.

By far the greatest part of the available water on the earth, over 1.35×10^9 km^3, or 97.8 per cent of the total, is in the oceans, and only 1700 km^3 (one ten thousandth of a per cent) in the streams (Table 10.1). The importance of the streams in the mean annual water budget is, however, much greater. Of the 104 000 km^3 of water that fall on average annually on the land surface of the earth, 71 000 km^3 evaporate and about 29 000 km^3 return

TABLE 10.1 Water balance of the earth

Total water available ~ 1.4×10^9 km^3, of which		
	($10^3 km^3$)	(%)
In the oceans	1 350 400	97.8
On the land:		
rivers	1.7	0.0001
lakes	230	0.00
groundwater	7 000	0.5
glacier ice	26 000	1.88
In the atmosphere	13	0.001

Source: After Wilhelm (1966) and Baumgartner and Liebscher (1990).

to the sea in the streams. Part of the remaining 4000 km^3 returns as glacier ice and part as direct groundwater discharge (Table 10.2).

10.2 COMPONENTS OF THE LOCAL WATER BUDGET

Some of the precipitation that falls lands on plants before it reaches the soil (**interception**), and some seeps into the soil (**infiltration**). If the subsoil is saturated or large pores or pipes are present in the soil, the infiltrated water flows as **interflow** (section 8.6.5) down the slope within the soil. Water that penetrates more deeply is stored as **groundwater** in the pores and joints of

TABLE 10.2 Annual water budget of the earth

	Volume 10^3 km^3
Evaporation:	
from the oceans	445
from the land	71
Total	516
Precipitation:	
on the oceans	412
on the land	104
Total	516
Discharge into the oceans:	
by streams	29
by glaciers	2.5
directly from groundwater outflow	1.5
Total	33

Source: After Wilhelm (1966) and Baumgartner and Liebscher (1990).

the underlying rock and reappears at the surface after a lapse of time as springs or seepage outflow.

There is a temporary storage of water at the surface in puddles when the topmost soil layer is saturated or impervious. If the precipitation continues, unconcentrated **surface runoff** may follow (section 8.6.4). Winter **snow cover** is a special form of surface storage.

A considerable volume of the water that falls is returned to the air by **evapotranspiration**. It evaporates either directly from the vegetation or the soil surface or after infiltration by absorption into plant roots by transpiration of the plants. The non-biogenic capillary rise of soil water in dry climates also contributes to evaporation.

The components of stream discharge are:

- **direct runoff** during a precipitation event which reaches the stream directly from the slopes without infiltration;
- **interflow**, which flows more slowly and continues to reach the stream after the end of the precipitation event;
- **base flow** which is supplied by the groundwater and the delayed remains of the interflow and is independent of specific precipitation events.

The **discharge quotient**, the relationship of discharge to precipitation, is a function of the amount of precipitation and of evaporation. Table 10.3 shows stream discharge height in various climates.

The water budget on the land surface can be expressed as follows for a section of the catchment area:

$$A = N + Z_o + Z_u - V - I - S_o \qquad (10.1)$$

where (per time unit)

TABLE 10.3 Mean annual precipitation and discharge height of various rivers

River system	Precipitation N (mm/year)	Discharge A (mm/year)	Discharge quotient (A/N)×100 (%)
Tji Anten (Java)	4935	3747	76
Tji Kapundung (Java)	2650	1580	60
Amazon	1900	655	34
Rhine	911	472	52
Danube	749	243	32
Seine	715	231	32
Elbe	601	158	26
Dniester	548	107	20
Dnieper	547	125	23
Red River (N. Dakota)	532	32	6
Mina (Algeria)	476	53	11
Don	403	66	16

Source: Keller (1962, p. 353) and Wechmann (1964, p. 490).

A is the discharge height;

N is the precipitation;

Z_o is the surface inflow;

Z_u is the supply from groundwater or inter-flow;

V is the evapotranspiration;

I is the infiltration;

S_o is the storage in hollow spaces on the surface;

Interception is not included in equation (10.1) because it occurs only during a precipitation event and because interception water drops from the vegetation and either becomes part of the rainfall reaching the soil or evaporates, both from the leaf directly and during transpiration.

Equation (10.1) shows a short-term, although not momentary, water budget in a small area that receives inflow from outside (Z_o). In catchment areas from which all the water flows out and none flows in, the component Z_o disappears. The storage component S_o is effective for only a short period and the water loss from infiltration *I* is equalized in the long term by the supply of groundwater and soil water to the streams by means of interflow and springs and by the share of infiltrated water to evapotranspiration.

The long-term annual mean for the water budget in an entire catchment area can, therefore, be simplified to

$$A = N - V \qquad (10.2)$$

10.3 GROUNDWATER AND SPRINGS

10.3.1 Groundwater movement

The groundwater fills the interconnecting hollow spaces, the pores, fissures and cavities below the **water table**. The height of the water table in the rock varies in relation to the rate of **groundwater recharge**, which is the net of supply from precipitation and infiltration and the loss through outflow from wells and springs. The rock in which the groundwater is stored is the **aquifer**. Below the aquifer is the **aquiclude**, an impervious rock in which there is no groundwater. The movement of groundwater flow follows Darcy's law

$$Q = kAH/L \qquad (10.3)$$

where

Q is the flow through rate in litres per time unit;

k is the **hydraulic conductivity** of the rock through which the water flows;

A is the flow through the cross-section, that is the sum of the cross-sectional spaces through which the water flows (pores, fissures, cavities);

H is the pressure difference between two cross-sections which are separated by the length *L* from one another. *H/L* is the pressure gradient or **hydraulic gradient.**

The hydraulic conductivity *k* increases with increasing size of pores and other hollow spaces in the rock. Clay, for example, has a very high pore volume but very small pore size and is practically impervious. Its value for *k* is consequently very low, in contrast to gravel which is highly pervious (Table 10.4). Clay or siltstone are usually aquicludes, and sandstone and conglomerate, which generally have large pores, are aquifers.

The hydraulic conductivity of a rock depends also on the jointedness of the rock. In rocks that are petrographically dense but crossed by tectonic joints and other fissures, groundwater can collect in these linear spaces. The joints intersect with one another and an interconnected linear groundwater storage, a joint aquifer, develops which has a common water table level. Plutonic rocks (granite, diorite and gabbro), most metamorphic rocks and limestones are typically dense petrographically but usually well jointed. Claystone and siltstone contain few joints, or no joints at all. Sandstones and many volcanic rocks have a large number of pores and if, in addition, they are well jointed, the joints provide more storage space and easier pathways for the

TABLE 10.4 Hydraulic conductivity of clastic sediments

Material	k (cm/s)
Gravel	> 0.1
Sand	0.1 to 0.001
Silt	0.001 to 0.00001
Clay	< 0.00001

Source: Herrmann (1977, p. 32).

groundwater to move through. The perviousness of the rock is usually greater along joints than through the pores.

Where the land surface cuts across the water table the water reappears at the surface in the form of **springs** or **seepage outflow**. Springs indicate a local concentration of the water in the aquifer, caused perhaps by the presence of joints or joint sets that are larger than in other areas nearby or by the convergence of a groundwater flow into a hollow such as a dell, gully or slope ravine that reaches down to the water table.

The **spring discharge**, the flow of water from the spring, causes the water table to dip down to the spring. A pressure gradient then develops in the aquifer and the groundwater flows to this exit point. The groundwater catchment area of a spring is the area of the aquifer in which the water table gradient is directed towards the spring.

Groundwater is important morphologically because, during dry periods, streams are supplied exclusively from groundwater storage by springs and seepage outflow. Springs also supply streams with the dissolved products of chemical weathering, thereby contributing to regional denudation. At the spring itself, constant saturation increases chemical weathering and cryostatic pressure during frost increases mechanical weathering. Positive pore water pressure in the water-saturated ground at the spring may also cause denudative instability (section 8.2.3).

10.3.2 Types of springs

Waste cover springs (Fig. 10.1) develop in dells and other hollow forms whose long axes are directed downslope and where the regolith is pervious but overlies impervious rock. The regolith serves as a shallow aquifer inclined downslope in which the infiltrated precipitation follows the gradient direction (Darcy equation (10.1)) and the flow of water converges towards the long axis of the hollow. Along the axis and downslope the water table lies nearer the surface than on the flanks of the slope or in the upper part of the hollow. The waste cover spring is the point at which the regolith is saturated up to the surface and where the water flows out

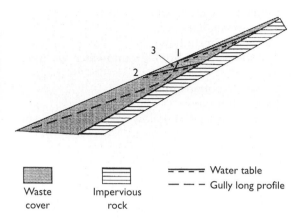

Waste cover   Impervious rock   ==== Water table
— — – Gully long profile

FIGURE 10.1 Waste cover spring (schematic), showing position of the spring: (1) during periods of substantial rainfall; (2) during periods of low rainfall; (3) after the erosion of a gully in the waste cover.

onto the slope. Because of the aquifer's shallowness, the spring outflow shifts upslope along the hollow axis during rainy periods and downslope when water supply is low. During dry periods it may stop flowing altogether.

If the stream from a spring flows intensively enough to erode a gully, the spring outflow becomes more localized. In Figure 10.1, the gully has a steeper gradient at its upper end than the slope so that with any change of the water table in the regolith, the groundwater outflow remains longer at this point than before. Positive pore water pressure can also cause spring erosion at this location, especially if pipes at the upper end of the channel are exposed (section 8.6.5). The spring erosion undercuts the upper end of the erosion gully and steepens it to form a gully head that is incised into the slope (Fig.10.2).

Contact springs occur in horizontally bedded or gently dipping rocks with differing perviousness. The groundwater flows out at the boundary between an aquifer and an underlying aquiclude (Fig. 10.3) on a valley slope. The aquifer is usually well jointed but the aquiclude is not. The joints in the aquifer concentrate the groundwater flow and the springs usually develop where large joints or densely spaced joint sets reach the slope surface. Because the springs are controlled by the almost horizontal

Figure 10.2 Waste cover spring with increased flow due to pipes in the Black Mountains, Wales.

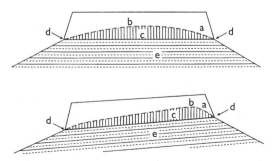

Figure 10.3 Contact springs (schematic). *Above*: in horizontal strata; *below*: in dipping strata. (a) Aquifer; (b) groundwater divide; (c) zone of aquifer saturated by groundwater; (d) contact springs; (e) aquiclude.

boundary between aquifer and aquiclude, they lie spaced along the slope as a **spring horizon**, or **spring line**, at about the same height. Contact springs have a more constant flow than waste cover springs because of the aquifer's larger storge capacity.

When the boundary of a horizontal aquiclude is exposed on opposite sides of a tableland by downcutting, the groundwater divide, or divide in the water table (b in Fig.10.3), lies more or less in the centre of the tableland. If the strata dip gently to one side, the groundwater divide is closer to the upper margin of the tableland, and the springs on the side of the dip direction have a larger groundwater catchment area and stronger flow than the higher-lying springs on the other side of the tableland, where the strata dip away from the outflow point. The latter are known as up-dip springs.

Fault springs occur on fault boundaries between pervious and impervious rocks. The rocks do not lie above one another but next to one another. The friction related to the crustal movement along a fault line shatters the rock. Stream erosion is more effective in this fault zone than in the undisturbed areas of rock on either side and a valley becomes incised in the more porous shatterbelt in which groundwater also moves more easily than in the neighbouring rocks. The contiguity of impervious and pervious rocks causes a lateral damming of the groundwater flow in the pervious rock. The valley development along the fault lowers the land surface to the level of the groundwater and after incision, produces a groundwater gradient in the aquifer directed towards the valley. The groundwater is highly mobile within the shatterbelt. Together, these factors have the effect that in a fault line valley, there is a series of strongly flowing fault springs aligned along the valley floor. Many thermal springs are fault springs, bringing water with high temperatures to the surface from considerable depth.

The aquifer in **artesian springs**, named after the Artois region in northern France, is an extensive area of rock lying between two aquicludes. Generally the strata form a large, gently sloping synclinal basin on whose uptilted margins the aquifer is exposed and into which the precipitation infiltrates. A free water table is present only on the margins of the basin, where it lies above the upper limit of the aquifer strata in the centre of the basin. The pores and joints in this central part of the aquifer are entirely filled with groundwater which, because of the height of the water table, is under hydrostatic pressure. At any point in the upper aquiclude that is breached, either by valley deepening or because it is in a fault shatter zone, the pressure pushes the water up to the surface, in some cases so strongly that it shoots up in the air. The covering aquicludes are often bored through artificially and the groundwater in the aquifer tapped. The Great Artesian Basin in Australia is so named because of its artesian aquifers.

The **karst springs** in areas of karst (Chapter 21) differ basically from other spring types in that the groundwater is stored in joints that are widened by solution weathering and eventually become karst caves. The caves and other joints form interconnecting systems within the rock which are able to absorb and store large quantities of precipitation very rapidly.

In contrast to the water movement through fine pores and narrow cracks in other aquifers, the water in karst caves flows and exits at the karst springs as a stream. The Orbe spring near Vallorbe in the Swiss Jura is an example. Karst water systems often contain U-shaped pipes like siphons in which the water is under pressure. If one of these pipes is cut by the land surface, the water rises to the surface and forms a spring lake or **vauclusian spring**, named after the spring lake at Vaucluse in France made famous by the poet Petrarch. The Blautopf near Blaubeuren in Germany and the Fonti di Clitunno near Foligno in Italy are other well-known examples of vauclusian springs.

Geysers are hot springs that shoot into the air at intervals for a few seconds or, in the case of large geysers, for a few minutes, as fountains of steam and water. The prerequisites are hot volcanic rock below the surface, a vertical pipe-like opening on to the land surface, such as the crossing of two large joints, and a lateral supply of initially cold groundwater.

The water fills the pipe and is heated from below by the hot volcanic rock. The boiling point on the bottom of the pipe is high because of the pressure of the water column. The water column expands as a result of heating and rises in the pipe until it reaches the land surface where it runs over. The water mass is then reduced because some water has flowed out at the surface, the pressure at the bottom of the pipe decreases accordingly and the boiling point is lowered there also. At the same time the water continues to be heated further. The lowering of the boiling point, because of the reduction in pressure, and the progressive increase of the water temperature as a result of further heating, continue until eventually the water at the bottom of the pipe reaches its boiling point and is abruptly transformed into steam. The steam, together with the remaining water in the pipe, then shoots out of the pipe, propelled by the pressure of the steam itself. The pipe is now empty and is refilled by lateral groundwater flow and the cycle begins again.

Providing the groundwater supply is constant, the geyser erupts at regular intervals because the heat supply from the volcanic rock remains the same. The regularity of Old Faithful in Yellowstone Park in Wyoming, USA (Fig. 10.4) originally gave it its name, but since an earthquake altered its water supply it erupts at intervals of 45–105 min. and the duration of the individual eruptions ranges from 2–5 min. The Yellowstone Park area contains more than 80 geysers and numerous hot springs with constant flows. Other geyser fields are located in southwest Iceland, the Kamchatka Peninsula, Russia, the North Island of New Zealand and Japan.

The silica in the water of geysers is deposited at the geyser outflow point and a crust of silica or **geyserite** develops which gradually forms a low cone around the mouth of the geyser funnel.

10.4 DISCHARGE, DISCHARGE REGIMES AND FLUVIAL MORPHOCLIMATE

10.4.1 The hydrograph and its components

The discharge of water has the dimension volume per time unit. It is usually expressed in litres per second (l/s), cubic metres per second

Figure 10.4 Eruption of the geyser Old Faithful in the Yellowstone National Park, Wyoming, USA.

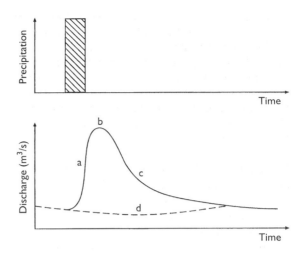

Figure 10.5 Hydrograph for isolated precipitation events (after Rodda, 1969, p. 405). (a) Rise of flood; (b) flood crest; (c) falling of flood; (d) base flow.

(m^3/s) and in the USA also in cubic feet per second (cfs; 1 cubic foot = approximately $0.0284\,m^3$). That part of the precipitation that becomes runoff and does not infiltrate or evaporate and is not taken up by plants or stored in puddles is termed the **effective precipitation**.

Figure 10.5 shows the discharge curve, or **hydrograph**, of a flood produced in a stream by a single, sufficiently intensive precipitation event. During the rise of the water level, the water in the stream is raised more rapidly than the neighbouring groundwater table and there is a flow of water from the stream to the groundwater due to infiltration in the banks of the stream. When the stream water level falls below the neighbouring water table, the movement direction is reversed and the exfiltration from the groundwater to the stream contributes to the discharge.

The falling limb of the hydrograph, or falling flood water level, is less steep than the rising limb because the contributions from surface runoff and interflow decrease only gradually. The hydrograph curve continues to flatten until the dry weather discharge, or base discharge, has been reached in which the stream is supplied only by groundwater.

Groundwater is significant for streams for another reason. Many stream channels lie on pervious rocks or on coarse-grained and, therefore, pervious stream deposits. Such pervious stream beds become impervious when their pore spaces are filled by groundwater. The stream then flows on a bed of impervious groundwater. When, during a dry period, the water table sinks beneath the stream channel floor, the channel bed itself dries up. Instead of a surface stream, there is only groundwater movement at depth in a downstream direction. The inhabitants of sub-humid and semi-arid tropical regions dig for water in the stream channel floor during the dry season because there the groundwater is nearer the surface than elsewhere. When the rainy season begins the water table rises above the stream channel floor and the stream can flow again.

Even at this time, the groundwater often supplies more to the stream flow than the runoff.

In humid climates and during the rainy season in humid subtropical climates, the hydrographs of successive rainfall events intersect. The flood water level may begin to rise before the falling flood of the previous phase has sunk to the level of the base flow (Fig. 10.6).

10.4.2 Discharge regimes and characteristics of discharge fluctuations

The seasonal fluctuations in the discharge of a stream reflect the hydro-climatic conditions, especially the distribution of precipitation and, where snow falls, the period of snow melt. The characteristic mean annual hydrograph for these conditions is termed the **discharge regime**. Pardé (1964) distinguished between a **simple regime** for a stream with only one pronounced maximum, a **first degree complex regime** for a stream with a main maximum and a lesser maximum, and a **second degree complex regime** for large streams that flow through more than one climatic region and are subject to different influences in different parts of their courses.

Figure 10.7 shows several types of discharge regime using the relative monthly mean of the discharge, expressed as a quotient of the mean monthly discharge divided by the mean annual discharge. **Nival** indicates a predominance of snow melt in the regime, **nivo-pluvial** that the snow melt discharge maximum is higher than that from rainfall and **pluvio-nival**, that the reverse is the case.

The Rhine is an example of a complex second-degree regime. The upper Rhine at Rheinfelden is strongly influenced by the glacial regime of the Alps with its summer maximum. At Kaub on the middle Rhine, the Alpine maximum is weakened but the snow melt and rainfall of the central uplands produces a weak second maximum between December and April. At Rees, on the lower Rhine, only this second maximum is clearly recognizable. The relative fluctuations of the Rhine are much less than those of simple regimes and first-degree complex regimes.

Stream classification on the basis of discharge regimes is useful but very rough and limited for geomorphological purposes because it is based on monthly means. The geomorphological work

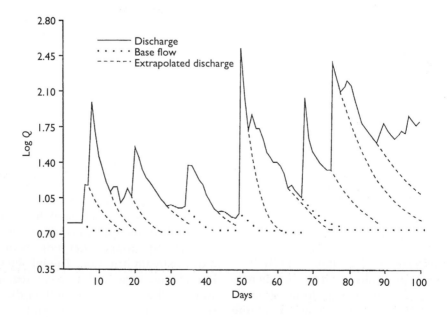

FIGURE 10.6 Discharge in a small stream (Bixier Run, Pennsylvania, USA) during successive rainfalls in the period from 20 December 1964 to 29 March 1965. Discharge Q in cubic metres per minute (after Ratzlaff, 1974).

of streams is accomplished by individual events which are insufficiently expressed by monthly means. Nevertheless, the discharge regimes indicate the seasonal distribution of the discharge events and as such are a useful addition to other hydrological information.

In many dry regions the evaporation rate is so high that the streams do not reach the sea but flow into a basin with no external drainage, where they either end in a terminal lake or evaporate and seep into the basin floor. These are **endorheic streams** and include the Volga in Russia, the Jordan and its terminal lake the Dead Sea and the Humboldt River in Nevada, USA. **Exorheic streams** flow into the sea and **exotic** streams are those, such as the Nile, that rise in an area of relatively high rainfall, cross a desert and flow into the sea.

Magnitude–frequency analysis can be used to examine discharge regimes that are geomorphologically important for erosion and transport. Fluvial hydrological magnitude–frequency anal-

ysis is usually limited to investigations of the frequency of annual runoff maxima of various sizes whose statistical recurrence rate, such as a hundred-year flood, is used in hydraulic engineering and as a basis for protective measures for settlements. In geomorphology, all daily discharge values above a threshold of effectiveness have to be included because erosion and transport take place not only during the annual maximum flood but also at other high-water events. Also of importance is the duration of the flood event.

Figure 10.8 shows two applications. The magnitude–frequency of the discharge is an important quantitative expression of the fluvial morphoclimate and its morphological effectiveness. Nevertheless, if shown as a single regression line of all discharge values (Fig. 10.8a), it does not give any information about the seasonal distribution of stream work. This can be overcome if magnitude analyses are made for each season or each month (Fig. 10.8b) and by linking

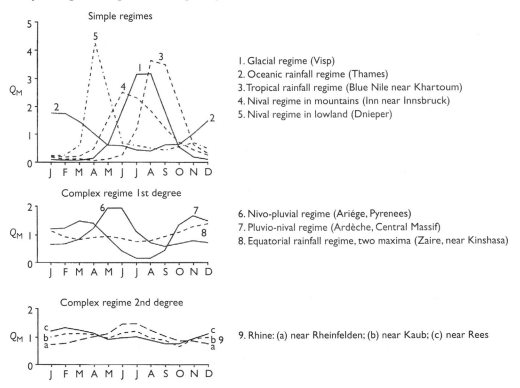

1. Glacial regime (Visp)
2. Oceanic rainfall regime (Thames)
3. Tropical rainfall regime (Blue Nile near Khartoum)
4. Nival regime in mountains (Inn near Innsbruck)
5. Nival regime in lowland (Dnieper)

6. Nivo-pluvial regime (Ariége, Pyrenees)
7. Pluvio-nival regime (Ardèche, Central Massif)
8. Equatorial rainfall regime, two maxima (Zaire, near Kinshasa)

9. Rhine: (a) near Rheinfelden; (b) near Kaub; (c) near Rees

FIGURE 10.7 Annual fluctuation of relative monthly discharge means in different discharge regimes (after Keller, 1961, pp. 279–91).

the magnitude-frequency analysis for the entire year with the determination of the discharge regime.

10.5 FLUVIAL HYDRAULICS

The term **hydraulics** refers to the fluid dynamics of liquids, especially water. Hydraulics form part of physics, of hydrology and hydraulic engineering. Most of the important hydraulic concepts used in functional geomorphology have been taken from these fields.

10.5.1 Laminar and turbulent flow

All flow of liquids is either **laminar** or **turbulent**. In laminar flow the particles move on parallel

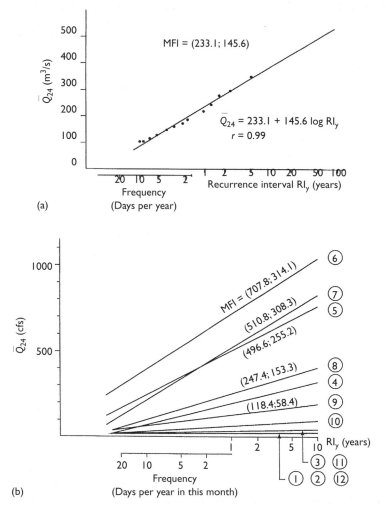

FIGURE 10.8 (a) Magnitude–frequency diagram of the discharge of the Kinzig (Black Forest). (Source: *Deutsches Gewässerkundliches Jahrbuch*, 1983–1987). (b) Magnitude–frequency lines for months of the year (1 = January, 2 = February etc.) for the Animas River near Howardsville, Colorado, USA. Notice the dry period in the months from October to March. (Source: US Geological Survey, Water Supply Papers 1960–1969). MFI: magnitude frequency index (section 5.2.1).

TABLE 10.5 Kinematic viscosity of water as a function of temperature

Water temperature (°C)	Kinematic viscosity ν (cm²/s)
0	0.018
10	0.013
20	0.010
30	0.009

Source: After Chow (1964, p. 1–4).

paths. Turbulent flow contains eddies. Whether flow is laminar or turbulent depends on three variables: the flow velocity V, the water depth D and the kinematic viscosity ν of the water. The higher the water temperature the lower the kinematic viscosity (Table 10.5).

The three variables form the Reynolds number Re whereby

$$Re = VD/\nu \qquad (10.4)$$

The critical threshold value between laminar flow and turbulent flow for natural stream channels is at about Re < 500. In reality, laminar flow is limited to such a small depth and low flow velocity that it seldom occurs in creeks or streams. With a relatively low velocity of $V = 20$ cm/s at a depth of 50 cm and with a water temperature of 20°C, that is, $\nu = 0.01$, for example, Re = (20 × 50)/0.01 = 100 000. The threshold value Re = 500 would be reached at this water temperature when $D = 1$ cm and $V = 5$ cm/s or when $D = 10$ cm and $V = 0.5$ cm/s.

10.5.2 Types of turbulent flow

There are three types of turbulent flow:

- streaming flow
- shooting flow
- plunging flow.

Plunging flow, which occurs at waterfalls, is fundamentally different from the other two because the water plunges in free fall over very steep, often perpendicular or overhanging rocks. Depending on the volume of the runoff and the height, the water falls either as a coherent mass of water or in individual water strands. If the falls are particularly high and the discharge low, the water is dissolved into drops and falls as mist.

Streaming flow is the least turbulent of the three. The water surface is fairly smooth, with small eddies that move downstream. The surface of **shooting flow** has numerous standing, often foaming waves, even if there are no bedrock humps or large boulders in the way (Fig. 10.9).

The standing waves in shooting flow indicate that the flow velocity of the water is greater than the propagation speed of the waves. In streaming flow the water moves more slowly than the propagation of the waves. A stone thrown in to the water confirms this: if the flow is streaming, a ring wave spreads in all directions from the impact point of the stone; if the flow is shooting, the stone splashes into the water without generating a ring wave. The closer the speed of flow in streaming flow is to the wave speed, the more the ring wave is distorted into an elliptical form in a downstream direction.

The propagation speed of a ring wave is the square root of the product of the gravitational acceleration g (981 cm/s²) and the water depth D (expressed in the same length unit, in this case cm). The **Froude number F**, which is the ratio between the flow velocity V (in cm/s) and this wave propagation speed gives a quantitative measure for the state of the flow:

$$F = V/(gD)^{0.5} \qquad (10.5)$$

$F < 1.0$ indicates streaming flow and $F > 1.0$ shooting flow. The critical threshold between the two types of flow is $F = 1.0$.

At the same flow velocity, there can be streaming flow in a deep section of the stream channel and shooting flow where the water is shallower. If the depth is the same, shooting flow can occur at higher velocities and streaming flow at lower velocities.

The flow velocity in the cross-section of a straight reach of a stream has its highest value in the middle of the stream at the surface or, if there is a wind blowing upstream, close to the surface. The zone of maximum flow velocity is a line on the water surface, the **line of highest velocity**. In a bend the centrifugal force shifts the line of highest velocity towards the outer margin of the bend.

The velocity decreases from the line of highest velocity towards the banks and the channel floor

FIGURE 10.9 Wurm River near Aachen, Germany. Standing waves on the surface of the water indicate stretches of shooting flow; where the water is smooth, streaming flow dominates.

because of internal friction in the water itself and bank and bottom friction in the channel. The greatest decrease takes place near the banks and the channel floor. In general, flow velocity at a depth 60 per cent of the total depth below the surface is approximately equal to the mean flow velocity of the vertical water column.

The mean flow velocity can be estimated as a function of the water depth, of the gradient and of the bottom roughness, using the Chezy equation or the Manning equation (section 8.6.2, especially equations (8.14) and (8.15)).

10.5.3 Hydraulic geometry of the stream channel

The introduction of the concept of hydraulic geometry by Leopold and Maddock (1953) brought new investigation methods to functional fluvial geomorphology. Central to this are the components of the discharge equation

$$Q = WDV \qquad (10.6)$$

where Q is the discharge (m^3/s), W is the width of the stream (m), D is the mean depth of the stream in a cross-section (m) and V is the mean flow velocity in this cross-section (m/s). W and D are geometric measures. The product $W \times D$

describes the cross-section area of the stream. V also contains as a geometric measure the length of the path the water passes through in a second.

In a stream reach without any additional supply of water or loss of water, the discharge Q and, therefore, also the product WDV remain constant. Any change in one of the components W, D or V causes a compensating change in the other components. When, for example, the stream width is reduced, the water depth increases. This leads, in keeping with the Manning equation (8.14), to an increase in the velocity which counteracts the increase in water depth. The greater flow velocity can simultaneously lead to bank erosion and to an increase in the width, the original decrease of which had generated the adjustment of the other two components. An initial change in D or V would be compensated for by the other two components. Thus all three components are linked by negative feedbacks (section 2.2.3).

With a constant discharge Q, therefore, the following is valid for successive channel cross-sections 1,2 and 3.

$$Q = W_1 D_1 V_1 = W_2 D_2 V_2 = W_3 D_3 V_3 \quad (10.7)$$

The product of the three components remains constant despite changes in the individual parameters. For this reason the equation (10.7) is known as the **continuity equation**. Because the components influence one another, with negative feedbacks to the component that caused the changes, the effect of a change in W, D or V on other components is not predictable quantitatively with any precision. For example, a change in the velocity can also result in an additional change in the erosion or sedimentation which, in turn, influences the channel cross-section and may cause changes in the form of the channel and the grain size of gravel on the channel floor. A basic **indeterminacy** exists as a result of these interactions (Leopold *et al.* 1964, p. 274). The quantitative relationships and changes of these components cannot be predicted but only estimated statistically on the basis of empirical investigation.

The changes that occur in the individual components W, D and V when the discharge Q changes, either at a particular point during a flood or along the stream as a result of increasing discharge downstream, are very complex and vary from case to case (Leopold and Maddock, 1953; Richards, 1982, pp. 142–61; and Knighton, 1984, pp. 99–114). A short summary follows.

The relationships in the equation (10.6) can be divided into three partial equations, in each case as $f(Q)$:

$$W = aQ^b \qquad (10.8)$$

$$D = cQ^f \qquad (10.9)$$

$$V = kQ^m \qquad (10.10)$$

The product WDV is then, and only then, equal to Q as required by equation (10.6), if the product of the coefficients

$$ack = 1.0 \qquad (10.11)$$

and the sum of the exponents

$$b + f + m = 1.0 \qquad (10.12)$$

The coefficients give the value of W, D or V for the discharge $Q = 1.0\,\mathrm{m^3/s}$; their numerical

quantity is determined by local conditions and of little interest here. More important is the size of each of the exponents. For the change in Q in a local channel cross-section, for example, at flood, Leopold and Maddock (1953, p. 26), after taking a great many measurements in the Midwest of the USA, obtained the mean values $b = 0.26$, $f = 0.40$ and $m = 0.34$. In the same study area the increase in Q downstream along the stream course during the mean annual high water produced mean values of $b = 0.50$, $f = 0.40$ and $m = 0.10$.

In general, therefore, the stream width increases downstream more or less proportionally to the square root ($b = 0.50$) of the discharge, but locally, at high water, by much less. The flow velocity at high water increases locally approximately proportionally to the cube root ($m = 0.34$) of the discharge, which anyone living on a river bank will be aware of. By comparison, the downstream increase in flow velocity is very small ($m = 0.1$). This is because the flow velocity is dependent not only on the water depth but also on the gradient which usually decreases downstream. It should be remembered that the increase in flow velocity is a mean increase, not related to individual values which in the upper stretches of a stream with gradient steps and deep pools can be above or below the mean value locally. In contrast to these variations, the exponent f for the change in water depth as a function of the discharge Q has, in both cases, local and downstream, the same mean value $f = 0.40$.

The values confirm a geometric law of change in the outline of the channel cross-section from the upper course downstream in streams whose discharge increases from the source to the mouth, namely, that the channel width increases downstream more rapidly with increasing discharge than the channel depth, that is, the relationship between width and depth increases downstream, which agrees with general experience in humid climates.

11

STREAM EROSION AND STREAM TRANSPORT

11.1 TYPES OF STREAM LOAD

The stream load is all the material carried by streams. It is made up of the **dissolved load**, the **suspended load** and the **bed load**. The load is supplied to the stream by the weathering and denudation processes on the slopes, by the groundwater in the catchment area, and by material from the stream channel itself.

Dissolved load is primarily a product of chemical weathering, although dissolved organic substances may also be present. The dissolved substances are ionic or molecular constituents of the water and move with it and are transported by it as long as the water can flow at all, even if the gradient is very low. The denudation of the catchment area continues by means of the dissolved load therefore after other types of denudation and transport processes have ceased.

In order to measure the dissolved load in a stream, the concentration of the dissolved material is usually determined in parts per million (ppm) which, in the case of water, is the same as milligrams per litre. The rate of dissolved load transport is obtained by multiplying the ppm by the discharge in litres per second. The amount that passes a measuring point during a time period is obtained by multiplying the rate by the number of seconds in the time period.

The suspended load is made up of solid particles that are small enough and light enough to be kept afloat by their buoyancy and the turbulence and to be carried along in the water. Most of the

particles are silt or clay (Table 6.3); where the current is strong, sand particles can be lifted from the stream channel floor and transported further. Very strong and turbulent currents, for example, shooting flow during floods, may move small gravels as suspended load for short distances.

Most of the clay and silt particles originate in the soils and are transported to the stream by wash denudation during heavy rains. They cause the muddy colour of streams flooding after rain storms. Suspended load is also eroded from the banks of the stream. Measurement of the suspended load is made by determining the suspended load concentration and multiplying it by the discharge. The concentration is obtained by weighing the insoluble matter or by measuring the turbidity of the water. In the stream cross-section, the concentration of suspended load is usually higher near the channel floor than near the surface, and higher near the banks than in the middle of the stream (Fig. 11.1). Any

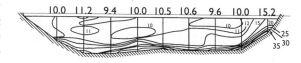

FIGURE 11.1 Distribution of suspended load concentration (mg/L) in the Danube River near Engelhardtszell on 15 April 1957 (Mangelsdorf *et al.*, 1990, p. 90).

determination of the suspended load must take this into account.

The bed load is composed of gravels (pebbles, cobbles and boulders) that are rolled or pushed along the floor of the channel or, if the current is very strong, moved in short jumps. Sand can be a component of either the bed load or the suspended load.

Most of the bed load reaches the stream from the valley slopes as a result of mass movement; a smaller share is derived from the erosion of the channel bed itself. In areas of intensive mechanical weathering, a larger proportion of coarse debris, and therefore also bed load, is produced than in areas of intensive chemical weathering, where only very resistant rocks, such as quartz dykes, deliver coarse debris. Transport by the stream rounds the fragments and gives them the typical form of fluvial gravel. During the development of the valley there can be a long-term deposition of bed load and accumulation of a valley floor. Lateral shift of the stream channel by lateral erosion, or renewed vertical erosion takes up some of the gravel accumulation again as bed load and transports it further downstream.

A determination of the volume of bed load is more difficult than for dissolved or suspended load. Usually a bed load trap is placed on the channel floor. The presence of the trap does, however, change the local current and disturb the bed load transport that is to be measured (Mangelsdorf *et al.* 1990, pp. 28–32). There have also been attempts to measure movement by dyeing pebbles or using radioactive tracers. More recent and more successful is the tracking of individual pebbles that have been magnetized or fitted with small radio transmitters (Ergenzinger and Conrady, 1982; Schmidt and Ergenzinger, 1990). During an individual flood event, bed load gravels are transported in small thrusts for distances of a few meters to tens of meters from one gravel bar to the next. Greater distances are limited to extreme flood conditions, unless the stream has a very large volume of water and rapid flow.

The relative importance of different types of stream load on the development of the stream channel varies greatly. The channel is most affected by the bed load because the gravels are deposited in the channel bed itself after being carried a short distance until moved on again, probably at the next flood. Suspended load remains in suspension until the stream enters a lake or the sea where the deposition of the fine material may lead to the development of a delta. If a stream inundates a flood plain during a phase of high water, suspended load is deposited on the flood plain because the water is shallower and the flow velocity much less than in the stream channel. Any lateral shift of the stream channel cuts into the fine-grained alluvial deposits in the stream bank. Compared to banks of gravel, banks of silt and clay are more resistant to erosion because of their greater cohesion.

The dissolved load has no influence on stream channel form. As long as the stream flows, the load is moved further. Only in very dry areas, where evaporation is high and salts are precipitated, is it possible for the former dissolved load to become a solid component of the dried-out stream bed.

11.2 EROSION AND TRANSPORT

Erosion is generally defined as the linear downwearing of the earth's surface, in particular by streams, in contrast to the areal downwearing by denudation. The term erosion is, however, sometimes applied more widely to include denudation processes.

Vertical erosion is the lowering of the stream channel floor. On a channel floor of unconsolidated material, vertical erosion is the net removal of sands and gravels whereas on bedrock, it is caused by the **abrasion** of the channel by the bed load. **Lateral erosion** is the wearing away of the stream banks by erosion.

11.2.1 Stream mechanics

The erosion and transport by flowing water is a function of the kinetic energy

$$E_k = mV^2/2 \qquad (11.1)$$

where m is the mass of the water and V is the flow velocity.

The Chezy equation (8.15),

$$V = CD^{0.5}S^{0.5} \qquad (11.2)$$

is also valid, where C is the Chezy coefficient, D is the water depth and S is the gradient.

The substitution of equation (11.2) for V in equation (11.1) shows that the kinetic energy of flowing water is directly proportional to the product of depth and gradient (DS). This relationship conforms to the definition of shear tension τ on the floor of the stream channel in the Du Boys (1879) equation

$$\tau = \gamma DS \qquad (11.3)$$

where γ is the specific weight of the water (grams per cm³), D is the water depth (cm) and S is the gradient expressed as the tangent of the slope angle (therefore dimensionless).

The depth slope product is, therefore, the controlling parameter for the stream's erosion competence, that is, its ability to set in motion pebbles of a particular size. As shown in equation (11.2), it is also proportional to the square of the flow velocity. A pebble with a mass m is moved in the stream when the shear force present in the stream is at least equal to the **critical shear force** necessary for the movement of this pebble. The critical shear force depends on the pebble's mass, its shape and its position in the stream channel in relation to the current.

In a gravel bar, the longest (A) axis of the pebbles usually lies transversal to the flow direction. The second longest (B) axis lies in the flow direction and at an angle inclined upstream, so that towards the current, it has an inclined surface whose upstream margin is covered by the neighbouring pebble. This **imbricated structure** of pebbles is particularly resistant to erosion (Fig. 11.2). It develops by the movement of pebbles until they have attained this optimal stable position. Only if the entire imbricated structure is destroyed by a renewed high discharge are the pebbles set in motion again. Flat pebbles offer greater resistance to renewed erosion than pebbles with a relatively large C-axis.

The accumulation of pebbles in a stream bed increases its roughness and reduces the flow velocity (Manning equation (8.14)). However, each pebble that protrudes above the floor of the stream channel bed causes the line of the current to be bent around it and especially over it. Above the pebble the current is narrowed resulting in a

FIGURE 11.2 Imbricated deposit of gravel, Northwest Branch of the Anacostia River near Adelphi, Maryland, USA. View upstream.

local increase in the flow velocity that works in the same way as the narrowing of the flow in a water suction pump: it produces a suction that is directed upwards, a lifting force which reduces the critical shear force necessary to set the pebble in motion and so increase its mobility. The relationship between current and pebble is not a one-sided causal relationship but a complex of hydrodynamic interactions that is still being investigated both empirically and theoretically (Richards, 1982, pp. 110–19; Knighton, 1984, pp. 56–60).

11.2.2 Erosion of different grain sizes

Based on empirical measurements Hjulström (1935) found a clear relationship between the flow velocity of the water and the eroded grain size, that is, the rock material set in motion by the water (Fig. 11.3). Both axes of the Hjulström diagram are logarithmic and it includes a large range of grain sizes and velocities.

The upper field of the diagram shows the velocities at which particles of a given size are eroded. It is limited below by a narrow band that marks the critical velocity at which the erosion of a particular grain size begins. This is a band rather than a line because the erodibility of particles of a given size also depends on their position and the way they lie in the stream bed. Single particles that lie on a sandy floor are more

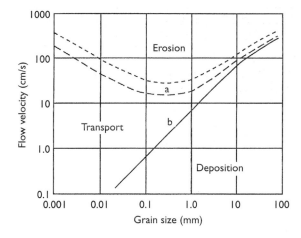

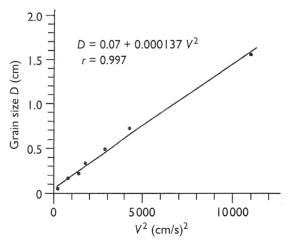

FIGURE 11.3 The relationship between grain size and critical flow velocity for erosion and transport (after Hjulström, 1935). (a) critical velocity for erosion; (b) critical velocity for the commencement of accumulation.

FIGURE 11.4 Grain size (cm) of fluvial bed load as a function of the square of the flow velocity (m/s) (after Wolman and Brush, 1961; *see also* Morisawa, 1968, p. 49).

easily set in motion than particles packed in an imbricated structure in a gravel bar. A lower flow velocity is required to erode fine sand and silt than that needed for still finer-grained clay particles. There is no cohesion in fine sand and silt, but clays are held together by an electrostatic force that also has to be overcome.

The lower curve in the diagram shows the flow velocity at which a particle of a particular size already in motion is transported no further and is deposited. Clay and fine silt do not sink at velocities of 1–2 cm/s, which explains the absence of suspended load deposits on the stream bed. Between the lower curve and the erosion curve lies the range of velocities within which particles of different grain sizes are transported. The greater the distance between the curves, the more continuous the transport. The flow velocity limit for the transport of grain sizes of more than 2 mm lies close to that for erosion, which means that if a piece of gravel that has been removed at one velocity arrives in a zone of sightly lower flow velocity, it is deposited after only a short distance.

The flow velocity at which erosion is initiated for grain sizes >0.5 mm is approximately proportional to the square root of the grain size so

that the maximum grain size eroded is approximately proportional to the square of the flow velocity. Wolman and Brush (1961) have confirmed this relationship in a comparison of data from various sources (Fig. 11.4).

The Hjulström diagram is only valid for unconsolidated material, the rock fragments transported to the stream by denudation processes or produced by weathering on the floor or

FIGURE 11.5 Potholes in the bed of the Mwania River near Machakos, Kenya, photographed during the dry season.

sides of the stream channel. On rock-floored stream beds surface friction by the bed load abrades the bedrock and causes vertical erosion. Where a stationary eddy rotates a pebble, a small hollow is produced. Others fall into the hollow, rotate and deepen it until eventually a **pothole** develops (Fig. 11.5). The pebbles within the hollow are also abraded and reduced in size. Potholes are formed in a similar way by meltwater flowing beneath glaciers.

11.2.3 Lateral erosion

Unlike vertical erosion which is produced by the immediate effect of the water flowing over the channel floor, lateral erosion involves additional processes. It is effective mainly in the area of the bank lying below the water level. The bank above is thereby undercut and steepened. Instability, slumping and the collapse of the bank into the water follows and the stream disintegrates the material and transports it away (Fig. 11.6).

Bank resistance varies with the season and discharge. If there is ground frost and ice formation in the pores, the bank material breaks up during the thaw that follows making it more easily erodible. Positive pore water pressure (section 8.2.3), where groundwater or interflow seep out in the bank area above the water level, also reduces the resistance of the bank material and leads to earthflows on the bank slopes making

FIGURE 11.6 Sliding as a result of lateral erosion in a creek in New Zealand. An old slide scar can be seen left of the recently developed slide.

them unstable, especially during the falling phases of floods.

11.3 DISCHARGE AND TRANSPORT RATES

The load transport rate, expressed as weight per time unit or volume per time unit, is an important measure of the total denudation rate for the stream catchment area above the point of measurement. Weight and volume are related by the specific gravity (g/cm^3) of the rock being transported and are reciprocally calculable. If the volume of the transport rate is divided by the surface area of the catchment area, the result is the mean denudation rate (section 3.3).

Because more load is transported during flood than during low discharge, there would seem to be a relationship between discharge and the rate of transport. This relationship does in fact vary not only according to the various types of load but also in different climates and seasons.

11.3.1 Dissolved load transport rates

The amount of dissolved load transported by the stream per time unit is the product of discharge and of the concentration of dissolved substances (section 11.1). Concentrations change in the different discharge phases depending on whether the stream is fed only by groundwater (dry weather discharge), from a combination of groundwater, overland flow and interflow (the rising limb of the hydrograph) or from a combination of groundwater and interflow (the falling limb of the hydrograph). The concentration also depends on whether the bedrock is soluble enough for the groundwater to deliver a major share of the total dissolved load or whether the soluble salts must first be produced by chemical weathering and are primarily present in the regolith. In the latter case, the transport of dissolved load to the stream would take place mainly in the interflow. In semi-arid climates, salt accumulates at the surface as a result of capillary movement and evaporation and after a dry period the overland flow in these salt areas is, at least temporarily, the main supplier of dissolved load.

11.3.2 Suspended load transport rates

Because the suspended load is primarily brought to the stream by the overland flow, the concentration of suspended load, measured in ppm, and the suspended load transport rate usually increase with increasing discharge. How much suspended load material reaches the stream depends on the intensity and duration of the overland flow, the availability of fine-grained unconsolidated material on the land surface and the vegetation cover. A dense cover of low vegetation protects the surface from wash denudation because it intercepts precipitation near the ground surface, reduces the flow velocity of the overland flow and impedes, with its densely spaced stems and its root network, the transport of soil particles. Where plants grow widely apart the overland flow tends to concentrate in narrow paths, which become rills and gullies and which increase delivery of fine material to the stream.

In the hydrograph of a flood event, the overland flow contributes particularly to the rise of the flood and to the flood peak. The maximum suspended load lies temporally close to, or earlier than, the discharge maximum (Fig. 11.7).

The relationship between the suspended load and the discharge changes if there is a change in

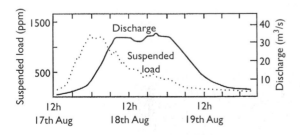

FIGURE 11.7 Suspended load concentration and discharge in the Enoree River, South Carolina, USA (after Einstein *et al.*, 1940).

the seasonal vegetation cover. For this reason estimates of the mean annual suspended load as a function of the mean annual discharge or of annual precipitation are not very useful (Langbein and Schumm, 1958). Instead, the **sedihydrograph** (SHG) is used to show the relationship between the mean monthly value of the suspended load concentration in tons of load per square mile of the catchment surface area and the monthly discharge contribution in tons of water per square mile during the course of the year (Wilson, 1972). Figure 11.8 shows two examples from the southwest of the USA. In the Rio Puerco the suspended load concentration changes little during the year and the mean monthly transport rate can be estimated reasonably accurately from the more easily available discharge data. The monthly discharge of the Gila River varies much less than the Rio Puerco but, despite almost constant discharge, the seasonal variations in the suspended load concentration range over more than two orders of magnitude so that an estimate of suspended load in this stream on the basis of discharge is not possible.

11.3.3 Bed load transport rates

Although many investigations have shown the relationship between grain size and discharge parameters, especially water depth and flow velocity, it is generally not possible to make reliable estimates of the total rate or **capacity** of bed load transport as a function of the discharge which includes all grain sizes of bed load (section 11.2.1 and Richards, 1982, pp. 110–19). This is because particles of a particular size are transported only above specific threshold values of the flow velocity and usually only for short distances in separate movements. Suspended and dissolved loads, once they are in motion, move continually until they are finally deposited or reach the sea. The availability of bed load is largely determined by factors that are unrelated to the discharge. These include the resistance of the rock, the type and rate of the dominating weathering and the denudation processes that predominate.

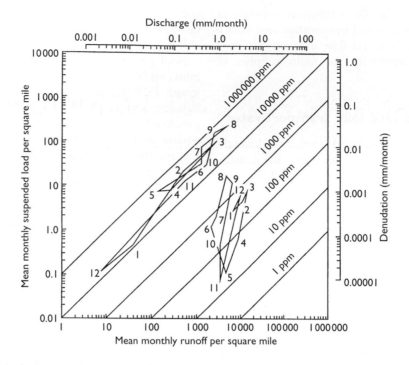

FIGURE 11.8 Sedihydrograph of the Rio Puerco River, New Mexico (stretched polygon) and the Gila River, Arizona (wider polygon) after Wilson (1972). The numbers on the curves indicate the month (January = 1 etc.). 1 ton of water per square mile = 0.386 m^3/km^2; 1 ton of suspended load per square mile = 0.155 m^3 (based on a specific weight of 2.5)/km^2.

The relative proportions of the bed load eroded and deposited locally are the main determining factors in the form changes of the stream channel. In a flood, the erosion takes place largely during the rise in water level and the deposition when the water level falls. The volume of the accumulation usually is similar to the amount of erosion immediately preceding it, so that, despite the large quantity of bed load transported, the form of the stream channel often is not very different after a flood than before (Schick *et al.*, 1987).

Bed load transport has been estimated to be roughly 10 per cent of the suspended load transport (Judson and Ritter, 1964). This is useful as an order of magnitude but does not apply in many regions of the earth and is anyway not particlularly important for an estimate of regional denudation rates (section 3.3) because, in general, suspended load and dissolved load make up by far the greatest part of the total stream load and, therefore, the total of rock material transported out of the catchment area.

12

STREAM CHANNEL FORMATION

12.1 THE RATIO BETWEEN WIDTH AND DEPTH

In humid climates, the width W of the stream generally increases more rapidly in the downstream direction than the depth D (section 10.5.3), as does, therefore, the ratio W/D (Fig. 12.1). The distance downstream, L in Fig. 12.1, is not the determining factor for this ratio but is used as a substitute variable for the downstream increase in discharge because detailed data were not available.

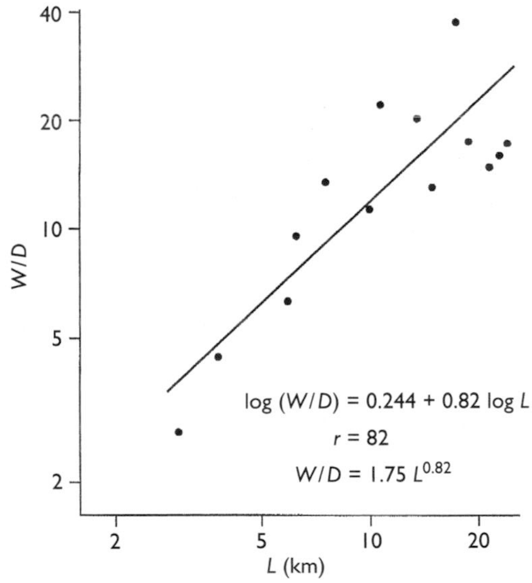

In stream channels of unconsolidated material, the ratio of width to depth is almost exactly inversely proportional to the percentage share of silt and clay in the bank material (Schumm, 1960; Fig. 12.2). The smaller the share of silt and clay, the less resistant the stream bank and the easier it is for the stream to widen both sides of its bed by lateral erosion. As the banks are eroded laterally, the depth of the stream channel in the stream cross-section is reduced. The simultaneous widening and shallowing of the stream is a major cause of stream braiding (section 13.2.2).

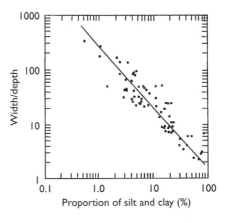

FIGURE 12.1 The ratio of width to depth (W/D) as a function of stream length L (distance from the source), in the stream channel of the Kall, Eifel, Germany. Based on data from Kim (1989).

FIGURE 12.2 The ratio of width to depth as a function of the share of silt and clay in the material of the banks of stream channels in unconsolidated material (after Schumm, 1960).

The figure equations read:

$$\log (W/D) = 0.244 + 0.82 \log L$$
$$r = 82$$
$$W/D = 1.75\, L^{0.82}$$

12.2 BEDROCK CHANNELS AND GRAVEL STREAM BEDS – EROSION-LIMITED AND TRANSPORT-LIMITED STRETCHES

A stream flowing in a **bedrock channel** that has no continuous gravel cover has more energy than is required to carry the bed load. Any isolated gravel beds in bedrock channels are temporary accumulations and are transported away at the next flood event. Gravel in the channel is itself worn down by friction and it also erodes the bedrock floor. Part of the stream's energy is used for downcutting. The remainder is dissipated by the high turbulence in the shooting flow of rapids and in waterfalls and is largely ineffective morphologically. Bedrock channels are, therefore, always **resistance-limited channels** and an indication of the high resistance of the channel bed and that the stream has an energy surplus for downcutting.

The stream channel in a **transport-limited stream** stretch is made wholly of sand or gravel that has been transported and deposited by the stream and which may be moved again during a flood event. There is always load available in a transport-limited stream channel for the energy of the stream to be fully used for transport. When more material is transported downstream than arrives from upstream in these stretches, the stream may eventually incise the bedrock and the stretch become resistance-limited (Hormann, 1963, 1965). The terms describe the fluvial equivalent of weathering-limited and transport-limited denudation (section 9.3).

A stream channel of unconsolidated material becomes resistance-limited if it contains blocks from the bordering slopes that are too large to be removed by the current. The blocks have a resistance similar to a bedrock channel, especially if they cover the stream bed without erodible spaces in between.

Streams moving a gravel bedload can alternate between being transport-limited and resistance-limited if all grain sizes are transported at high water; but at low water the erosion competence is insufficient to move the coarser components. Only bedrock channels are always resistance-limited.

12.3 GRAVEL BARS IN THE STREAM CHANNEL

The movement of the bed load in the stream is largely a function of **bed shear** on the stream bed (section 11.2.1). The bed shear is approximately proportional to the square of the flow velocity. It varies greatly within the stream's cross-section. A pebble that moves near the line of highest velocity ceases to move when it is transferred by turbulence in the water from the high-velocity zone to the neighbouring area where the bed shear is less than the critical value necessary for the movement of the pebble.

Repetition of this process with large quantities of pebbles leads to the accumulation of a **gravel bar**. Some sorting by size takes place because only those pebbles accumulate that cannot be transported further by the bed shear present in the current in the area of the gravel bar. The sorting is incomplete because the gravel bar itself is an obstacle and its surface roughness and the reduced depth of the water flowing over it cause a further decrease in velocity and bed shear (section 11.2.1).

The gravel bar is usually aligned in the flow direction. Many streams have gravel bars alternately on their left and right banks, even where the channel is almost straight which is an indication that the line of high velocity swings from side to side. At bends in the stream course, the line of high velocity is pushed to the outside of the bend and **point bars** are deposited on the inside of the bend where velocity is low (Fig. 12.3). The gravel deposits in gravel bars are imbricated (section 11.2.1). They also catch smaller pebbles and sand in the large pores and depressions on their surfaces.

12.4 RIPPLES, DUNES AND ANTIDUNES ON SANDY CHANNEL BEDS

Sand banks form in a similar way to gravel bars but their shape is changed more easily and small forms that are especially characteristic for sand bars develop on the channel floor. A stream channel floor is, physically, the interface between

Figure 12.3 Point bars near Kamihori, Japan.

sand and water. Normally, water movement in streams is turbulent (section 10.5.1) and the flow velocity on the bottom of the stream bed varies spatially as does the movement of the sand so that eroded depressions and elevated accumulations of deposited material occur close to one another on the channel floor. The small-scale relief on the stream bed created by turbulent flow influences the flow of the current itself in a feedback: the channel floor, altered by the current, and the change in the spatial differentiation of the turbulent flow, altered by the form change on the floor, adjust to one another.

The ripples, dunes and antidunes on the channel floor are an indication of the adjustment. They change the surface roughness of the channel floor and therefore the mean velocity of flow, in accordance with the Manning equation (equation (8.14) and section 8.6.2).

Ripples develop during quiet streaming flow (Froude number $F < 1.0$, section 10.5.2) as small asymmetrical ridges across the flow direction, with their steep sides directed downstream. A prerequisite seems to be that the grain size of the sand is less than 0.7 mm (Simons and Richardson, 1963; *see also* Simons, 1969). Ripples under water are in their form and the mechanism of their development very similar to wind ripples (section 8.7.3). Where the water is deeper and the stream flow more rapid, similarly asymmetric but much larger **dunes** develop on whose surface ripples may also form.

Antidunes develop in standing waves in shooting flow (Froude number $F > 1.0$). Their steeper side faces upstream. The sand waves of the antidunes and the standing waves on the surface of the water are in phase. At very high flow velocities the crests of the antidunes are eroded and the channel floor levelled.

In principle, the same mechanisms affect gravel beds but because of the low mobility of coarse material, their forms are less well developed.

12.5 RIFFLES AND POOLS

Short stretches of shooting flow and streaming flow alternate along the courses of many, particularly smaller, streams. The stretches of shooting flow, **riffles**, are more rapid and shallower than the intervening streaming flow stretches, **pools** (Fig. 12.4).

Flow dynamics determines the succession of riffles and pools in a transport-limited stream stretch. In addition to the eddies around a perpendicular axis and the turbulence around a horizontal axis transversal to the general flow direction, there is a considerable rotational helical movement of the water along the long axis of the stream. This **secondary circulation** overlies the general downstream direction of the current and produces a more or less rhythmically repeated local strengthening and weakening of the flow velocity and also of the energy available for the erosion and transport of bed load.

Negative feedbacks develop as a result that limit the effectiveness of these increases and decreases in flow. A pool that initially is eroded by the strong current causes, because of its greater depth, a local increase in the stream cross-section and, in accordance with the continuity equation (10.4), a reduction in flow velocity. Eventually the velocity decrease will mean that the pool is no longer deepened.

The development of pools and riffles in the stream channel is not only the geomorphological result of variations in the flow but also a component of the hydrodynamic flow equilibrium and belongs, like many other spatially recurring landform components, to the **rhythmical phenomena** of the earth's surface (Kaufmann, 1929). The mean distance between sets of riffles is roughly five times the stream channel width.

Creeks and streams with steep mean gradients and a coarse gravel load tend to develop sequences of **steps and pools** rather than riffle and pool sequences. The gravels often accumulate in the riffles, forming steps in the channel bed. The largest gravels are seldom moved and catch the smaller grain sizes which are then deposited in imbricated layering on the upstream side. Between the steps built of coarse gravels over which the water falls or shoots, are the pools. Drifting branches increase the ability of the steps to catch gravel.

Steps and pools sometimes occur in bedrock channels but, in this case, the steps are often

FIGURE 12.4 Sequence of riffles and pools in the Royal Gorge of the Arkansas River near Pueblo, Colorado, USA.

related to resistant rocks or dykes and the intervals between the steps are determined less by flow dynamics than by rock structure.

12.6 VALLEY FLOORS, NATURAL LEVEES AND ALLUVIAL LOAM

The floor of a **flat-floored valley** is almost horizontal and often wide (Fig. 12.5). The **floodplain** is that part of the valley floor flooded frequently at high water.

Valley floors develop in two ways: as a rock-floored valley bottom or as an accumulation valley floor. A **rock-floored valley** is formed by a stream that no longer incises but erodes laterally in a course that swings from side to side across the valley floor. The valley slopes are undercut and steepened by the lateral erosion and retreat by denudation. The channel floor lies in the bedrock and is covered by only a thin layer of gravel and sand beyond the channel.

As the stream swings across the valley floor, it deposits material on the insides of the bends, which covers the rock floor beyond the current position of the channel. The thickness approximates the height of the stream bank above the channel bottom. Subsequently, the stream shifts within the valley floor into the deposited material, erodes it on the outside of its stream bend and redeposits it on the inside of the next bend

FIGURE 12.5 Flat-floored valley of the Perlenbach near Kalterherberg, Eifel, Germany.

downstream. The deposition is not permanent therefore and the material is in a state of intermittent long-term transport.

An **accumulation valley floor** cannot be distinguished easily from a rock-floored valley from its surface. It is produced by the continuous deposition of gravels and sands in an existing incised valley where deposition has replaced incision. Both the channel floor and floodplain are composed entirely of these alluvial deposits.

The difference between a rock-floored valley and an accumulation valley floor is relevant in relation to the groundwater budget, particularly if the bedrock beneath the gravels is impervious and only the gravels are available as an aquifer (section 10.3). An accumulation valley floor is also much less resistant to erosion than a rock floor since the sands and gravels of its channel bed have already been transported and may well be removed again during the next flood.

During flood, the stream covers the floodplain. However, the flow velocity decreases rapidly from the immediate area of the stream bank out into the floodplain because the depth of water on the floodplain is much less than in the stream channel. If there is enough turbulence in the stream at the flood stage, sand and gravel may be carried over the bank and deposited in a strip close to the bank on the valley floor. Repetition of this process leads to the formation of a **natural levee** as the valley floor near the stream bank becomes more elevated than the area further from the stream (Fig. 12.6).

The higher the natural levee, the less often it is flooded by subsequent high-water events. Its possible height above the channel floor is, therefore, limited. The deposition of bed load during falling water in later floods takes place to a greater extent in the channel floor itself which also is raised in height, thus reducing the difference between the natural levee and the channel floor and increasing the probability that water will again flow over the stream banks during floods.

In some rivers, such as the lower Po River in northern Italy, the raising of the natural levees and the channel floor has progressed so far that the water level of the river lies at a higher altitude than the valley floor beyond the levee.

FIGURE 12.6 Freshly accumulated natural levee, photographed shortly after a flood in the North-west Branch of the Anacostia River near Adelphi, Maryland, USA.

The first European settlers on the Mississippi River built houses on the natural levee, not only because the river provided the best transport route, but because the levees were least likely to be flooded and because the land surface lies higher above the local water table there than further from the river. Natural levees do not have the same height along their entire length; during floods the water flows over on to the valley floor beyond at low points in the levee.

Natural levees slope only very gradually away from the river onto the valley floor. In the higher areas the levee can be used for the cultivation of plants that would not survive if the water table were high. On those parts of the valley floor that lie further from the river, the water table is closer to the surface and a **back swamp** may develop.

Even if the stream does not flow over the levee during high-water events, the flood water seeps through the porous banks and into the groundwater in the valley floor area. The water table behind the levee may rise and temporarily cover the valley floor until the flood falls again and the groundwater can drain back through the pores of the levee sediments to the stream.

Natural levees develop only if there is a large enough proportion of coarse bed load (sand and gravel) present and enough turbulence during floods for part of the load to be carried over the bank. Streams primarily transporting suspended

load have, at most, only very minor natural levees.

On the areas of floodplain lying further from the stream, the deposits during floods are mainly the sands, silt and clay of the suspended load. They form **alluvial loam**. Alluvial loam ranging from tens of centimetres to a few metres in thickness was deposited on valley floors in the central uplands of Germany. It originates from a period of forest clearing in the Middle Ages that caused wash denudation on the hill slopes (Mensching, 1949, 1951).

12.7 THE TENDENCY TOWARDS A LOCAL DYNAMIC EQUILIBRIUM IN A STREAM BED

A similar tendency towards equilibrium for the mass balance of the bed load exists on transport-limited streams as exists for the denudative mass balance on slopes (section 9.2). The height of the channel floor in the cross-section is determined by the balance equation

$$H = H' + (A - R)\Delta T \qquad (12.1)$$

where H is the height after the timespan ΔT, H' is the height at the beginning of ΔT, A is the bed load arriving from upstream per time unit, expressed as thickness of the layer that this bed load will form if all of it is deposited, and R is the amount of bed load transported downstream per time unit, including any additional bed load from local erosion expressed in the same way as A.

Initially, it can be assumed that the stream sections immediately upstream and downstream from the cross-section maintain their heights unchanged. When more bed load arrives from upstream than is transported away downstream, part of the load is accumulated and the height of the channel floor increases. As a result, the gradient downstream is increased but the gradient upstream is reduced. The stream is now able to transport downstream more of the bed load than before and less material arrives from upstream.

The accumulation of the bed load that is not transported and the accompanying local raising of the channel floor continues, until all the bed load that arrives is also transported further downstream, that is, until $R = A$ and, therefore, also $H = H'$. The supply and removal at the observed cross-section have reached a dynamic equilibrium.

Conversely, an excess in the removal over supply ($R > A$) lowers the stream channel. The gradient downstream is reduced and upstream increased. The removal is reduced and the supply A increases until $R = A$ and $H = H'$.

The equilibrium is dynamic in both cases. H, R and A together form a **process response system** in which they steer one another by means of negative feedbacks (section 2.2.3).

In reality, the actual attainment of the equilibrium is possible only if the period of uniform stream flow and therefore transport potential lasts longer than the **relaxation time** necessary until A and R are equal. Normally, however, the discharge, which primarily affects the bed load transport, changes continuously, especially during floods, when there is net erosion, a surplus of R over A, during the rising of the water level, and net deposition, with a surplus of A over R, during the falling of the water level. Net supply and net removal of the bed load usually occur at different times in the same flood event. But, as described in section 12.5, if there is little difference in the form of the stream channel before and after a flood, then over a longer period A and R cannot be far from a dynamic equilibrium.

Also, the stream channel is not a purely linear form with a longitudinal gradient but has, in addition, a variable width. Lateral erosion, particularly when it takes place on both sides of the stream and widens the stream channel, can contribute to the equilibrium state. Increased removal R in equation (12.1) depends, partially or wholly, on this type of lateral erosion. The water depth and thereby the flow velocity and bed shear are reduced as the stream widens and its ability to move bed load decreases.

The bed load arriving in a resistance-limited channel stretch from upstream is usually transported through the stretch without local deposition. Accumulation does not take place because the potential for bed load to be removed is larger than the actual supply. Lowering of the channel is only possible if the rock floor of the channel is eroded but, because of the high resistance of the floor, this takes place much more slowly than on unconsolidated material. As long as the potential for removal remains higher than the quantity of bed load that arrives in the channel stretch, plus the amount delivered locally by erosion, there cannot be a local dynamic equilibrium of the mass balance.

The way a stream responds locally to changes in its discharge by changing its gradient, its bed load supply or removal, how, that is, these components and the form of the cross-section change under the influence of their feedbacks, is not deterministically predictable but subject to the same indeterminacy principle as the components of hydraulic geometry in the continuity equation (10.6).

STREAM CHANNEL PATTERNS

13.1 VALLEY FORM AND STREAM CHANNEL PATTERN

The two main types of fluvial valley forms are the **V-shaped valley** and the **flat-floored valley**. The side slopes of a V-shaped valley border immediately on the channel. The orientation of the stream and of the valley is the same and the length of the stream and of the valley are similar. In a flat-floored valley, the valley floor lies between the stream and the foot of the valley side slopes. It is produced by lateral erosion, accumulation or a combination of the two. The width of the flat valley floor allows the stream direction to diverge from that of the valley, either by swinging from one side of the valley to the other or by dividing its channel into several arms. In a flat-floored valley, the stream can be considerably longer than the valley and on a depositional plain unconfined by valley sides, the course direction may change even more. There is therefore a relationship between the possible channel patterns of a stream and the type of valley in which it flows.

Three patterns are usually distinguished: straight streams, braided streams and meandering streams. Most **straight stream** stretches are in narrow V-shaped valleys that are themselves straight and prevent any lateral swinging by the stream. Their alignment is generally a consequence of linear tectonic structures in the bedrock, such as faults or dense sets of joints. The rocks along these structures are disrupted and weather more rapidly, forming zones of weakness in which the stream can erode more easily. Where linear zones of weakness of this type

intersect both the valley and the stream may make a sharp bend, not to be confused with the curving bends of meanders.

Straight streams in flat-floored valleys have usually been artificially straightened. Natural streams flowing on valley floors tend to braid or meander.

13.2 STREAM BRAIDING

There are three main types of **stream braiding**. **Erosional braiding** usually develops in stretches of resistant rock. Streams with a large bed load and unstable banks develop **in-channel braiding. New channel braiding** occurs if the bed load contains not only coarse components but also a large proportion of silt and clay particles.

13.2.1 Erosional braiding in a bedrock channel

Streams transport their loads through stretches of resistant rock without any significant local deposition. The bed load is a tool for vertical erosion in the bedrock channel of the stream bed. The bedrock is generally not equally resistant to erosion in all parts of the stream bed. The wider the stream bed the greater the chance that there are differences in resistance due to variations in rock type or the spacing of joints (Fig. 13.1).

Because of the differences, parts of the stream bed are more strongly incised than others and form local rock channels within the main channel floor in which the water flows at greater

FIGURE 13.1 Erosional braiding at the Great Falls of the Potomac near Washington DC.

depth and higher velocity (*see* Manning equation, section 8.14) than in the neighbouring channels. The higher flow velocity means that the stream's capacity to cut down is increased in these rock channels and a local positive feedback (section 2.2.3) develops within the erosion process. The duration of the self-reinforcement is limited because the deepening of the channel leads in the long term to a local reduction in the gradient.

Several such bedrock floor channels can incise next to one another in the resistant stretch of a wide stream. The greater their total cross-section, the greater the amount of water concentrated in them and the smaller the amount of water flowing on the shallower channel floor between them. If erosion of the deeper channels progresses, the uneroded part of the stream's rock floor eventually lies above the water level and the stream is divided into several arms separated by ridge-shaped bedrock islands. Erosional braiding usually occurs at waterfalls and rapids (section 14.5) because of the large current energy required for this type of development.

13.2.2 In-channel braiding

Fluvial gravel and sand deposits are not cohesive unless they contain a large proportion of clay and silt particles. Stream banks formed of non-cohesive, unconsolidated material are very easily eroded and during a flood both banks can be shifted back by lateral erosion. The stream is widened as a result and becomes shallower; its velocity is lower and its transport power thereby diminished. The sand and gravel eroded from the banks is transported only a short distance and redeposited as sand and gravel bars. The surface roughness and height of the bars reduces the velocity of the water flowing over them and they become obstacles in the stream which catch the transported bed load during flood, further increasing their height. Ultimately, they are so high that at times of low and average discharge the water does not cover them but flows around them in separate channels.

Because of differences in the amount of accumulation in the two arms of the stream on either side of a gravel bar, the water level of one arm often lies higher than that of the other. If it rises enough, cross-currents flow across the bar and incise small transversal channels. Some of these widen to form new larger channels.

The formation of gravel banks in the stream bed by local redeposition has the effect that the divided flow channels are forced laterally to the banks which have already been shifted back.

The unstable non-cohesive bank material is again eroded and redeposited as sand or gravel bars in the stream bed. The term in-channel braiding indicates that the actual cause of braiding is the erosion of both banks accompanied by an increase in width of the stream bed. The entire cross-section of the flowing water does not necessarily have to change because the accumulation of the bars within the stream bed can compensate for that part of the cross-section that is added by the widening of the bed.

In-channel braiding can develop without any net accumulation and the stream remain in an equilibrium of its mass balance. Only the lack of cohesiveness in the banks and the local erosion and redeposition are decisive. Nevertheless, there are many streams, particularly in glaciated high mountain areas, where in-channel braiding takes place simultaneously with the net accumulation of large amounts of material. If the meltwater is loaded with moraine debris, the braided bed of a stream may cover the entire floor of a valley (Fig. 13.2).

With every additional branching, the water available in the individual channel branch for bed load transport, and therefore the transport power, is reduced. In any stream, the total number of branches cannot exceed a maximum which is dependent on the total discharge and the quantity of bed load to be transported. Because the discharge varies, only the deeper branches are active at low water, the shallower dry up. At high water all the gravel bars may be flooded, in which case the stream has a unified water surface.

The ability of the stream branches to transport bed load varies. Coarse bed load that is moved from a large channel branch to a smaller may become deposited because the transport power of the latter is less. Repetition of this process gradually clogs the smaller branch until it becomes inactive.

The more new branches are developed, the greater is the probability that old ones become filled because the total number of branches in this type of stream usually remains more or less constant, an expression of an inherent tendency to dynamic equilibrium between the two processes in the system.

Many of the debris-carrying streams of the Alps have in-channel braiding if they have not, like the Rhine, the Inn and others, been forced

FIGURE 13.2 In-channel braiding in the Rosegg Valley, Grisons, Switzerland. In the background, the end moraine of the Tschierva Glacier reaches into the valley.

into a single fixed bed and transformed essentially into drainage canals. Until it was regulated in the 19th century, the upper Rhine was braided between Basel and Karlsruhe and had a meandering course from Karlsruhe northwards as far as Mainz. The change from braiding to meandering was probably due to the progressive downriver decrease in the grain size of Rhine sediments and a corresponding increase in bank resistance. Traces of the former course showing the natural pattern of the braided river in the upper half and the meandering river in the lower half of the river stretch between Basel and Mainz are visible on present day topographic maps.

Wadis, stream channels with rare ephemeral flows that occur in deserts, also usually have coarse bed load with low bank stability and tend therefore to be braided.

13.2.3 New channel braiding

New channel braiding, also known as **avulsion**, is related to the formation of natural levees, described in section 12.6, and occurs when the water level in the stream lies above the land surface in the back swamp because of an increase in the height of the natural levee and the channel floor. The levee is not the same height along its entire length. During a flood, the stream flows over the levee at its lowest point. The concentration of the overflow at this location produces a strong local current that erodes the levee and cuts out a flood channel on the levee slope directed away from the stream. On the lower Mississippi these overflow channels are called **crevasse channels**. The water flows through the channels to the back swamp where it follows the gradient behind the levee, more or less parallel to the stream course.

Further downstream, where the levee joins the valley side, the water flowing in the back swamp is dammed between the levee and valley side. Because of their longitudinal gradient, the levee and the river at this point are lower than at the overflow channel further upstream. If the level of the dammed water becomes high enough, part of it flows back to the main river over the levee as a return-flow channel which is then also lowered at this point by erosion.

The entire process is repeated at the next flood. The overflow and return-flow channels are now lower than the rest of the levee and are lowered further with each following flood event. The lower the overflow channel the larger the amount of water that flows along the back swamp channel and the more frequently flooding of the new channel occurs.

The new channel in the back swamp area follows the direction of the valley gradient and is, therefore, usually straight and shorter than the corresponding stretch of the main river course. Its gradient is also greater and with each flood there is additional erosion in the channel which gradually takes on the character of a river bed. Eventually water also flows through at low water as well as during floods.

During subsequent floods, natural levees form along the banks of the new channel so that the river is divided into two branches, both with natural levees and with a back swamp between them.

An important difference between in-channel and new channel braiding is that the former produces a large number of new branches but the latter only two, or occasionally three. Also, new channel braiding develops only during floods but in-channel braiding can occur without bank flooding. Bank stability is a prerequisite for new channel braiding. It is maintained if there is a sufficiently high proportion of silt and clay in the bank material. If the banks are eroded, in-channel braiding may be initiated. One type does not exclude the other and there are transitional and mixed types. The Yukon River in Alaska is a stream with both in-channel and new channel braiding (Figs. 13.3 and 13.4).

13.3 RIVER MEANDERS

There are two types of meanders: **free meanders** and **incised meanders**. Free meanders lie entirely in a valley floor or a plain that has been formed of material deposited by the river. The river can erode laterally into its bank, remove this material and change its course within the valley floor. Incised meanders develop when a meandering river incises its course by vertical erosion. The deepened valley follows the curves of the river and the valley itself meanders.

13.3.1 Free meanders

The course of a freely meandering river swings in a series of curves. It is distinguished from other types of stream course, especially braided streams, not only by the succession of bends, but, in general, also by an absence of braiding. Exceptionally, some freely meandering rivers have a temporary phase of new channel braiding (Fig. 13.4).

A simple measure of the intensity of meandering is the **sinuosity** of a river. It is expressed as the ratio of the length L of a river stretch, with all its bends, to the straight distance D between the two end points of the stretch, therefore

$$P = L/D \qquad (13.1)$$

If a river stretch is completely straight, then $P = 1$; for a meandering stretch it can be twice as large or more. The sinuosity is also an expression of the ratio between the gradient of the straight valley floor and the longitudinal gradient of the stream.

Schumm (1960; 197, p. 113) examined the relationship between sinuosity and the composition of the stream bed (Fig. 13.5). He showed that all the stream beds in which $P < 1.5$ have less than 8 per cent silt and clay. The shortage of fine material means that their banks are made of incohesive sands and gravels and are easily attacked by lateral erosion, so that in-channel braiding is more likely to develop than meanders.

Streams with a significantly higher proportion of silt and clay in their bed material have a

FIGURE 13.3 New channel braiding in the Yukon catchment area, Alaska.

FIGURE 13.4 Free meanders in the Yukon catchment area, Alaska with cutoffs and new channel braiding.

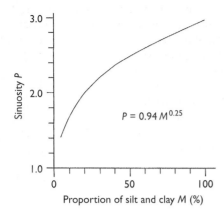

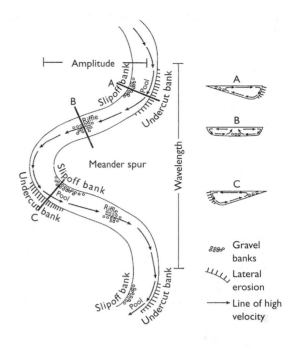

FIGURE 13.5 The relationship between the sinuosity and the proportion of silt and clay in the stream bed material (after Schumm, 1977, p. 117).

higher sinuosity. The silt and clay content is particularly important in the banks of the streams that were originally formed of alluvial loam, deposited from the suspended load on the valley floor, which has been exposed again laterally by the stream. The higher the silt and clay content, the greater the resistance to erosion and the smaller the width-to-depth ratio of the stream bed (section 12.1); also, the helical secondary circulation of the water remains single or double celled (section 12.5).

The characteristics of channel form and current are the most important prerequisites for the development of meanders. Because of the relatively limited width of the stream bed there is only one single line of high velocity, instead of several next to one another as at the beginning of in-channel braiding. The stability of the banks has the effect that lateral erosion takes place only where the line of highest velocity is near one of the banks. The opposite bank is in a zone of low velocity where accumulation predominates. The stream bed shifts laterally in the direction of the eroded bank without, however, changing its width.

The helical secondary circulation causes the line of highest velocity to swing from one side of the stream bed to the other. The stream bed is also shifted spatially to the right and left of the original flow line, so that it forms a succession of

FIGURE 13.6 Plan view and cross-sections of meanders (schematic).

bends and begins to meander (Fig. 13.6). Exceptionally, the initial swing of the line of highest velocity can be caused by an obstacle in the stream such as a tree trunk or large rock. Usually, though, it is the dynamics of the turbulent flow that start the swinging.

The **wave length** of a meandering stream course is expressed by the mean distance between two curves bending in the same direction, from one right curve to the next, for example. In general the wave length is equal to about 10–14 times the width of the stream bed. The **amplitude** of the meanders is the width of the swing between the extreme points of the bends.

The wavelength is influenced mainly by two factors: the size of the mean discharge Q_m and the proportion of clay and silt M in the banks of the stream bed. The dependence on discharge seems clear: large rivers have large meanders. However, a higher proportion of clay and silt in the banks increases the sinuosity. With a given discharge, a higher sinuosity means, in general, not only a greater amplitude, but also a reduction in the wavelength. Schumm (1977, p, 115) has quantified this relationship (shown here for metric units):

$$\lambda = 1936 Q_m^{0.34} M^{-0.74} \qquad (13.2)$$

where λ is the wavelength, Q_m is the mean discharge (m^3/s) and M is the clay and silt share of the bank material (per cent).

Figure 13.7, a nomograph, shows the relationship graphically. For $M = 40$ per cent, the wavelength of a meander with a discharge of $Q_m = 2 m^3/s$ is 160 m, with $Q_m = 2000$ m^3/s it is 765 m and with $Q_m = 2000$ m^3/s, 1674 m. For a given discharge of 200 m^3/s the wavelength varies, however, between 388 m at $M = 100$ per cent and 2134 m at $M = 10$ per cent. The numbers are only approximate. Schumm's equation is the result of a statistical analysis of data with a considerable scatter.

There is a clear relationship between meanders and riffle–pool sequences (section 12.5). Both are determined by the helical secondary current whereby a meander wavelength is equivalent to two riffle-pool sequences. The riffle–pool sequences of a previously straight stream course can gradually change to meanders if the erosion

power of the stream and the bank erodibility allow this. The pools are in the meander bends close to the outer bank of the curves and the riffles in the more or less straight stretches between the bends (Fig. 13.6).

The stream channel cross-section in the bend is asymmetric with the deep water and steeper slope on the outside and shallow water and flatter underwater slope on the inside of the bend. The line of highest velocity is closer to the outer, **undercut bank** and the high flow velocity results in lateral erosion there. On the **slipoff bank** on the inside of the bend, away from the line of highest velocity, the current is weak and sedimentation predominates. With further erosion on the outside of the bend and sedimentation on the inside, the bend shifts laterally and the meander continues to develop.

The line of highest velocity in the bend is curved and is pushed against the outer bank by centrifugal force. Also, the general downvalley direction of the current affects the position of the line of highest velocity which is not closest to the bank at the apex of the bend but a little downstream from it. It is here that the most intensive

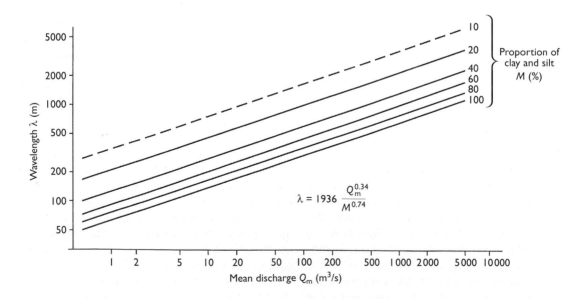

FIGURE 13.7 Nomograph to determine the meander wavelength as a function of the mean discharge Q_m and of the silt and clay content M of the bank material. Based on the functional equation (13.2) from Schumm (1960, 1977).

lateral erosion takes place, with the result that the meander bend shifts gradually down valley. The rate of lateral erosion on the bank can be measured by placing long thin erosion pins (section 8.8.2) in horizontal rows in the slope of the undercut bank and measuring how much they project from the bank after a given time period.

With an increase in amplitude, the stream length increases, the gradient is reduced and the lateral erosion also decreases progressively. Meander growth is, therefore, limited. Erosion can be renewed locally by neck or chute cutoffs of the meander bends (Fig. 13.4). A **neck cutoff** occurs when a meander bend is shifted down valley more rapidly than the bend downstream from it. The **meander spur** between the bends narrows until it is cut through by lateral erosion. This can take place without the stream flooding the valley floor. A **chute cutoff** develops if the valley floor is covered during flood to a sufficient depth and the general current follows the valley gradient rather than that of the meandering stream channel. The gradient of the valley floor is steeper than along the stream channel of the meanders because the straight distance is shorter. If the flood water erodes, it will create a straight channel between the two neighbouring bends. Initially, such a channel is so shallow that it remains dry after the flood has subsided. At the next flood it is again used as an overflow channel and further eroded until the stream can flow through also in times of normal flow discharge. Because its gradient is higher, compared to the longer course of the older meander bend, the new chute cutoff eventually becomes the main channel.

At first, the **cutoff bend** remains open at both ends and there is some flow through it but at lower velocity and with less transport power than in the new channel. Part of the sediment load is, therefore, deposited at the channel entrance which becomes narrower and is finally obstructed during normal flow; water and sediment can then only reach the cutoff bend during flood. Because of the direction of the flow, the upper end of the cutoff is blocked by sediment earlier than the lower end, where sediment is transported in mainly as a result of turbulence at

flood stage. Once both ends are blocked, the cutoff bend becomes an **ox-bow lake**. Sediment still arrives during floods in the lake, which gradually fills and is converted into a swampy depression. A former meander bend is visible in the valley floor after it has been filled in because of its lower position, gley soils (section 7.5.3) and its hygrophytic vegetation.

13.2.2 Incised meanders

Incised meanders develop when a meandering stream incises its bed so that the stream's course and the valley in which it flows are meandering. If there is an increase in the amplitude of the meander bends during the downcutting phase, it is termed an **ingrown meander**, but if there has been little change in its plan form after incision, it is an **inherited meander**.

Incised meanders cannot change their position as easily as free meanders because the valley slope above must also retreat with each lateral shift of the meander bend which requires much more energy.

Ingrown meanders are more frequent than inherited meanders. As in free meanders, centrifugal force pushes the line of high velocity outwards and, at the same time as it is downcutting, the stream erodes laterally into the outside of the bend. The valley slope is undercut and steepened and becomes the **undercut slope** of the meander. On the opposite side is the **slipoff slope** on which the river has shifted laterally outwards during the incision.

The steepest part of the undercut slope is where the line of highest velocity is closest to the outer bank, a little downstream from the apex of the bend (Fig. 13.6). How steep the slope is depends on the intensity of the lateral erosion and the rock resistance (section 8.2.5). The steepness of the slipoff slope is a function of the relationship between the rate of vertical downcutting and the rate of the lateral erosion directed outwards, away from the slope. The slipoff slope is usually much less steep than the undercut slope, which gives ingrown meanders their characteristic asymmetrical valley cross-section. The Mosel valley in Germany between

Koblenz and Trier has many well-developed ingrown meanders (Fig. 13.8). Ingrown meanders can also become cut off, although this happens much more rarely than in the case of free meanders. Because of the relatively high ridges of the meander spurs between the valley bends, only neck cutoffs are possible. Lateral erosion narrows the neck of the meander spur, lowers it and eventually removes it. The incision of the stream progresses simultaneously and the new cut through the spur is deepened further. The old meander bend dries up rapidly and becomes a **cut off valley**. The **cut off spur** forms an isolated hill with the cut off valley on one side and the new channel through the spur on the other (Fig. 13.9). A tributary that originally joined the main stream in the cutoff meander stretch now flows as an **underfit stream** in the cutoff valley, underfit because it flows in the valley that was created by the larger main river and because it is undersized compared to the valley it now uses.

There is little difference in the inclination of the slipoff and undercut slopes in **inherited meanders**. They occur where the rock structure encourages vertical erosion but impedes lateral erosion. The effect of centrifugal force on lateral erosion is, however, not entirely neutralized and there is a gradual rather than principal difference between ingrown and inherited meanders. The Goosenecks of the San Juan River in southern Utah in the USA are inherited meanders (Fig. 13.10). The river is incised in a series of limestones. In the upper, thinly bedded and largely unjointed strata, there is a marked contrast between the inner and outer slopes of the valley bends. Below in the thickly bedded densely jointed rock, both valley sides are almost perpendicular.

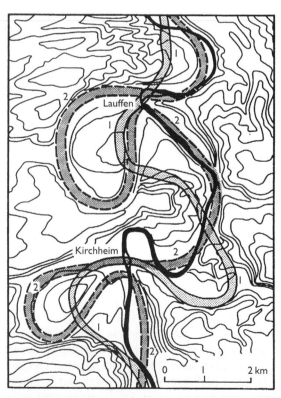

FIGURE 13.9 Cutoff incised meander on the Neckar River near Lauffen, Germany. (1) Early Pleistocene course; (2) Young Pleistocene/postglacial course. Present-day course marked in black (after Wagner, 1960, p. 90).

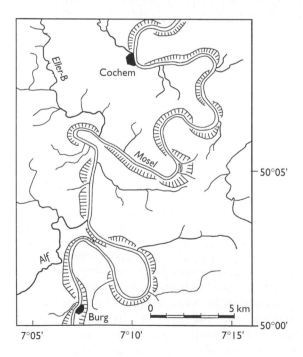

FIGURE 13.8 Incised meanders on the Mosel River between Burg and Cochem, Germany. The positions of the steep slopes are shown by hatching.

FIGURE 13.10 Inherited meanders. Goosenecks of the San Juan River, Utah, USA.

A clear relationship has been found between wavelength and mean discharge of free meanders which indicates that they can adjust relatively rapidly to changes in mean discharge that may have come about as a result of a change in climate or for other reasons. The wavelength of incised meanders, by contrast, is determined by the meander wavelength at the beginning of incision. Incision generally continues for tens of thousands of years, longer, for example, than the succession of climatic periods in the Quaternary. Also, the volume of the spur that would have to be removed by lateral erosion for the wavelength to be changed, is too large. Even when single neck cutoffs locally modify the number of incised meander bends, their wavelength remains more or less constant. Only after incision has ceased can lateral erosion gradually form a flat valley floor in which a more rapid shifting of the stream course, and a change of its

wavelength, is possible. But then the incised meanders have become free meanders.

13.3.3 Other types of asymmetric valleys

In addition to the **valley asymmetry** resulting from the contrast between undercut and slipoff slopes, **structurally determined asymmetry** can occur in valleys incised along a fault so that slopes are of different rock resistance. Structure also determines the asymmetry of **isoclinal valleys**. In this case, the strata dip across the valley direction and the slope on the valley side towards which the layers dip is steeper than on the opposite side where the slope is influenced more by the bedding planes of the strata.

It has been suggested that asymmetry in some central European valleys is due to climate. Causes that have been cited include loess deposition on one side of a valley or differences in the intensity of solifluction during the last ice age on slopes facing in different directions and therefore receiving different amounts of solar radiation. These interpretations are not generally convincing, even though they may be correct in some cases.

13.4 ASYMMETRY AT STREAM MOUTHS: ENTRANCE ANGLES AND THE DEFERMENT OF JUNCTIONS

When a tributary stream enters a main river, its line of highest velocity is shifted to the downstream side of its entrance by the main river's current. The cross-section of the tributary stream channel becomes asymmetrical, as does the distribution of its erosion and transport capacity. There is a tendency, therefore, for the tributary stream to erode laterally on its bank that is downstream in relation to the main river and to deposit bed load on the side that is upstream, so that the tributary **entrance angle** is not a right angle but is slanted in the downstream direction of the main river. The entrance angle of the Mosel into the Rhine at Koblenz is an example of this type of asymmetry.

At a tributary valley's junction with a V-shaped valley there can also be asymmetry if the tributary valley slope on the downstream side relative to the main valley is undercut by one-sided lateral erosion and becomes steeper than the upstream slope.

Tributary stream junctions can be shifted laterally for some distance on depositional plains. The junction of the Ill in Alsace, France, for example, is deferred for about 100 km parallel to the Rhine in the upper Rhine plain before entering the Rhine near Strasbourg. The deferred junction of the Yazoo River into the Mississippi is also of this type.

THE LONGITUDINAL STREAM PROFILE AND ITS DEVELOPMENT

14.1 THE LONGITUDINAL PROFILE

The **longitudinal profile** or long profile of a stream is the gradient line of its water surface from the source to the mouth. In a diagram, it is shown as a curve in a two-dimensional rectangular coordinate system with the distance from the source, measured along the stream course on the horizontal axis, and the elevation of the water surface on the vertical axis. The stream profile is usually longer than the long profile of the valley in which the stream flows because the valley gradient is measured along the valley axis, the centre line of the valley, and does not take account of bends in the stream.

The ideal form of the longitudinal profile of streams whose discharge increases from source to mouth is concave. It is related to the concave profile of wash denudation slopes (section 8.6) and reflects similar hydraulic conditions. Because the drag force of flowing water is a function of the product of gradient and water depth (section 8.6.3) and depth increases with increasing discharge (section 10.5.3), a progressively lower gradient in a downstream direction is sufficient to transport the bed load. Many longitudinal stream profiles do, however, contain steeper stretches and waterfalls and deviate from the ideal form.

An important function of a graphic representation of the longitudinal profile is the identification of **knick points**, those locations in the profile where there is a pronounced local increase or decrease in the gradient. Knick points can be an indication of differences in the rock, the presence of local tectonic disturbances, abrupt changes in discharge or critical phases in valley development such as active headward erosion.

The elevation difference between the source and the mouth is so small in comparison to the length of the stream, that any profile in the diagram must have an exaggerated vertical scale if significant differences in the gradient are not to disappear in the profile curve. It is convenient to have a vertical exaggeration in which the elevation difference between source and mouth appears to be about half as high as the horizontal length of the long profile. A 5 km long creek, for example, whose source lies 200 m above the altitude of the mouth, would, on a scale of 1:25 000, have a length of 20 cm and an elevation difference of 8 mm.

Figure 14.1 shows the longitudinal profile of the Rhine. The 1236 km long river has its source at about 3000 m so that the vertical distance from source to mouth is only 0.24 per cent of the length. In the profile curve (a), the vertical scale of the diagram has been increased 250-fold. Because of the large range of natural gradients in the profile, from 4 per cent in the upper course to almost nothing in the lower course, local differences in gradient cannot be clearly identified in either the upper or the lower part of the profile at this vertical exaggeration. Only in the middle course from Lake Constance to the upper Rhine

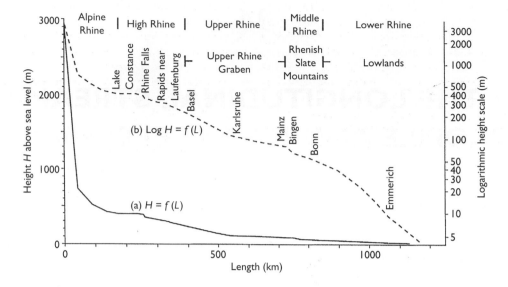

Figure 14.1 Longitudinal profile of the Rhine River. (a) linear height scale; (b) logarithmic height scale.

Graben are knick points apparent, particularly those at the Rhine Falls near Schaffhausen.

To overcome this problem the profile can either be divided into three parts, each with its own vertical scale, or a **logarithmic vertical scale** can be used (curve (b) in Fig. 14.1) which has the effect of enlarging the height differences in the lower part of the elevation scale and reducing them in the upper part. With the logarithmic scale, the interpretation of all parts of the profile line is easier since both the local steeper stretches and also local decreases of the gradient become visible.

14.2 BASE LEVEL AND PROFILE DEVELOPMENT

14.2.1 Base level

A stream cannot erode its bed to an altitude that is lower than the altitude of its mouth. The main **base level** of entire river systems, together with all their tributaries, is sea level. Endorheic streams in dry areas flow into basins with no external drainage and their base level is the basin floor or a terminal lake, such as the Great Salt Lake in the western USA, Lake Chad in Africa and the Caspian Sea, the Aral Sea, the Dead Sea and Lake Balkhash in Asia.

The base level has a controlling effect on the intensity of form development in the entire catchment area because the stream transports away the unconsolidated material it produces by downcutting as well as the products of weathering and denudation from the slopes.

Within a stream system there are also local and regional base levels although in principle every point in the profile controls the vertical erosion upstream of it and can act as the temporary local base level. The regional base level for every tributary stream system is the mouth of the tributary to the main stream. Base levels also exist within a stream course at the points at which a stretch dominated by erosion changes to one in which accumulation predominates. The complex profile of the Rhine has at least three successive segments, each with its regional base level. Lake Constance (395 m above sea level, Fig. 14.1) forms the first noticeable level in the profile and is the base level for the entire catchment area of the Alpine Rhine. At Stein am Rhein, where the river leaves the lake, the erosion stretch of the High Rhine begins, with the

Rhine Falls at Schaffhausen and the rapids at Laufenburg. At Basel, the Rhine turns north and enters the accumulation area of the upper Rhine Plain (the Rhine Graben). Although there has been minor erosion by the river in the plain, it is primarily an area of sedimentation and serves as the base level for the stretch upstream. The river leaves the plain at Bingen and enters another erosion stretch in the narrow valley cut through the Rhenish Slate Mountains which continues past Koblenz to Bonn, interrupted only by the Neuwied Basin, a downwarp. Below Bonn the river enters the Lower Rhine embayment, the last regional base level above its mouth in the North Sea.

14.2.2 Changes in base level and headward erosion, denudation and sedimentation

Base level is the controlling influence on vertical erosion and any change in base level height affects the work of the stream. There are several causes for height changes in the base level:

1. long term progressive lowering or rising of the sea level;
2. crustal movement which raises or lowers parts of the catchment area;
3. accumulation in interior basins of endorheic streams;
4. vertical erosion or accumulation of a main stream that affects the local or regional base level of the tributaries;
5. capture which results in a stream becoming tributary to another main stream lying at a different altitude.

If the base level is lowered, the relief is increased in the catchment area upstream. Initially, only that part of the stream immediately above the base level is affected by the change. The lowering produces a steeper gradient locally and a new impulse for erosion in the stretch adjacent to the base level which is then also lowered. In turn, this newly lowered stretch acts as the local base level for the stretch directly upstream, which, because of the now steeper gradient, also erodes. In this way, the erosion impulse is transferred upstream, headward, not only in the main stream but also into any tributaries in the system. The process of **headward erosion** is the most important mechanism for stream incision and for valley development (Fig. 14.2).

Stream incision increases the height and steepness of the valley slopes. Denudation processes taking place on the slope are intensified,

FIGURE 14.2 Headward erosion of a knick point in the headwaters of the North Saskatchewan River, Canada.

initially on the stream banks and adjacent foot-slope and then, as **headward denudation** (Ahnert, 1954) progressively upslope, in most cases as far as the divide, the boundary of the catchment area.

An increase in the height of the base level reduces the gradient immediately upstream. The transport power of the stream is diminished and load deposited.The accumulation in this stretch of the stream produces a temporary higher base level for the next stretch upstream. Like headward erosion, the process works its way upstream, in this case, as **headward sedimentation**.

14.3 EQUILIBRIUM TENDENCY IN PROFILE DEVELOPMENT

The raising or lowering of the regional or local base level influences the fluvial processes in both the upstream and the downstream direction. Figure 14.3 shows the temporal change, the spatial differentiation and shifting of vertical erosion and the supply and removal of bed load in a

longitudinal profile of a stream during and after a tectonically caused change in the local base level. The tendency to equilibrium discussed in section 12.7 in relation to the cross-section of the stream bed can also be applied to the development of a longitudinal profile.

As in equation (12.1), A is the rate of bed load supply, the amount of bed load transported per time unit from upstream. R_p, is, unlike in equation (12.1), the potential rate of bed load removal. The actual rate can be smaller than this potential. E shows the rate of local vertical erosion. A, R_p and E have the same dimensions: height change of the stream bed per time unit.

Fig. 14.3 (a) shows a smooth concave longitudinal profile in which the **transport equilibrium**

$$R_p = A \qquad (14.1)$$

dominates. The actual rate of transport equals the potential rate at all points on the profile.

In Fig. 14.3(b) a vertical movement has taken place on the previously inactive fault as a result of which the upstream stretch of the stream has been raised and the stretch downstream lowered.

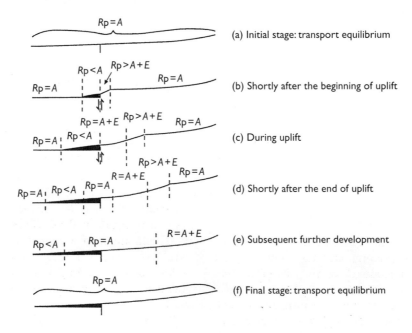

FIGURE 14.3 Changes in a long profile during and after vertical tectonic movement (after Ahnert, 1967).

The gradient at the fault is steepened and a knick point develops in the profile. The steepening increases the removal potential R_p so that there is energy available for local vertical erosion over and above that required to transport the bed load from upstream. Not all of this energy is used for the actual downcutting. Immediately upstream from the fault there is a **surplus of removal potential**

$$R_p > A + E \qquad (14.2)$$

that is used up by the greater turbulence in the stream (shooting flow). Downstream from the fault, the gradient is lower and the removal potential is too small for the increased bed load supply arriving from the steepened stretch above to be carried away. A **transport deficit** predominates:

$$R_p < A \qquad (14.3)$$

and part of the bed load is deposited.

The originally uniform longitudinal profile is now subdivided into four sections. In the lower-most and uppermost sections, the former transport equilibrium still dominates. In the upper of the two middle sections, there is erosion and surplus removal potential and in the lower section a transport deficit and accumulation is taking place.

As long as the rate of vertical erosion at the fault remains smaller than the rate of relative uplift of the upper block, the erosion stretch becomes steeper and the erosion more intensive. Eventually the erosion rate catches up with the rate of uplift so that immediately above the fault a new state of dynamic **erosion equilibrium** develops (Fig. 14.3 (c)):

$$R_p = A + E \qquad (14.4)$$

Like the transport equilibrium it is dynamic because its components and the local gradient are linked in a feedback relationship. The two profile sections, one with a surplus of removal potential and the other with a transport deficit, which had bordered one another, are now separated by this new fifth section. The section with the surplus of removal potential and erosion is

shifted upstream as a consequence of headward erosion, together with the knick point at the upper end of the erosion stretch (Seidl and Dietrich, 1992).

Shortly after the tectonic movement has ceased (Fig. 14.3(d)), erosion at the fault also stops because the uplift, the motor of the system differentiation, is no longer present. The removal potential at the fault decreases but adjusts itself at this location to a new dynamic transport equilibrium (equation (14.1)). This state expands both upstream and downstream from the fault line.

The profile has now attained its highest degree of spatial differentiation. The knick point at the upper end of the erosion stretch moves further upstream and the stretch in which there is net accumulation, downstream. The remains of the initial transport equilibrium stretches at the upper and lower ends of the profile are progressively replaced.

In Figure 14.3(e), the erosion that started at the fault line has reached the upper end of the profile and the net accumulation, the lower end. The new transport equilibrium spreads at the expense of the neighbouring sections until the entire profile is dominated by this state (Fig. 14.3(f)).

If there is a new uplift phase, the cycle begins again; but, if the uplift phase lasts long enough and the uplift stays at more or less the same rate, erosion equilibrium (equation 14.4) eventually develops in the entire area above the fault and transport equilibrium below it.

The very long period of time required for this form of development means that verification in the field is impossible. Only the states in the differentiation that are simultaneously present in the different sections of the long profile can be observed in the field. Moreover the development of a stream course is three-dimensional, not only the development of a profile; in some streams, there are changes taking place in the cross-section, such as braiding, which contribute to the establishment of an equilibrium (sections 12.7 and 13.2).

Changes in grain sizes of the bed load along the profile also cause divergences from the basic longitudinal profile form. In an investigation of the effect of grain size upon the channel gradients of numerous small streams in the Appalachians, Hack (1957) found the following relationship (converted here for metric units):

$$S = 0.006(D_{50}/A)^{0.6} \qquad (14.5)$$

where is the S local stream channel gradient, expressed as tangent, D_{50} is the median value of bed load grain sizes (mm) and A is the size of the drainage basin (km²). This shows that the drainage basin (or catchment) size and the bed load grain size are of equal importance for the gradient and that a change in the bed load grain size when a coarser bed load is brought into the stream by a tributary, for example, would cause a change in local gradient that would appear as a deviation from the ideal longitudinal profile form. The same relationship seems also to be valid for other upland areas in the middle latitudes. Parameter values similar to Hack's have been obtained in a study of streams in the Eifel upland in Germany (Schröder, 1991).

14.4 CAUSES OF KNICK POINTS IN LONGITUDINAL PROFILES

Knick points appear as a clear break in the longitudinal profile at which the gradient increases or decreases. They deviate from the normal profile and, for this reason, are of diagnostic value for relief interpretation.

Knick points are concave or convex. **Concave knick points** generally only become apparent in a longitudinal profile drawn on a logarithmic scale (profile (b), Fig. 14.1). The following are frequent causes of concave knick points.

1. The entry of a tributary of similar size to the main stream so that there is a considerable increase in the discharge. The resulting increase in water depth allows the transport of the grain sizes present at a lower gradient, as long as the product of depth and gradient, and therefore the competence or drag force, remains large enough for the bed load to be transported (Du Boys equation (11.3)).
2. The entry of a stream into an area of easily eroded rock. The bed is lowered more rapidly

and becomes flatter than the upstream stretch in the more resistant rock.
3. The entry of a stream from an upland area into a lowland, such as an area of tectonic down faulting. This type of concave profile disappears relatively rapidly following the deposition of an alluvial fan (Fig. 14.3).

Convex knick points indicate that vertical erosion at this point is not able to develop a smooth longitudinal profile or that the knick point has been created only recently and has not yet been worn away. The following are possible causes for the development of convex knick points.

1. An active fault crossing the stream, as shown in Fig. 14.3. Knick points above a fault are often an indication of recent movement. Once crustal movement begins, the knick point shifts upstream by headward erosion. Its distance from the fault is a relative measure of the time that has lapsed since the beginning of the crustal movement.
2. The stream is crossed by resistant rock strata. If the strata of the resistant rock are more or less horizontal or dipping gently, headward erosion shifts the knick point gradually upstream. In the case of very steeply dipping rock or a dyke, the knick point does not shift.
3. Where a small tributary enters a large stream that has cut down its bed, a convex knick point may form if the smaller stream cannot incise at the same rate as the large stream and enters it with a steepened stretch at its mouth. Many of the small streams entering the Rhine between Bingen and Koblenz have this type of knick point.There is a major knick point at Harper's Ferry in Virginia, USA in the Shenandoah at its junction with the Potomac. When there is a large height difference between the main valley and the tributary, the tributary valley is a **hanging valley**. They are common in the formerly glaciated regions of high mountains.

Convex knick points develop more often in small rather than large streams. The riffle–pool sequence that occurs in small streams (section 12.5) is a more or less rythmic sequence of knick points. Resistant stretches and rapids also tend to be associated with convex knick points.

14.5 WATERFALLS

Waterfalls are the most conspicuous knick points. There is no sharp dividing line between rapids and waterfalls. A convex knick point over which the water moves in free fall at low water, can become an area of rapids at high water if the water level below the break of gradient is close to the upper edge of the knick. Waterfalls can best be defined as knick points over which the water usually flows in free fall.

The bed above the waterfall of a large river, such as the Zambezi, the Niagara or the Potomac, is usually wide and the water shallow. The stream flows over resistant rock in this stretch, which usually limits vertical erosion, in contrast to the regolith of the banks which is not very resistant to lateral erosion. To a varying extent erosional braiding divides the area of the falls themselves (section 13.2.1). Below the falls is a gorge which is the result of intensive headward erosion. Based on their lithology, structure and morphogenesis, waterfalls can be subdivided into at least three types: Niagara, cascade and hanging valley waterfalls.

14.5.1 The Niagara type of waterfall

The **Niagara type of waterfall**, named after the Niagara Falls on the border of Canada and the USA, is characterized by a more or less horizontal layer of resistant rock underlain by weaker rock. The resistant rocks form a bedrock scarp over which the stream plunges. The turbulence at the foot of the falls produces a **plunge pool** which hollows out the rock beneath the resistant layer. Eventually, parts of the resistant layer break off and the waterfall retreats upstream without any lessening in steepness (Figs 14.4 and 14.5).

The steep scarp in the Niagara Falls is formed of the very resistant Silurian Lockport dolomite. The less resistant rocks are claystones, limestones and sandstone. The Niagara River, which connects Lake Erie and Lake Ontario, plunges about

FIGURE 14.4 Niagara type of waterfall – schematic.

55 m at the falls. The falls are divided into three branches by two flat, rocky islands, which were formerly under water but are now partially wooded. On the Canadian side are the large Horseshoe Falls and in the USA the similarly wide American Falls. Between the two are the narrower Bridal Veil Falls. This division is the result of erosional braiding.

The Niagara Falls developed at the end of the last ice age, about 10 000 years ago, after the ice in the region had melted and stream work had begun. At first, the Niagara River flowed over the edge of the Lockport dolomite cuesta scarp near Lake Ontario. Headward erosion has caused the falls to retreat 10 km since then and to erode a deep gorge into the dolomite scarp. The mean rate of retreat is 1 m per year; in reality the retreat is in the form of major rock fall events that are separated by periods of several years. Large dolomite blocks, angular because of jointing, lie at the foot of the falls where they are gradually broken down and removed. The upper edge of the American Falls projects at right angles in several places due to the influence of the joint boundaries (Fig. 14.5).

Other Niagara type falls include the Kaieteur Falls on the Potaro River in Guyana where hard conglomerate and sandstone layers above not very resistant claystone have formed a 226 m high knick point, the over 100 m high Victoria Falls on the Zambezi in Africa and the 30 m high Gulfoss Falls in Iceland. In both the latter, basalt and not sedimentary strata form the resistant rock. Gulfoss flows over two basalt layers between which lies easily eroded sedimentary material.

FIGURE 14.5 Niagara Falls, USA/Canada.

14.5.2 Cascade waterfalls

Cascade in this case means that the water falls over several, usually small, steps rather than one large one. **Cascade waterfalls** occur on homogeneous, mainly plutonic or metamorphic rocks. The rock is divided by a fissure system of almost horizontal pressure release joints and almost vertical tectonic joints that provide the surfaces on which weathering and erosion can take place and create the step-like form of the cascade. The densities of the joints vary and therefore also the resistance of the rock to erosion. Because several tectonic joint directions are present, one of them is often more or less the same as the flow direction of the river and, if the joints are densely spaced in zones along which the river erodes more rapidly, it becomes divided by erosional braiding into several channels, each with its own falls. The number of the channels is usually greater than in the waterfalls of the Niagara type.

Upstream of the falls, the river flows in an unbraided bed. The headward erosion of the individual waterfall channels, and consequently the retreat of their falls, occurs at different rates, depending on the size of the discharge and the resistance in the joint zone in which the channel has developed. A channel that is more rapidly incised than its neighbour receives an increasing proportion of the stream flow and gains erosional power. This is a positive feedback process. The neighbouring channels receive less and less water and their erosional power is reduced. Eventually the water flows almost entirely in the active channel and the neighbouring channels become dry, except at high water.

The active channel also widens as more water is concentrated in it and the waterfall at its upper end retreats into other zones of varyingly resistant joints, where renewed braiding into several channels takes place. The drying up of former channels and the erosion of new channels approaches a state of equilibrium in which the total number of braided erosion channels remains more or less constant as the cascade waterfall retreats.

The waterfalls of the Fall Line, the border between the largely crystalline rocks of the Appalachian Piedmont and the younger, for the most part unconsolidated, sediments of the Atlantic Coastal Plain in the eastern United States, are cascade falls. They include the Great Falls of the Potomac at Washington DC (Fig. 14.6) where the total height of the several individual steps is 21 m. The Great Falls are also

FIGURE 14.6 The Great Falls of the Potomac near Washington, DC, an example of a cascade waterfall.

strongly braided with several flow channels and numerous dry channels.

14.5.3 Hanging valley waterfalls

The high mountains were much more glaciated during the Pleistocene ice age than today. Glaciers reached into many of the main valleys and were joined by glaciers from tributary valleys. The ice followed the preglacial river valleys but modified their forms by glacial erosion and tended also to deepen the main valleys more than the tributary valleys which then entered higher up on the slope rather than at the level of the main valley floor. Once the ice melted, the tributary valleys became hanging valleys and

their streams form **hanging valley waterfalls** which plunge down on the sides of the main valley they enter.

Many of the hanging valley waterfalls in the Alps have retreated by headward erosion from the immediate area of the main valley slope since the end of the ice age about 10 000 years ago (Fig. 14.7). A narrow **saw cut gorge** between the main valley and the present position of the falls marks the path of retreat. The distance of this retreat depends on the discharge of the stream and the resistance of the rock. The Staubbach Falls in the Bernese Oberland in Switzerland, for example, has not cut into the slope at all and falls

FIGURE 14.7 Waterfall of the Tana near Grandvillard in the Canton Fribourg, Switzerland, an example of a retreating hanging valley waterfall and the erosion of a saw cut gorge.

FIGURE 14.8 Staubbach Falls near Lauterbrunnen, Switzerland, unlike many other hanging valley waterfalls in the Alps, no saw cut gorge has been eroded.

as cloud of moisture droplets into the Lauterbrunnen valley (Fig. 14.8). A short distance upstream, the Trümmelbach Falls, which receive their water from the Eiger Glacier and have a greater discharge, have cut a short, very steep gorge into the valley side.

15

RIVER TERRACES

A terrace is a more or less level area that is limited both above and below by inclined surfaces. **River terraces** are remnants of former valley floors that remain on the slope after further downcutting. They are created directly by the stream and indicate that vertical erosion was interrupted by a phase of lateral erosion or of accumulation during the valley's development. It is not always easy to distinguish river terraces from other levels on the valley slopes. Artificial terraces no longer used for agriculture, vineyards or orchards look increasingly natural after some years. Other level areas on slopes are denudation terraces, which follow the outcrop of a resistant rock in the slope (section 20.3.9) and have been formed solely by differential denudation.

Before a river terrace can develop a valley floor must exist. There are two types of valley floor and two different types of river terraces: a **rock-floored terrace** which develops from a rock-floored valley bottom and an **accumulation terrace** which develops from an accumulation valley floor (Fig. 15.1). The term terrace refers only to the landform and not to the material; the landform remains a river terrace even after the original gravel cover has been removed and as long as the level created by the stream can be identified.

15.1 ROCK-FLOORED TERRACE

The gravel cover of a rock-floored terrace is relatively thin, normally no thicker than the depth of the stream bed. It lies on the flat, laterally eroded surface of the bedrock (Fig. 15.2). The lateral erosion followed a phase of downcutting during which the valley had been incised down to the bedrock floor. A renewed phase of downcutting cut into this **rock-floored valley bottom** (section 12.6), remnants of which were then preserved as rock floored terraces on the slopes of the deepened valley (Fig. 15.1, A3). Rock floored terraces have, therefore, three phases of development:

1. vertical erosion
2. lateral erosion and valley widening
3. vertical erosion.

The presence of rock-floored terraces is an indication of long-term downcutting, usually due to tectonic uplift, that was interrupted. During a pause in uplift, lateral erosion took place. Because the terraces are formed in the resistant bedrock floor, they retain their form for a long time, even on steep slopes.

15.2 ACCUMULATION TERRACES

The level surface of an **accumulation terrace** is a remnant of an accumulation valley floor (section 12.6). The development of the terrace has at least three phases:

1. vertical erosion
2. accumulation
3. vertical erosion.

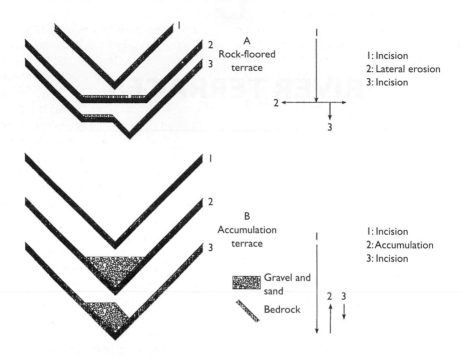

FIGURE 15.1 Development phases of the two types of river terraces shown in valley cross-sections (schematic). Note that the form of the valley cross-section of the rock-floored terrace (phase A3) is the same as that of the accumulation terrace (phase B3).

FIGURE 15.2 Rock-floored terrace in the Auvergne, France.

The valley is formed in the first phase of vertical erosion. During the accumulation phase, downcutting ceases and the level of the stream bed is raised by the deposition of gravels which form a wide accumulation valley floor. When vertical erosion is renewed the stream cuts down into this surface of unconsolidated fluvial material and remnants of the former valley floor are left as an accumulation terrace (Fig. 15.1,B3 and Fig. 15.3).

This simple scheme is complicated by the fact that the sequence of accumulation and vertical erosion is often repeated several times, and may be interrupted by phases of lateral erosion. Deposits from different accumulation phases can cover one another so that the incisions of differ-

FIGURE 15.3 Large accumulation terrace in the Kumaon District in the Himalayan foothills of India.

ent downcutting phases intersect (Fig. 15.4). It is important, therefore, to limit the term terrace to the form; the gravels, as such, are not a terrace.

15.3 LOCATION AND PRESERVATION OF TERRACES IN THE STREAM VALLEY

If there are terraces at different levels on the side of a valley, the highest are the oldest. In general, older terraces are less well preserved than younger terraces, although there are exceptions. The broad main terrace on the middle Rhine, which lies on the slopes above the narrow gorge, is much better preserved than the younger middle terrace on the steep slopes of the gorge (Fig. 17.3).

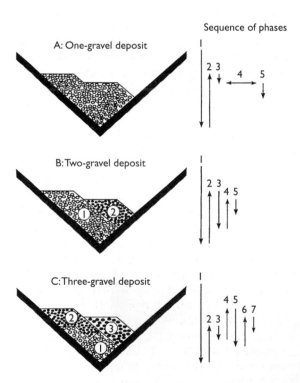

FIGURE 15.4 Three valley cross-sections, each with two accumulation terraces of identical forms but of different origin. The numbers indicate the age sequence of the deposits.

Terraces are often preserved on the slip-off slope of a valley meander. The stream is directed away from the slipoff slope in a meander while it downcuts and a valley-floor remnant on this slope is not undercut by stream erosion. It is also flatter than other slopes in the same valley and wearing down by denudation is therefore less intensive.

The **confluence spurs** of slopes on the upstream side of a tributary valley are not eroded laterally or undercut at its confluence with the main valley so that any terrace located on the spur is more likely to be preserved than one on the downstream side of the tributary mouth, to where the line of highest velocity of the tributary stream is pulled by the current of the main stream.

Some of the medieval castles on the steep valley slopes of the middle Rhine, the castles of Gutenfels, Katz, Maus and Rheinfels for example, have been built on the small rock-floored terraces of protected confluence spurs on the upstream side of tributary valleys.

15.4 THE CAUSES OF TERRACE FORMATION

There are four significant complexes of causes for the formation of river terraces:

1. crustal movement, especially tectonic or isostatic movement
2. eustatic change in sea level
3. climatic fluctuations
4. stream capture.

The most important factor in the formation of a rock-floored terrace is a long period of **crustal uplift** with vertical erosion interrupted by a pause for lateral erosion during which a rock-floored valley floor is created. With renewed uplift and incision the rock-floored valley floor is dissected and becomes a terrace. The pause in vertical erosion during which lateral erosion dominates is not necessarily due to an interruption of crustal uplift. Long term fluctuations in discharge and bed load that reflect climatic changes can also cause the downcutting rate to vary. The interpretation of terrace formation becomes complicated when the effects of crustal movements and climatic fluctuations overlie one another.

Eustatic changes also have a direct effect on the erosion and transport behaviour of all exorheic rivers. If the sea floor has a steeper gradient than the river, a fall of sea level produces headward erosion from the coast inland. A stationary sea level generally causes lateral erosion and valley widening and a rise in sea level, deposition, initially at the coast and then extending inland.

The sea level fell more than 100 m during the last ice age. Similar changes took place during and after each of the preceding ice ages. Very probably the fall in sea level was in phases and produced terraces that were later flooded by the post glacial sea level rise. On many coasts, including the east coast of North America, there are terraces near the coast which correspond to the highest sea level of the interglacials. Elsewhere, eustatic changes in sea level and local or regional crustal movements are superimposed on one another in such a complex way that it is very difficult to separate their effects. Usually, the effect of eustatic changes on stream erosion and terrace formation is limited to areas near the coasts.

Climatic fluctuations primarily affect stream discharge, especially the magnitude–frequency of their effective runoff events, and the grain size and the volume of the transported load.The terrace systems of the Iller and Lech rivers in the Swabian–Bavarian Alpine foreland in Germany are climatically controlled terraces (A. Penck and E. Brückner 1901–1909; Habbe and Rögner, 1989; Fischer, 1989). During the ice ages the rivers in the area formed large gravel levels into which they incised in the intervening warmer interglacials. The glaciers reached into the Alpine foreland in the ice ages and large quantities of bed load were available from their moraines for the rivers to remove and redeposit on the gravel levels. During the interglacials vegetation covered the moraines, as it does today, and the glaciers retreated to higher altitudes. The streams' bed load was also less, the gravel surfaces were incised and terraces formed. This cycle of deposition and incision was completed

at least four times during the Pleistocene. Each new gravel surface was lower than the previous one, indicating that the change from glacial accumulation to glacial erosion was apparently overlain by a tendency for the rivers to be progressively lowered. For this reason, the Pleistocene terraces of the Iller-Lech are rock-floored terraces although the 'rock floor' is formed of the Tertiary molasse sediments of the Alpine foreland, a largely unconsolidated rock.

Where there is no progressive net downcutting and terraces are formed only as the result of alternating accumulation and downcutting, climatic fluctuations generally produce accumulation terraces.

The development of the terraces in the Rhine gorge between Bingen and Bonn where the river cuts through the Rhenish Slate Mountains is produced by a complex of causes. Tectonic uplift of the mountains began in the Tertiary and is still continuing. Since the beginning of the Pleistocene, it has totalled 250 m. The Rhine has eroded its valley down by about the same amount, more or less maintaining the altitude of its course. Most of the incision, about 190 m, has taken place in the last 6–700 000 years (Bibus 1980, p. 252). During interruptions in the downcutting the valley was widened and gravels and sands deposited. The subsequent renewal of downcutting caused the formation of terraces. Depending on the interpretation, there are about a dozen terrace levels in the gorge (Fig. 15.5).

The Pleistocene Rhine terraces are divided into high, middle and low series of terraces. The most prominent of the high terraces is the broad area of the **main terraces** at 130–200 m above the river on which there are large cultivated areas and villages. The lower edge of the younger main terrace forms the upper edge of the narrow valley. Between Bingen and Koblenz it is particularly well defined, especially at the Loreley (Fig. 17.3). The **middle terraces** are usually in the form of narrow level areas on the slopes of the gorge, noticable often only because of the castles built on them. The **lower terraces** are used by both road and rail. Settlements tend to be where the deposits of tributaries entering the Rhine have widened the area of the lower terraces.

The high and middle terraces are both rock-floored terraces. The lower terraces are rock-floored between Bingen and Koblenz but downstream from Koblenz are accumulation terraces with nested gravel deposits (Fig. 15.5) because uplift in this area was less in the late Pleistocene than upstream and replaced by subsidence in the Neuwied Basin between Koblenz and Andernach.

It is probable that the pauses in downcutting that led to the development of the middle and high terraces were caused by pauses in the uplift or at least by a reduction in its intensity. In the case of the accumulation terraces among the lower terrace levels, it is more likely that climate-related changes in the bed load were of greater significance, in a way similar to the bed load changes during the glacials and interglacials in the Alpine foreland. During the last two glacials the Rhine deposited part of its bed load and formed a valley floor. In the intervening warm period, there was less bed load and the river incised into the former valley floor, which then became an accumulation terrace (Bibus, 1980).

The climate-determined alternation between accumulation and downcutting occurred several times. Signs of cryoturbation and ice wedges found in gravels of the middle and the main terraces suggest that, during the formation of these older terraces, the alternation between cold and warm climates probably also resulted in an alternation between gravel accumulation and erosion, although indications of colder climates in the gravels could, in part, have developed after the gravels were deposited.

River terraces formed as the result of **stream capture** (sections 17.1.1 and 17.1.2) are a special case. If a stream with a relatively high base level is captured by the upper course of a lower lying stream, the captured stream has a new, lower base level and cuts down into its former valley floor. This is a one-time process that is limited to a captured stream and produces only one terrace level.

In general, it is possible to determine the origin of a terrace level only by observing the

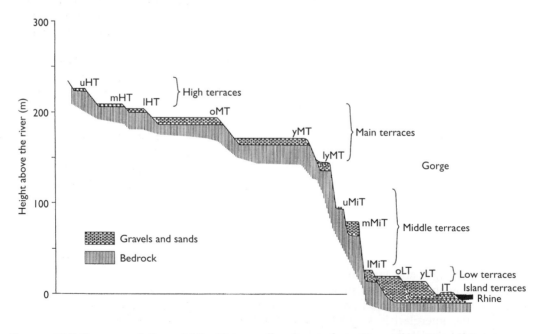

FIGURE 15.5 Terraces of the middle Rhine valley in a schematic profile. Not all terraces are necessarily present on any individual profile in the Rhine valley. uHT, mHT, lHT: upper, middle and low high terrace; oMT, yMT, lyMT: older, younger and lower younger main terrace; uMiT, mMiT, lMiT: upper, middle and lower middle terrace; oLT, yLT: older and younger low terrace; IT: island terrace (simplified from Bibus, 1980; nomenclature partly after Quitzow, 1974).

regional distribution of the terrace system. River terraces caused by crustal movements are limited to the zone of uplift of the crustal block. Eustatic terraces occur at similar altitudes on all tectonically stable sea coasts but at different levels on unstable coasts. Terraces resulting from climatic change can often be recognized by particular types of stream deposits or fossil soils. Many terraces are, however, the result of several different processes and it is often difficult to separate their relative influences in any one terrace.

15.5 THE DIAGNOSTIC SIGNIFICANCE OF RIVER TERRACES

Although terraces are relatively inconspicuous forms in the landscape, they are of great importance as indicators of the endogenic and exogenic morphological development of an area. Because all river terraces are remnants of valley floors, their original gradient is the gradient of the valley at the time the valley floor was formed. The **correlation of terraces** involves the identification of terraces which belong to the same former valley floor and their reconstruction by comparing altitudes of terrace segments and analysing the grain size, rock type and heavy mineral content of the gravels and sands found on the terraces, and also the extent to which the grains have been weathered. If datable material is available, it is possible to determine the absolute age of the deposits.

The rate of downcutting in the period between the formation of two terrace levels is estimated from the difference in their heights and, by correlating and dating the terraces of its streams, the uplift sequence in a mountain range can be determined. In the same way, sea level changes can be interpreted from eustatic river terraces.

The original gradient of a valley floor is inclined in a uniform direction. Any deviation of the terrace gradients is an indication of later crustal warping, whose age and extent can be determined if datable material is available. Figure 15.6 shows the height of the main terrace and the upper and lower middle terrace from the Mainz Basin through the Rhenish Slate Mountains to the lower Rhine beyond Duisburg (Quitzow, 1974). Between Bingen and Koblenz the main terrace is strongly upwarped. The middle Rhine terraces are also upwarped in this area but less strongly, showing that the upwarping began after the formation of the main terrace and a considerable period before the formation of the upper middle terrace and continued after the formation of the lower middle terrace.

The lower position of the terraces in the Neuwied Basin between two faults resulted from the subsidence of this area relative to its neighbouring blocks. Downstream from the Basin the altitudes of the terraces converge, indicating that the rate and total amount of uplift was progressively smaller in this direction. Between Duisburg and Nijmegen, there are **terrace crossings**. Because of the subsidence in the German and Dutch lowlands, the main and middle terrace gravels of the Rhine were lowered below the level of the river and disappear as a morphological form. From here to the North Sea coast their gravels form layers in normal deposits with the older gravels

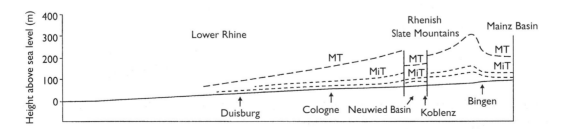

FIGURE 15.6 Long profile of the main terrace level (MT) and the middle terrace levels (MiT) of the Rhine between the Mainz Basin and the North German Lowland (after Quitzow, 1974).

from the terraces lying below the younger (Woldstedt and Duphorn, 1974, pp. 261–73).

Schematic diagrams of valley cross-sections often show entire flights of terraces with identical levels on either side of the valley. This would mean that each of the successive valley floors, whose remains form the terraces, was in every phase of the downcutting narrower than the previous one. This rarely happens. Usually the original valley bottom had the same width at the different levels and the terraces are preserved in those locations where they are protected from denudation (section 15.3). In reality, a particular terrace level is more likely to be preserved on only one side in a given cross-section and up valley or down valley on the other side.

It has sometimes been maintained that only a symmetrical occurrence of the same terrace level on both sides of the valley, known as paired terraces, is evidence of alternating downcutting and valley floor formation, while the asymmetrical distribution of terraces, or unpaired terraces, is the result of the river's course swinging from side to side of the valley during continuous downcutting. This distinction is incorrect. Most river terraces are asymmetrically distributed in the valley cross-section and unpaired terraces are the rule. Some gravel-covered slipoff slopes with a low angle, which have been formed by the incision of valley meanders, may resemble terraces but are parts of slipoff slopes and not terraces. Whether or not a terrace is a remnant of a former valley floor does not depend on its presence or absence in a particular cross-section, but that it appears along the valley at a comparable height in several locations and with similar deposits so that reconstruction of the former valley floor is possible.

16

ALLUVIAL FANS AND DELTAS

16.1 ALLUVIAL FANS

An **alluvial fan** is the characteristic accumulation form of a stream that flows from the mountains on to a plain and loses so much gradient that part of its load must be deposited. The accumulation occurs in the same way as valley floor accumulation, with the difference that accumulation on the valley floor is limited laterally by the valley's sides, whereas an alluvial fan accumulates on the plain from the stream's exit point from the mountains in a more or less semicircular deposit (Fig. 16.1).

The term **alluvial cone** is often used to describe this form because its highest point is at the valley exit to the plain and its surface slopes into the plain in a semicircle, covering almost

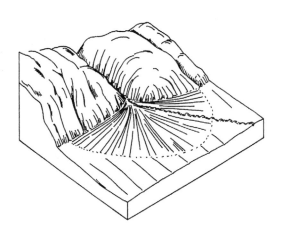

FIGURE 16.1 Alluvial fan (after Strahler, 1975, p. 443).

180°. The term cone is sometimes limited to alluvial fans with a gradient of more than 20° (Bull, 1977; *see also* Lecce, 1990, p. 4) and sometimes used as a synonym for alluvial fan. Rachocki and Church (1990) have published a collection of papers that covers recent research on alluvial fans.

The coarser the load and the smaller the stream, the steeper and more noticeable the fan. Large fans form at the foot of mountains in the catchment basins of arid and semi-arid areas in which a large amount of mechanical weathering occurs so that coarse bed load is produced and transported. In humid regions the proportion of bed load is relatively small and is more likely to consist of smaller grain sizes. The alluvial fans in these areas have very gentle slopes and are noticeable mainly because of their relatively good soil and the land use rather than their relief. The fan of the Neckar River near Heidelberg in Germany extends with a radius of about 10 km on to the upper Rhine plain from its valley exit and is clearly visible even on satellite photographs because of its distinct land-use pattern.

16.1.1 Form and development

Alluvial fans develop similarly to natural levees. The sudden reduction of gradient of the stream at its valley exit causes accumulation and thereby a local increase in gradient, in keeping with the natural tendency of a stream to establish a dynamic equilibrium between supply and removal of its load at all points along its long profile. The accumulation zone downstream from the fault in Fig. 14.3 can also be regarded as a section through an alluvial fan at its exit point from an uplifted area.

The accumulation in the stream bed itself raises the height of the bed above the level of the land on either side. This process is particularly rapid during flood discharges when natural levees develop on the stream's bank. The lateral gradient from the levee to the plain is, as a result, steeper than the gradient along the stream. Sooner or later the stream flows over the levee during a flood or a break may be eroded in the levee and the stream flows out laterally on to the plain.

The same development takes place in this new branch of the stream: reduction in gradient at the exit to the plain where there has not yet been any accumulation, followed by accumulation and the local development of natural levees and the breaking through of the new levee during a flood. Whether, with the development of a new branch of the stream, the existing stream bed continues to contain water or dries up, depends on the total flow, the local gradient of the various branches and their ability to erode. The deeper a lateral flow incises a break in the natural levee, the more water it can divert from its previous channel. The total number of branches is obviously limited. Increasing branching reduces the transport power of the individual branches and thereby the effectiveness of the entire system. In principle, it is not necessary for several branches to exist simultaneously, it is enough that a stream flows over its natural levee at various times at various locations, successively building up the semicircular surface of the fan in all directions. Fans built up of coarse sediment have a greater tendency to develop several simultaneously active branches than fans of fine-grained sediment.

The process also has much in common with the processes of new channel braiding and in the case of a coarse cohesionless load the same laws apply. An alluvial fan is, in effect, a system of accumulating braided channels without lateral limits.

Some alluvial fans develop as a result of a single catastrophic runoff event. In this case, sheet floods and mudflow types of mass movement, as well as fluvial transport, contribute to the building up of the fan. The Roaring River

alluvial fan in the Rocky Mountain National Park, Colorado is a fan of this type (Blair, 1987). On the margins of mountains in dry areas where there are infrequent but sometimes intensive precipitation events or periods of rapid snow melt, fans have been formed almost completely of the deposits of successive mudflows (section 8.3.6). They are in fact **mudflow cones** rather than alluvial fans. Examples from the White Mountains on the border between California and Nevada have been described by Beaty (1963, 1990).

16.1.2 Size, gradient and growth of alluvial fans

The empirical function

$$A_f = cA_d{}^b \qquad (16.1)$$

where A_f is the surface of the alluvial fan and A_d is the surface of the catchment area, was formulated by Bull (1964) to estimate the surface area of alluvial fans.

Kesel (1985, quoted by Graf, 1988, p. 187) found that in various areas in North America, the value given for the coefficient c varies between 0.15 in the interior basins of the southwestern USA and 3.84 in the California coastal ranges which are still being uplifted strongly. In 77 per cent of the cases examined, it was less than 1.0. The exponents b were mostly between 0.8 and 1.0, with only a very few lower and none higher.

It seems that the surface area of the alluvial fan is almost linearly proportional to the size of the catchment area. The differences in the proportionality factor c are due to local differences in rock type, relief, vegetation cover and the intensity and frequency of precipitation events.

The gradient of the alluvial fan is a function of a number of factors of which the grain size and the discharge during floods are of particular importance. The coarsest bed load is usually deposited in the upper part of the fan close to the valley exit and finer material further out so that the long profile of many alluvial fans is concave. Basu and Sarkar (1990) have measured gradient values on alluvial fans at the foot of the Himalayas near Darjeeling, along profiles approximately 8 km long, of 3.5°–10° (60–175–m/km) in the upper parts of the fans, 1.5°–3.5° (27–60 m/km) in the central areas and less than 1.5° in the lower parts. These are high gradients for fans of this size and are due mainly to the high relief of the catchment area where slopes are steep and intensive denudation delivers coarse debris which is transported to the alluvial fans by the floods of the monsoon season. By contrast, the Leba Fan in Poland, an alluvial fan of similar size, has a mean gradient of only about 13' of arc or 3.7 m/km (Rachocki, 1990).

Because of its semicircular form, the area of the alluvial fan increases with the square of the radius. If the rate of its surface area growth remains constant, the rate of radial growth gradually decreases and the alluvial fan extends at an increasingly slower rate into the plain. Also, if the load supply remains the same, the rate of increase in the surface area declines because part of the load is not deposited on the outer margin but on the existing alluvial fan.

The more the surface of the alluvial fan grows, the greater is the proportion of the bed load that contributes to an increase in the height of the fan, rather than to its area. In the late phases of fan development bed load reaches the margins of the fan only rarely. Suspended load and dissolved load are regularly transported further than the bed load, but their contribution to the growth of the fan is negligible.

The growth of the fan gradually comes to an end when the upland part of its catchment is no longer uplifted in relation to its lowland part, and the gradient of the stream in the upland is gradually reduced. Much of the bed load is deposited instead at the upper end of the fan because the stream is less able to transport its bed load further over the fan surface and accumulation expands from the apex of the fan upstream back into the upland valley.

With this **headward accumulation**, which is the opposite of headward erosion, part of the former area of erosion and transportation becomes an additional area of accumulation and the fan extends headwards as an accumulation valley floor into the uplands. The size of the supply area and, therefore, the debris supply itself is thereby reduced and for this reason, too, the alluvial fan's growth slows down.

16.1.3 Dissection and terracing of alluvial fans

On many alluvial fans terraces have developed because streams have incised into their own deposits, in some cases in several phases. There are a number of causes for downcutting on a fan. One is an increase in the ratio of discharge to bed load following changes in the catchment area. Figure 16.2 shows a slide that was triggered by a heavy rain. At its foot, a small sandy alluvial fan developed, formed by water loaded with sand flowing from the slide mass. Towards the end of this development, which lasted only a few hours, a channel had been incised by the water in its own deposit. As long as the slide mass was still moving, the water pressed out of it could transport a large amount of sand to the foot of the slide where it accumulated. With the end of the movement, the available sand load was reduced but the water continued to flow out of the slide mass and eroded a channel in the fan.

On large alluvial fans whose development reaches back into the climatic changes of the Quaternary and, particularly, the Pleistocene, the changes in the load supply or the discharge have frequently been brought about by climatic factors. This is analogous to the development of climatically caused accumulation terraces. The availability of large quantities of material in the cold periods, or ice ages, resulted in accumulation and the decrease in bed load during the warm periods in erosion. The Pleistocene change from cold to warm periods in the humid mid-latitudes also coincided with a change from wetter **pluvial periods** to dry periods, similar to those present in the arid regions of North America, which lead to cycles of accumulation and erosion on alluvial fans in these areas as well (Lecce, 1990).

Tectonic activity can also cause terrace formation on fans. As long as the rate of uplift of the upland area exceeds the downcutting rate of its streams, that is, as long as the gradient of these streams continues to increase, the bed load increases and the alluvial fans on the upland margin continue to aggrade. There is a tendency for dynamic equilibrium to be established between supply and removal of the load. If the uplift rate of the upland slows down, the erosion

FIGURE 16.2 Small dissected alluvial fan at the foot of a slide in Aachen, Germany (*see* Fig. 8.9).

rate of the streams can catch up with, or overtake, the rate of uplift in which case the stream gradient declines. The transport rates and the grain size of the bed load also decrease towards the margin of the upland. A lower gradient than before is sufficient for transport on the alluvial fan and the streams incise into its upper part. Repeated decreases and increases in the rate of uplift in the upland lead to alternating phases of incision and accumulation on the alluvial fan and to the development of terraces.

Incision following a reduction in load also takes place on the fan if there is a long-term slowing down or cessation of uplift because the slopes in the catchment area are then less steep and the quantity and grain size of the load transported to the alluvial fan declines.

16.2 DELTAS

Deltas are branched river mouths. They are triangular in outline and are related to alluvial fans, with the difference that the sedimentation of a fan takes place on a plain and that of a delta in the sea or a lake.

16.2.1 Delta bedding structure

When a stream reaches standing water it loses its gradient completely and deposits its entire bed load. The suspended load can be transported further by turbulent currents which reach a little further out into the standing water, but then sinks to the bottom (Fig. 16.3).

The bedding structure of the delta sediments is a consequence of the behaviour of the stream load (Fig. 16.4). The suspended load is deposited in almost horizontal layers of fine sediment beyond the mouth of the stream, the **bottomset beds** (c in Fig. 16.4).

The bed load deposits are built out from the original mouth of the stream layer by layer into the standing water (b in Fig. 16.4), the **foreset beds**. Their angle of dip (inclination) is the angle of repose of material under water and remains constant. The direction of dip is forward away from the stream's mouth. The uniform angle and direction of dip distinguishes delta beds from cross-bedded sediments within stream channels and from dunes.

The development of foreset beds in the standing water immediately in front of the stream's mouth is possible only if the stream has extended its gradient forward into its new area

FIGURE 16.3 Delta of a small stream on Midsummer Sjö, Peary Land, North Greenland. The water of the delta arms shoots with its suspended load about 10–20m out into the lake. Because of its higher temperature, it causes an increased melting of the lake ice (photographed July 1960).

FIGURE 16.4 Delta bedding structure: (a) topset beds; (b) foreset beds; (c) bottomset beds.

of deposition. This takes place with the deposition of sediments on top of the foreset beds. The bedding of these **topset beds** (a in Fig.16.4) is similar to the stream gradient and therefore, nearly horizontal.

As the sedimentation progresses, the foreset beds cover the bottomset beds that were deposited in front of them, and the topset beds cover the foreset beds. This regular arrangement of the three types of beds distinguishes delta bedding from all other types of sedimentary bedding structures.

16.2.2 The development of the delta outline

The discharge of a stream depositing a delta in the sea or a lake varies. During floods the stream builds up natural levees on its banks that extend, together with the topset beds, into the standing water body. Eventually these deposits cause the level of the stream to be higher than the water level on either side of the stream beyond the levees. As in the case of natural levees on a floodplain or on an alluvial fan, flooding breaches the levee at some locations and by eroding these breaches creates a system of new braided channels which form a delta (section 13.2.2).

Each new channel builds up the characteristic delta bedding structure at its mouth and extends itself forward into the area of standing water. Floods transport suspended load into the area between the branches or arms of the delta, where this fine material together with organic matter is deposited.

In small lakes with no strong currents or waves, the delta grows outward unimpeded and may eventually fill the entire lake basin. A delta that extends into the sea is increasingly subject to attack by waves and currents the further it projects out from the coastline. At the same time the supply of sediment by the stream to its mouth decreases the further the mouth is from the mainland catchment area of the stream and source region of the sediment. These factors limit the growth of deltas. Ultimately, the forward edge of a delta lies where the supply rate of the sediment by the stream and the rate of removal by waves and currents are equal. The tendency for the establishment of a dynamic equilibrium of the mass balance plays a decisive role in determining the form of a delta. The outlines of the various types of deltas on sea coasts are largely an expression of the contrast between the relative strength of stream work and the action of the waves and currents in the sea.

16.2.3 Cuspate deltas

A small stream cannot create and keep active a large number of arms in its delta simultaneously because the individual arms would lose too much transport power. The development of a new arm usually leads therefore to the abandonment of an old one. If only one active arm is present, this builds itself out into the sea in the form of a natural levee until the equilibrium between sediment supply and removal is reached. From this point on, all the additional sediment that arrives is carried away by the waves and currents. The wave refraction at the mouth of the delta has the effect that the sediment is turned to the outsides of the levees and deposited. The levee projects into the sea as a triangular landform whose tip is located at the mouth, hence the name **cuspate delta**. The Tiber delta (Fig. 16.5a) and the Ombrone delta on the west coast of Italy are examples.

16.2.4 Winged deltas

If the stream of a cuspate delta builds its levee so far out into the sea that the lateral shifting and deposition of the sediment by the waves does not extend back to the shore, bars develop on either side of the mouth of the stream, forming a **winged delta**. The bars run more or less parallel to the coast and usually end in the sea in a hook that curves round towards the land. The Ebro delta belongs to this type (Fig. 16.5b). The Po delta on the Adriatic coast of northern Italy and the Danube delta in the Black Sea are complex

forms of a combination of cuspate, winged and arcuate deltas (section 16.2.6).

16.2.5 Bird's foot deltas

The **bird's foot delta** is made up of several natural levees built out into the sea so that its outline resembles the spread claws of a bird's foot (Fig. 16.5c). Unlike cuspate or winged deltas they are not widened much by lateral sedimentation or action by the sea. They develop if the stream load is very large and the effect of the waves and currents particularly weak. The work of the stream dominates and the delta is a long way from dynamic equilibrium. Bird's foot deltas are not common on sea coasts because stream action seldom dominates sufficiently for them to form. That part of the Missisippi delta active at the present time is the best example (Fig. 16.5c). Two of its arms, the Southwest Pass and the South Pass, push their natural levees forward

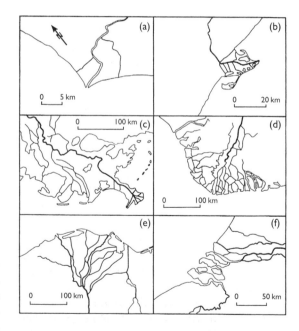

FIGURE 16.5 Delta types: (a) cuspate delta, Tiber, Italy (after Strahler, 1975, p. 430); (b) winged delta, Ebro, Spain; (c) bird's foot delta, Mississippi, USA; (d) arcuate delta, Niger, Nigeria; (e) complex arcuate delta, Nile, Egypt; (f) estuarine delta, Rhine–Maas, Netherlands (before artificial changes).

into the Gulf of Mexico. Two smaller arms are developing in an easterly direction, both of which have large areas of sedimentation on either side. The active part of the Mississippi delta is only a small part of a much larger delta with a radius of 120 km which extends in a semicircle from the mouth of the Atchafalaya River, an old delta arm in the west, to the barrier island chain of the Chandeleur Islands in the east. The delta has not always been a bird's foot delta but has had different outlines at other phases in its development.

16.2.6 Arcuate deltas

The **arcuate delta** is the most completely developed equilibrium form of a large delta. Ideally, it develops from a bird's foot delta when the individual delta arms have extended so far out that further forward growth is prevented by the action of waves and currents and the sediment arriving at the mouth is moved laterally by currents and deposited in the form of barriers (section 23.5.1). Initially, open lagoons remain behind the barriers and between the delta arms. Once the barriers between the arms are continuous, the lagoons are separated from the sea and become **delta lakes**. Flooding by the streams gradually desalinizes the water in the lakes, which are also progressively filled by sediments. The barriers connecting the mouths of the delta arms give the delta its arcuate form.

The Niger delta in the Gulf of Guinea has an almost perfect arcuate form (Fig. 16.5d). It is composed of a large number of mouths and is more than 200 km wide. The most easterly delta arms near Port Harcourt have developed as estuaries, an indication that here the influence of tidal currents is more important than fluvial transport.

The Nile delta is often described as an arcuate delta; it is arcuate in overall shape and has a number of delta lakes but the two largest mouths of the Nile, the Masabb Rashid near Rosetta and the Masabb Dumydt near Damietta, protrude as small local cuspate deltas from the arcuate form (Fig. l6.5e). This complexity is not unusual in large deltas.

16.2.7 Estuarine deltas

An **estuary** is a river mouth that has been widened towards its mouth by the alternation of tidal currents in a seaward and landward direction. The strength of the tidal currents in the estuary depends mainly on the tidal range (section 23.3.2). The mouths of the Elbe, Weser and Ems in Germany, the Schelde in Belgium and the Thames in England are typical esturaries. An **estuarine delta** combines the characteristics of the delta with those of an estuary. It has several arms which, however, do not, as in other deltas, reach the sea as streams with parallel banks but banks that are widened towards the sea by the effect of the tides. The Amazon has developed a large estuarine delta with arms that are more than 10 km in width. The combined delta of the Rhine and Maas is also of this type (Fig. 16.5f) although it has been considerably changed artificially. The estuarine delta at the mouth of the Colorado River at the northern end of the Gulf of California has a particularly large spring tidal range of more than 7 m.

Often some arms of a delta are developed as an estuary while others in the same delta have normal parallel banks all the way to their mouths. The Orinoco delta has well-developed estuary arms in its eastern part where it opens to the Atlantic, fully exposed to the waves of the northeast trades, and there is a tidal range of 2 m. Its central and western parts, on the Gulf of Paria, sheltered by the island of Trinidad, have developed normally. Many of the arms of the Ganges on the Bay of Bengal, which has a spring tidal range of up to 5 m, have an estuarine character. The Niger delta at Port Harcourt is another example.

Estuarine deltas can form in two ways, either through destruction of a prior delta when the delta arms are widened in a seawards direction by tidal current erosion or as a constructive landform through the deposition of bars in a wide, initially open esturary. Continued deposition on the bars raises their surface and reduces the frequency with which they are flooded by high tides. A vegetation cover develops and the bars eventually become islands which divide the estuary into several arms. This built-up form of estuarine deltas has, therefore, the elongated outline of the estuary and subparallel arms rather than the diverging arms and broader outline that generally indicate the destruction form of the estuarine delta.

Parts of an arcuate delta can also take on the shape of an estuarine delta if a delta arm that has been transporting a large load loses this function to another arm and becomes affected by increased tidal erosion which is no longer compensated for by the stream load.

16.2.8 Age and distribution of deltas

All deltas on sea coasts are very young forms. They can only have developed within the last 5000–6000 years after the end of the postglacial eustatic rise in sea level. Considering the size of the deltas of the Ganges, Nile, Niger, Mississippi and other large rivers, it is clear that deltas are some of the most active geomorphological forms. The mean rate of growth of many of the larger deltas is, even at the present time, several tens of meters a year (Kelletat, 1989, p. 125). The Hwangho has the highest rate with a growth of 100–268 m annually. In addition to extending forward, deltas also change the detail of their local forms rapidly in reaction to changes in the supply and removal of sediments.

Deltas in lakes are not affected by eustatic fluctuations and can, therefore, be older. The largest lacustrine delta is the delta of the Volga in the Caspian Sea with a surface area of over $27\,000$ km^2. It is the eighth largest delta in the world.

Virtually all large rivers that flow into seas with a small tidal range, such as the Baltic, the Black Sea, the Mediterranean, the Gulf of Mexico, the Caribbean Sea and the South China Sea have deltas because in these seas tidal currents are too weak to hinder delta growth significantly. Also, the relatively small surface areas of these seas compared to the oceans means that the height and possible erosive energy of the waves are also small.

Deltas can develop on open ocean coasts if the sediment load of the stream at the coast is large enough to offset the destructive effect of the tides and waves to remove it. This is not often the case. Some deltas have developed on such coasts where there is some protection within a bay or behind some islands. Without such shelter, river mouths develop as estuarine deltas or as pure estuaries. On the North Sea, for example, estuaries predominate because of the large tidal range, frequent storms and the relatively small loads of the streams arriving at the coast.

17

STREAM AND VALLEY NETWORKS

17.1 CHANGE AND INTEGRATION OF STREAM NETWORKS

Stream systems, with their valleys, create the texture of the relief in their catchment areas. Even on an atlas map, major differences in the basic patterns of streams and their valley networks are obvious. The southern tributaries of the Danube in the Swabian and Bavarian Alpine Foreland in Germany flow almost parallel to one another towards the Danube. In the Bohemian Basin of Czechia the rivers converge from the east, south and west on a section of the Elbe River that is only a few tens of kilometres long. The Elbe then leaves the wide Bohemian Basin, the only river draining the area, and flows northwards, crosses a sandstone upland in a narrow **transversal valley** and reaches the lowland beyond Dresden in Germany.

The difference between the largely integrated stream network of the Elbe in Bohemia and the lack of integration in the Alpine foreland rivers has to be explained. Also, how it is that a river, which has to follow the gradient that is present, can cross a mountain range? A significant process in the progressive integration of stream networks is the **capture** of one stream by another; the development of many transversal valleys is associated with this process.

17.1.1 Capture by lateral shifting of the divide

Two neighbouring streams compete for the watershed or divide that lies between them. If one of the streams erodes down more deeply,

then the slope on its side of the divide usually becomes steeper than the slope on the other side of the divide. Denudation is also more rapid on this steeper slope and it is worn back more rapidly. The divide then shifts in the direction of the weaker stream.

In the early stages of form development on an inclined surface that has not been eroded much, the streams flow almost parallel to one another, following the regional gradient. This process can be seen in miniature in the parallel flow of gullies on an artificial slope, such as a new road cut in unconsolidated material (Fig. 17.1a). A stream that cuts down deeper than the neighbouring stream initiates a self-reinforcing, positive feedback: its valley not only becomes deeper but also wider in cross-section, the upper edges of its slopes lie farther apart so that the stream receives more precipitation, which increases its discharge and its ability to erode.

The side slope of the widening valley intersects with that of the neighbouring valley and the divide between them becomes a ridge. The continued widening of the larger valley takes place at the cost of the smaller neighbour, the divide shifts further towards the latter and is gradually lowered as a result. Positive feedback also occurs now in the neighbouring valley, but working in the opposite direction: its catchment area becomes progressively smaller, its discharge and power to erode decline and its ability to compete with the larger neighbour decreases.

Eventually, some point on the lowered divide reaches the level of the stream in the smaller valley. The two valleys are no longer separated and the smaller stream, lying at a higher level, changes its direction and becomes a tributary of the larger stream at this point on the divide. The capture of the small stream by the larger stream is complete. The catchment areas are now combined and there is an increase in discharge below the point of capture. The **captured stream** has a new lower local base level, its erosion power is increased and it incises into its former valley floor, which becomes a terrace (section 15.4).

At the point of capture, the captured stream bends sharply in an **elbow of capture** from its previous course to the valley of the **capturing stream** (Fig. 17.1b). The valley stretch in which the continuation of the captured stream flows has lost the upper part of its catchment and becomes a **beheaded valley**. The valley is now too large for the **beheaded stream** that flows in it. It is an **underfit stream**, too small to have formed the valley in which it now flows (section

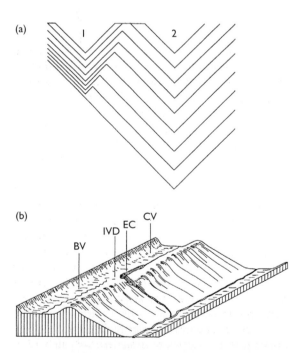

FIGURE 17.1 Capture of parallel neighbouring channels by the lateral shifting of the divide. (a) Development of the valley cross-profiles (schematic). (b) Valley system shortly after capture has taken place. CV: captured valley; EC: elbow of capture; IVD: in-valley divide; BV: beheaded valley.

13.3.2). Between the upper end of the beheaded stream and the elbow of capture there is an **in-valley divide**. With time all signs that the capture took place are removed from the landscape.

17.1.2 Capture by headward erosion of the valley head

In the Black Forest in Germany a head stream of the Danube was captured by headward erosion of the Wutach River (Fig. 17.2). The former Danube headwater flowed from the area of the Feldberg (1493 m) eastwards and originally joined the Danube above Tuttlingen at an altitude of 616 m. At this time the Wutach was a short stream flowing southwestwards to the Rhine, which it joined near Waldshut at about 314 m above sea level. It incised its valley by headward erosion and, in the sense of the positive feedback described in section 17.1.1, lengthened its course and thereby enlarged its catchment area. Because of its lower base level, the valley head and the upper course of the Wutach were at a lower altitude than the head stream of the Danube. Eventually the Wutach cut back far enough to capture the Danube head stream. The original valley floor today forms a terrace at the capture point about 350 m above the present level of the Wutach (section 15.4).The elbow of capture is particularly noticeable in this case because the stream bends from an easterly direction to a southwesterly direction, more than 90°. In the beheaded valley, the Aitrach flows as an underfit stream, which has its source in the area of the in-valley divide near Blumberg.

By the process of capture, a stream acquires an additional tributary and enlarges its catchment area. Repetition of the process leads to a progressive integration of streams which reduces their number and increases the catchment area size. A highly integrated stream system with a large main stream is usually an indication of a long period of development.

Capture can take place in several ways. In the case of the Wutach, it was the difference in local base level heights. The Danube lies at a greater altitude in this area because the distance to its mouth on the Black Sea is much greater than the distance from the mouth of the Wutach into the

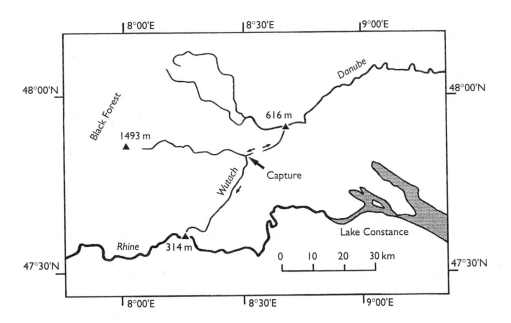

FIGURE 17.2 The stream systems of the Wutach capture.

Rhine to the North Sea. Captures also often take place when two streams flow next to one another on rocks of varying resistance and the stream on the less resistant rock incises more rapidly and gains an advantage that leads to capture. Similarly, differences in the discharge of neighbouring streams can result in capture.

17.2 TRANSVERSAL VALLEYS AND WATER GAPS

Transversal valleys are so termed because they appear to cut through a mountain range, connecting areas of lowland. The cause of this phenomenon puzzled observers of nature long before any scientific explanation had been put forward. In the 18th century Thomas Jefferson hypothesized about the transversal valley of the Potomac across the Blue Ridge in the eastern USA.

Generally, one of two sets of processes lead to the development of transversal valleys: **antecedence** and **superposition**. A third, **headward erosion**, is usually associated with capture. Antecedence means that the stream course already

existed before the upland through which the stream flows was uplifted. The stream eroded its course into the mass that was being uplifted, held its position and created the narrow valley in which it crosses the upland. Like the Elbe River, the valley of the Rhine through the Rhenish Slate Mountains (Fig. 17.3), the valley of the Danube through the southern Carpathians (Iron Gate) and the valley of the Columbia River in the Cascade Mountains in the western USA are all major antecedent river valleys. A stream must have a considerable erosive power in order to compensate for the uplift of the mountain by incision and for this reason antecedence is usually limited to large rivers. The courses of smaller streams are diverted by crustal movement or capture because they cannot keep pace with the uplift.

A stream in an antecedent valley is older than the structure it crosses. A **superimposed valley** is cut down into a structure that is older than the stream. A prerequisite for its development is the presence of a zone of resistant rock lying between two zones of less resistant (weak) rock, across which the stream flows. Initially the land surface has a low relief and can even be composed of sediments which cover the resistant and

FIGURE 17.3 Antecedent gorge of the middle Rhine. In the foreground on the right bank of the river are the steep Loreley rocks. The extensive level areas above the narrow valley belong mostly to the main terrace of the middle Rhine. (Photograph from the Loreley Communal Administration, St. Goarshausen).

weak rocks. Following a lowering of the base level the stream begins to incise its valley into the structure beneath. Its erosive power must be sufficient to erode down into the resistant rock as well as the weak rocks. On the slopes of its increasingly deepened valley, resistant and less resistant rocks alternate. The less resistant are more rapidly worn down by weathering and denudation. Also, tributaries that enter the main valley in a stretch of less resistant rock can erode their valleys more easily and gain an advantage over tributaries on the resistant rock. Captures take place and the tributary stream network becomes more and more concentrated in the zones of less resistant rock. Through the combined effect of erosion by the tributary streams and denudation on the slopes, the areas of less resistant rocks become lower than the areas of more resistant rock which remain as ridges between the neighbouring zones of weaker rock and

which are crossed by the main streams in narrow superimposed transversal valley stretches.

There are numerous transversal valleys in the Appalachian Mountains in the eastern United States. The Appalachians are Palaeozoic fold mountains that had been levelled by denudation by the end of the Palaeozoic. The capped fold structure of resistant quartzites and less resistant limestones and slates remained. The axis of the original folds was southwest–northeast, almost parallel to the present east coast of the USA. After the separation of North America from Europe and Africa during the plate tectonic opening of the Atlantic Ocean in the late Mesozoic, the drainage oriented itself towards the new ocean with the result that the main streams in the eastern USA flowed eastwards across the old Palaeozoic fold structure. As they began to incise their valleys headwards from the coast into the Appalachian fold structure, the zones of limestones and slates were lowered by stream erosion and slope denudation. Long steep quartzite ridges were left on the flanks of the folds, which were crossed only by the main streams (Fig. 17.4). The stretches of superimposed transversal valleys that cut across these ridges and through which streams flow, are known locally as **water gaps**. **Wind gaps** are former water gaps that lie above the present stream level, their streams having being captured or diverted. Water gaps and wind gaps played a major role in the history of the early settlement of the United States because they provided the only routes from the east coast

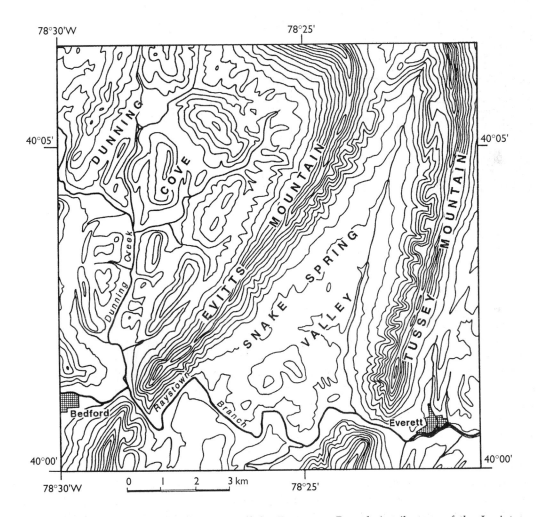

FIGURE 17.4 Superimposed water gaps of the Raystown Branch (a tributary of the Juniata River) in the Appalachians near Bedford and Everett, Pennsylvania. The mountain ridges Evitts Mountain and Tussey Mountain are hogbacks (section 20.3.10) formed of Silurian Tuscarora quartzite on both flanks of an anticline the crest of which has been worn down (now forming the Snake Spring Valley – a relief inversion). Dunning Cove is structurally a synclinal basin with low ridges made of moderately resistant rock. The valleys neighbouring Dunning Creek form a ring-shaped valley net with in-valley divides (section 17.4).

through the Appalachians to the interior of the continent.

Transversal valleys formed by headward erosion occur mainly on the flanks of the anticlines and upwarps where a resistant cover rock is underlain by a less resistant rock (Fig. 17.5). At first, the land surface corresponds approximately to the surface of the resistant cover rock so that the stream flows down the dip of the upwarped rock. If the stream erosion in the upper part of the stream near the crest of the upwarp cuts through the resistant rock into the weaker rocks beneath, these become exposed and are more rapidly lowered than the area where the resistant rock has not been removed. Headward erosion and denudation expand the areal lowering in the

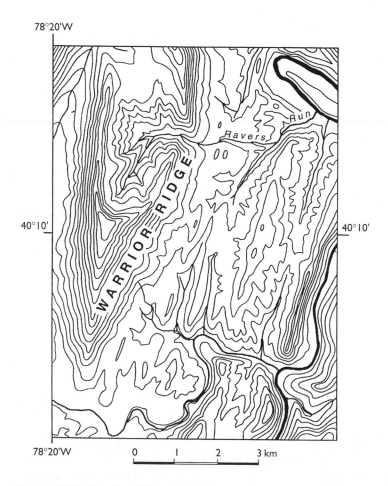

FIGURE 17.5 Warrior Ridge in the Pennsylvania Appalachians is structurally an anticline of resistant Tuscarora quartzite. In the centre of the upwarp this rock has already been removed and less resistant claystone outcrops at the surface. The eastwards-flowing Ravers Run has cut a transversal gorge through the eastern flank of the anticline by headward erosion and the less resistant rock is being removed by its headwaters.

core of the upwarp further. The effect can be increased by the capture of other streams by the main stream.

17.3 STREAM ORDER AND VALLEY ORDER SYSTEMS

In a paper that is regarded as the beginning of modern quantitative geomorphology, Horton (1945) developed a system to describe the internal organization of stream networks that became known as the **Horton stream order system**. A few years later, Strahler (1952, 1957) introduced a simple but useful modification which was generally adopted and is used here. The system can be applied equally well to stream networks and to the related valley networks.

Horton and Strahler assumed that stream networks have a recognizable inherent hierarchical order: small creeks combine to become small streams which in turn combine to become rivers and so on. The order system indicates the position of a particular stream section in the hierarchy of a stream network. Also, quantitative parameters can be derived from it which characterize the stream system.

In the order system every section of a stream network, from its source to its mouth in the sea or a lake, is classified by an order number. The streams flowing from the source are first-order streams. They are closest to the watershed and have no tributaries. A second-order stream is formed by the junction of two first-order streams, a third-order stream by the combination of two second-order streams and so on. When a stream of a higher order is joined by streams of a lower order, its order number does not change. The order number is increased only when a stream joins another stream of the same order; the order then changes by a value of one. Figure 17.6a illustrates the scheme.

The stream with the highest order in a stream network characterizes the order of the entire network. Figure 17.6a shows a fourth-order stream network. The system is also applied to catchment areas: a fourth-order catchment area is one drained by a fourth-order stream network. The **stream number** is the number of streams of

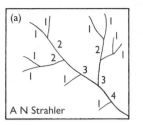
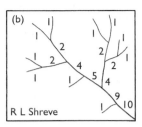

FIGURE 17.6 (a) Stream order system (after Strahler, 1952, 1957); (b) stream magnitude (after Shreve, 1966, 1967).

a particular order in a stream system. It declines more or less geometrically with increasing order size. This law of stream numbers is not a natural law but a logical consequence of the order system because the numerical relationship of the stream sections of the Kth order to the (K + 1)th order is always at least 2:1.

The parameter often used to characterize a stream network is the **bifurcation ratio**, R_b. This is the ratio of the stream numbers of two successive orders, n_k and n_{k+1}:

$$R_b = n_k / n_{k+1} \qquad (17.1)$$

Because, by definition, the smallest possible value is $R_b = 2.0$, it is useful to use a **net bifurction ratio** R'_b, instead of the Horton bifurction ratio; this is determined by subtracting the two obligatory tributaries from the stream number n_K:

$$R'_b = (n_k - 2)/n_{k-1} \qquad (17.2)$$

R'_b is more useful for comparing the characteristics of stream networks than R_b because it relates directly to the number of stream sections of order K that are above the defined minimum.

Like the stream number, the mean stream gradient declines with increasing stream order and the surface area of the catchment area increases with increasing stream order.

The significance of an analysis of the numerical relationships in Horton's scheme is not the normality of the relationships in themselves, since they are largely a function of the system's definition. More important is the possibility to identify and explain deviations from the norm.

Shreve (1966, 1967) introduced a classification scheme that used the term magnitude instead of order. The streams of first magnitude are identical with the Horton/Strahler first-order streams but at the mouth of every tributary, the magnitude of the stream increases by the magnitude value of the tributary stream. In Shreve's system, therefore, the magnitude of a stream is equal to the number of streams of first magnitude present in its catchment area (Fig. 17.6b).

The system has some advantages in applications to particular problems. For example, the magnitude of a stream depends to a greater extent on the size of the catchment area than does the stream order of Horton/Strahler. In an area that is hydrologically, geologically and topographically more or less uniform, the Shreve magnitude, which is quickly determined from maps, can be used as a substitute variable to estimate the approximate size of the catchment area.

Both methods do, however, usually base the determination of first order, or first magnitude, streams, on maps on which not all water courses are necessarily shown. Also, it has to be decided whether an erosion channel with periodic flow should be defined as a stream or not.

An analysis of the **valley network** is different from the stream network analysis in that the dry valleys are also included. The Horton/Strahler method can also be used to order the divides because their spatial system is essentially a counterpart of the valley system.

The **drainage density**, or **valley density**, is an important hydrological and fluvial morphological characteristic of stream networks. It is determined as

$$D = L/A \qquad (17.3)$$

where D is the drainage or valley network, L is the total length of all water courses or valleys (km) and A is the total area of the region (km^2). If miles are used, one mile per square mile is equal to 0.63 km/km^2.

Drainage density varies depending on the perviousness of the rock. It is usually higher on not very pervious clayey rocks than on sandstones or chalk. On a relatively young surface of uniform material, the stream density can also be an indication of the age of the drainage network

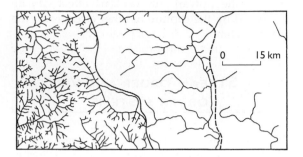

FIGURE 17.7 Varying stream density on Pleistocene glacial deposits of different ages in Iowa, USA: left, 17 000 years old; middle, 15 000 years old; right, 13 000 years old. The continuous line indicates the position of the ice margin 15 000 years ago, the broken line the position of the ice margin 13 000 years ago (after Gregory and Walling, 1973, p. 398).

development. Figure 17.7 shows drainage densities on late glacial deposits of different ages in Iowa, USA.

Strahler (1952, 1957, 1964) defined a number of other characteristic measures of stream and valley systems which are useful for quantitative morphographic descriptions of stream networks and catchment areas, including indices for the circumference of the catchment area and the ruggedness of the relief. The latter is defined as the product of valley density and relief. The disadvantage of indices of this type, which are the product of two variables, is that they can conceal an infinitely large variation in either component. Gregory and Walling (1973, pp.37–92) have discussed the Strahler and other stream network and catchment measures in some detail. Shorter surveys have been made by Schmidt (1984, pp.26–39) and Abrahams (1984).

17.4 STREAM AND VALLEY NETWORK PATTERNS

The pattern of streams and their valleys in a landscape can vary greatly. The first attempts to characterize patterns were made by Willis (1895) and Russell (1898) (*see* Feldman *et al.*, 1968). Zernitz (1932) distinguished seven main pattern types which are still largely used today.

Stream pattern types are, of course, fixed points in a continuous reality and there are transitions between all pattern types. They are essentially descriptions of characteristics. Some are fairly closely related to the causes of their development; the identification of causal relationships is, however, of secondary importance. The spatial pattern of stream networks applies as well to valley networks and the terms are used for either.

The **parallel stream network** is the simplest type of drainage (Fig.17.8a). It occurs primarily on young land surfaces with a uniform gradient, such as recently raised sea floors or young plains of subaerial deposits. The streams of the Swabian–Bavarian Alpine Foreland in Germany approach this type. With subsequent shifting of the divides between the parallel streams and capture, they evolve into more integrated larger stream systems.

A **radial stream network** (Fig.17.8b) develops when the original slopes radiate in all directions from a central summit or upwarped divide. Apart from the differences in its slope direction, it resembles a parallel network. Radial patterns are typical for large volcanic mountains that are still, or were recently, active and for young dome shaped upwarps such as young laccoliths and other intrusions with a circular outline. Radial networks also tend to become more complex later.

The **dendritic stream network** (Fig. 17.8c) is characteristic of areas where there are no strong structural or lithographic controls. The pattern resembles the even branching of a tree. The streams are usually highly integrated so that they form large systems and are, therefore, relatively old. Most have developed from simpler patterns.

The stream courses of **rectangular drainage patterns** (Fig. 17.8d) are made up of straight stretches and more or less right angled bends. They are usually controlled by a strongly developed joint system in which the joint sets are approximately at right angles to one another. Faults can also contribute to the development of this type of pattern. The joints and faults weaken

the rock resistance and become the controlling lines for erosion. Streams eroding along them have an advantage compared to their neighbours on more resistant rock and, with captures, determine the stream network pattern.

The **trellis drainage patterns** (Fig.17.8e) are influenced by the geological bedding structure. It is also known as the Appalachian drainage pattern because it is typically developed on the folded Appalachian Mountains. The folds have

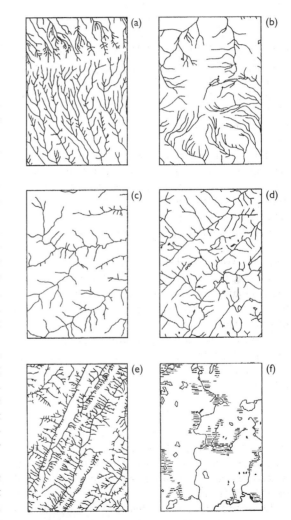

FIGURE 17.8 Stream network types: (a) parallel, (b) radial, (c) dendritic, (d) rectangular, (e) trellis, (f) deranged (after examples in Thornbury, 1954, pp. 121–2).

been capped by denudation so that the steeply dipping resistant and less resistant strata now outcrop next to one another parallel to the strike direction. The resistant rocks rise as long hog back ridges. In the less resistant rocks, longitudinal parallel valley zones have developed. In only a few places have major rivers cut through the ridges of resistant rock in superimposed transversal valleys (section 17.2). The trellis stream pattern is made up of three types of stream:

1. small creeks, mostly first or second order, which flow from the divide into the adjacent longitudinal valley (flow direction at right angles to the strike of the fold),
2. longitudinal valley streams, which have eroded the longitudinal valleys and receive the small streams flowing from the divide;
3. higher-order superimposed transversal valley streams, which run across the strike of the folds.

Within the longitudinal valleys are in-valley divides, an indication that the drainage pattern has been changed by capture, which has contributed to the stream's adjustment to the geological structure.

An **annular drainage pattern** develops if rock strata of differing resistance form a circular upwarp which has been capped so that the resistant and less resistant rocks outcrop concentrically. The hog backs in the resistant rocks and

the valley zones in the weaker rocks are also circular in ground plan and the resistant rocks, similar to the trellis pattern, of which it is a variation, are crossed in only a few places by the main streams.

Unlike other drainage patterns, the **deranged drainage pattern** is characterized by an absence of any apparent spatial organization. It is a young drainage pattern and is frequent on irregular relief in areas of accumulation such as the moraine deposits of the last continental glaciation which left behind a relief of moraine hills separated by basins. Precipitation collected in the basins to form lakes which then overflowed and the outlet stream followed the gradient to the next basin. The lakes in the young moraine landscape of northern Germany and the numerous small lakes in Michigan, Wisconsin and Minnesota in the USA have developed in this way. Because the streams follow the local, more or less random gradient direction, there is no dominating drainage direction. Only in a later phase, when headward erosion from the sea or from a large river begins to influence the moraine area, is a drainage direction gradually determined. The lakes then disappear, partly because the streams flowing in gradually fill them up with the easily eroded moraine material that surrounds them and partly because the streams flowing out of them incise their outlets and lower the water level in the basins.

STREAMS AND SLOPE DEVELOPMENT IN THE FLUVIAL SYSTEM

18.1 THE FLUVIAL PROCESS RESPONSE SYSTEM

Since most of the earth's land surface is formed by fluvially controlled processes, the interaction of the work of streams and of slope development is of fundamental geomorphological importance. Denudational slope forms can only develop if height differences are present. These are caused mainly by crustal movements and volcanism and, in fluvial landscapes, secondarily by stream incision which leads to the formation of valleys. A further prerequisite for slope development is that the bedrock is weathered sufficiently for the loose material to be removed by denudation processes so that the slope form is changed.

The most important components of the fluvial process response system, together with their interrelationships and feedbacks, are shown in a simplified form in Fig. 18.1. The system is driven by **eksystemic** endogenic and exogenic energy supplies and contains three categories of **ensystemic** components.

1. **Form components**:
   (a) regional relief
   (b) stream bed and valley floor and their gradient values
   (c) slope form and slope height.
2. **Process components**:
   (a) weathering
   (b) denudation
   (c) fluvial erosion
   (d) fluvial transport

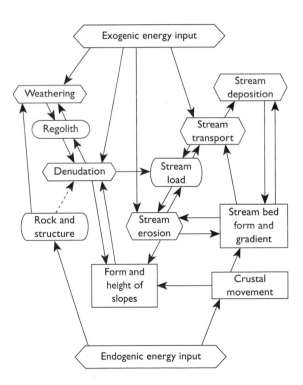

FIGURE 18.1 Components of the fluvial morphological process response system and their functional relationships.

(e) fluvial deposition.
3. **Material components**:
   (a) rocks and geological structure
   (b) regolith
   (c) stream load.

18.1.1 Eksystemic energy supplies

The effect of the initial height differences produced by endogenic processes depends on their spatial distribution. The endogenic forms of most volcanoes vary within narrow limits. By contrast, crustal movements create a wide variety of structural forms which can cause very different reactions in the exogenic processes, especially in stream erosion. If, for example, a crustal block is uplifted along a fault without tilting, the height differences created are first of consequence only immediately at the fault line. There the gradients become steeper and it is there that the stream erosion concentrates. In the central region of the block, the exogenic effect of the endogenic uplift impulse is felt much later when headward erosion of the streams reaches this area after a considerable time lag and forms deep valleys with steep slopes.

If the fault block is tilted or upwarped during uplift, the new gradient conditions are present everywhere from the beginning and the streams can react simultaneously in all parts of the block, although here too headward erosion is a determining factor in the development of the exogenic landforms.

The exogenic energy supply regulates the heat and water budgets of the earth's surface. The complex nature of their effects on the surface is mostly the result of the wide range in the possible magnitude of meteorological events, particularly of precipitation. The longer a landform develops, the less important become individual events or short-term clusters of such events and the more important are the long-term magnitude frequencies of events that are a characteristic of the particular climate, especially the long term cumulative effects of such events. Magnitude–frequency analysis (section 5.2.1) is useful not only for analysing meteorological and hydrological events, but also for the analysis of the magnitude and frequency of exogenic geomorphological process events of various types and intensities.

18.1.2 Form components

Regional relief results from the combination of relief-increasing crustal movements and relief-reducing denudation. At first, the work of the streams in the denudation process of down-wearing increases the relief locally and regionally, as long as the rate of fluvial downcutting is greater than the denudation rate on the divide. Exogenic processes reduce relief only when the denudation rate of the divides is greater than the downcutting rate of the streams or the uplift rate of the crust.

Slope form and height are influenced directly only by fluvial erosion and denudation. Downcutting at the slope foot by the stream increases the slope height and steepness. Denudation at the divide counteracts the increase in height and steepening. The form of the slope profile depends largely on the type and intensity of the denudation processes (section 9.4). If the material transported by the denudation processes to the slope foot is not, or only after long intervals, removed by the stream, it is stored as a local accumulation of **colluvium** at the slope foot. **Valley floor forms**, **stream bed forms** and **stream gradients** are mainly products of fluvial erosion and fluvial transport.

Depending on their size, form components require different lengths of time for their development (section 1.3). Regional relief is areally extensive and needs, in order of magnitude, from 100 000 to 10 million years to adjust to changes in the endogenic or exogenic energy supply. An approximate steady state of constant form of the regional relief as a whole can therefore be reached only when the endogenic and exogenic energy supplies remain at a nearly constant level for such a long period, excluding short-term oscillations.

The components of slope form and slope height are only local in extent and can adjust more quickly to changes in the energy supply, usually within 10 000 to 1 million years. Adjustments on the valley floor and in the stream bed can take place in an even shorter period, about

100 to 10 000 years, not only because of the limited area of these form components, but also because the greatest concentration of energy in the entire system is present in the stream. The chance for a form component to reach a steady state is therefore greater, the smaller its area, and the greater the concentration of exogenic energy present.

18.1.3 Material components

The **bedrock** with its **structure** provides the source material. Weathering alters it to **regolith**. Once the regolith is transported downslope by denudation processes, it is also termed **slope debris** as long as it remains on the slope. Once removed from the slope foot by the stream it becomes part of the **stream load**; the rest of the stream load is supplied by the erosion of the stream bed itself.

The properties of the regolith and the stream load change as a function of the particular processes at work and of time. A regolith that is worn down only very slowly continues to weather *in situ* and eventually develops into a mature soil. Both its grain size spectrum and its chemical composition change during this process. If denudation is more intensive, the regolith does not develop beyond the stage of a largely unweathered slope debris or skeletal soil.

The stream load, especially the bed load, changes depending on the intensity and on the frequency of the transport events. Frequent transport leads to the bed load being rounded more than when transport of the load is a rare event. Reduction in the size of the bed load components is caused not only by transport of the material but also by further weathering of the gravel and sand grains.

18.1.4 Process components

The preparation of the rock by **weathering** takes place over the entire land surface in direct response to the variations of the heat and water budgets. The intensity of this process is linked by feedbacks with the material that produces the regolith. **Denudation** also occurs over the entire surface but its intensity is strongly dependent on

slope form, especially on the inclination of the slope.

In Fig. 18.1 the areal processes of weathering and denudation are shown opposite the predominantly linear processes of stream action: **fluvial erosion**, **fluvial transport** and **fluvial deposition**.

There is a progressive increase in the energy concentration per unit of surface area from weathering, through denudation of the slopes to the work of the stream itself that is largely due to the spatial distribution of the availability of water. The soil moisture, which contributes to weathering, is spatially and temporally more evenly distributed over the land surface than the water that flows downslope into the valley, with an increasing volume, as interflow or overland flow which does denudational work and is areally effective and not usually spatially concentrated.

The water in the stream bed is concentrated in a narrow channel, eroded by the stream itself in order to transport away all the water and the sediment that is brought to it from the slopes. When more load arrives from the slopes than the stream can remove immediately, local deposition increases the stream's gradient and thereby improves its transport capability. If there is a surplus of energy beyond that required for transport of the load, the stream erodes its bed and reduces its gradient until it is only sufficient for the load transport. Should the stream not be able to lower its gradient because the channel floor is too resistant, then the surplus energy is dissipated in shooting or falling flow and by increased turbulence.

Net fluvial erosion can only take place when more energy is available than is needed to transport the material that arrives from upstream and from the slopes on either side of the stream and when this surplus is large enough to attack the stream bed. Usually lateral erosion requires less energy than the erosion of the channel floor.

Bed load transport is very closely related to the available energy, whereby the grain size of the material plays a significant role. By contrast, the transport of the suspended load and dissolved load depends very little on the stream's available energy and far more on the amounts of these types of load delivered to the stream.

Deposition of material occurs when the transport energy is no longer sufficient. This can apply either to the amount of bed load to be transported or to its grain size. How long does the sediment remain where it has been deposited? Earth history shows that all sediments are removed again sooner or later either directly as unconsolidated material or after they have been transformed by diagenesis, perhaps with the addition of metamorphosis, into a hard rock and uplifted again. In the latter case there is, of course, a different process response system at work than the one that prevailed when the original deposition occurred.

18.1.5 The continuity of natural process systems and the discontinuity of their representation

One disadvantage of the components of the fluvial process response system shown in Fig. 18.1 is the division of the system into discrete subsystems which in reality mesh with one another and in part superimpose one another. A fluvial morphological process response system such as a stream catchment area is a spatial, temporal and functional continuum. How it is subdivided depends on the purpose. The actual continuity of the functional structure, the spatial connections and the temporal development are not changed by the subdivisions. The system shown in Fig. 18.1 could be divided into a smaller or larger number of components. Each component is a unit only when seen from the outside; in itself it is a process response system that can be broken up into form, material and process components. Similarly the spatial unit of a stream catchment area can be subdivided into smaller areas and their development divided into time units.

The various degrees of subdivision of the continuum reflect a wide range of spatial and temporal scales in which the functioning of a process response system can be examined. Spatially they range from the relationship between components at a particular point on a slope or in a stream bed to those that are valid for an entire catchment area and temporally form the relationship between processes, forms and materials at a particular moment in time to those that are significant for the formation and development of catchment areas over millions of years.

18.2 The linking of processes with differing magnitude–frequencies

Compared to the spatial and temporal continuity of fluvial morphological process response systems, some of the processes within the system are not continuous but take place as discrete process events. Their effect on the shaping of the forms is the result of the magnitude–frequency of process events (section 5.2.1). The relationship between the magnitude–frequencies of different processes within a given system is of significance for their interaction.

In general, processes with a low-energy concentration occur more frequently or proceed more continuously than processes with a higher energy concentration. Granular disintegration takes place almost continuously on sandstone rock wall surfaces as a result of mechanical weathering, with intensity maxima during seasons with night frosts. Block disintegration and rock fall events are, by contrast, separated by periods of tens or hundreds of years. Chemical weathering of the bedrock and the subsequent weathering of the regolith require relatively little energy per unit of time, but are quasi-continuous as long as moisture is present in the soil.

Downslope movement of regolith of the slow, low-energy creep process is also more or less continuous. The higher-energy wash denudation of the regolith by overland flow and piping is limited to heavy rainfall events that are separated in time. Mudflows and landslides involve much higher amounts of energy and are consequently repeated even less frequently. The frequency of fluvial erosion and fluvial bed load transport events of different intensities varies greatly from one stream to another because they are strongly dependent on gradient and discharge and therefore size of the catchment area.

There are locations in the structure of a process response system where different process associations border on one another and where the material to be transported is transferred from one process association to another, the relationship between weathering and denudation at each individual point on the slope, for example (sections 9.2 and 9.3). In the case of transport-limited denudation, regolith cover is a type of intermediate deposition for the slope debris that is lying ready to be transported.

Reneau and Dietrich (1991) have described an intermediate deposition within a denudation system in the Pacific Coastal Ranges of the USA. Slow creep transports the slope debris to dell-like hollows in the slope where the material is stored. Infrequent extreme heavy rainfall events soak the debris in the hollows. Positive pore water pressure builds up and the debris flows as a mudflow into the valley, leaving a deep gully in the hollow. After the mudflow event, the gully is filled gradually by local collapsing and wash denudation of the regolith on the unstable walls of the gully and by continued creep movements. The slope hollow functions as a **coupling point** between the slow, almost continuous creep process and the intensive but rare removal of the debris by mudflows.

Coupling points also occur at the foot of slopes where the denudational transport on the slope is replaced by fluvial transport. Harvey (1977, 1991) has investigated this type of coupling point in a number of small valleys of the Howgill Fells in the uplands of Cumbria in England. The climate in the area is cool and moist. On an average of 25–30 days per year the rainfall exceeds about 14 mm, the critical value at which transport of regolith material to the slope foot is initiated. The transport is mostly by wash denudation in gullies. The transported material accumulates at the slope foot as alluvial cones which remain until the creek flowing along the valley removes them during a sufficiently high flood. In contrast to the high frequency of the denudation events on the slope, the removal of the cones by fluvial transport takes place only every two to five years. In the intervening period new alluvial cones develop at the slope foot.

In these examples, processes of differing magnitudes and frequencies border on one another. More important, at the coupling points or temporary deposition areas, there is a cyclical alternation of supply and removal neither of which, however, predominates in the long-term. The observed change from predominating supply to more or less sudden removal is only an oscillation in a long term state of approximate dynamic equilibrium of the local mass balance at the coupling point.

Temporary deposition also occurs in the fluvial system itself. A measure of this is the **sediment delivery ratio**, which expresses the relationship between the amount of sediment load that the stream transports out of its catchment area during a time period, annually, for example, and the amount transported from the slopes of the catchment area in the same time period. Gregory and Walling (1973, p.204) cite values from the USA that range from less than 30 per cent to more than 90 per cent. The size of the catchment area investigated plays a significant role in the size of the ratio. Large catchment areas contain a larger number of possible intermediate locations where sediment can be deposited temporarily. As a result, the ratio decreases with increasing catchment basin size. Also decisive is how far the catchment basin is from a dynamic equilibrium of its mass balance. In a perfect equilibrium, the stream transports the entire load that it receives out of the catchment area, a sediment delivery ratio of 100 per cent. A similar value would be expected when the catchment area is dominated by weathering-limited denudation and there is a deficit in the available sediment load in relation to the potential sediment removal (section 9.3). Low sediment delivery ratios indicate a surplus in the available sediment supply within the catchment area relative to sediment removal and deposition within the area by, for example, aggradation of valley floors.

There are also differences in the associated **spatial magnitude–frequency**. Many temporally rare events such as landslides are also rare spatially, while events that occur frequently also take place more or less simultaneously at many locations. This relationship might provide the basis for an analysis of the spatial frequency of different kinds of geomorphological process that would allow, by analogy, estimates of the temporal magnitude-frequencies of events that are too rare to be observed repeatedly in the same location.

18.3 VALLEY CROSS-SECTION FORMS AS AN EXPRESSION OF THE PROCESS STRUCTURE

The form of a valley cross-section, or valley cross-profile, is the result of stream work and slope development. A stream's line of incision is the line on either side of which the valley slopes develop. The relationship between the intensity of vertical erosion by the stream and the denudative retreat of the valley slopes determines the ratio between valley depth and valley width. Intense vertical erosion and little or no denudation on the slopes produces a **saw cut gorge** with almost perpendicular rock walls. The gorge is only as wide as the stream bed. Saw cut gorges are frequent in high mountain areas glaciated during the last ice ages. Glacial erosion produced steps in the long profiles of the valleys. At these steps, and at the junction of some hanging side valleys (section 14.4), there are gorges with steep gradients which meltwater streams had already begun to deepen under the glacier. The glaciers probably also protected the side walls of the gorge in the early phases of its development from denudation. There are numerous saw cut gorges in the valleys of the Alps including the Via Mala on the Hinterrhein in Switzerland (Fig. 18.2). The Trient gorge, also in Switzerland, on a tributary of the Rhône, is incised into a hanging valley. The waterfall at the back of the gorge is being extended by headward erosion. The gorge of the Trümmelbach near Lauterbrunnen in the Bernese Oberland has, by contrast, cut back only a few tens of metres into the steep flank of the main valley.

Gorges can only develop in resistant rock. Their possible depth in all cases is limited by the critical height of their rock walls (section 8.2.5). If the critical height is exceeded, rock falls and landslides follow and a **V-shaped valley** develops, at first with steep slopes and weathering-limited denudation (Fig. 18.3). This state lasts as long as the rate of downcutting is greater than the maximum possible rate of retreat of all parts of the slope by weathering and denudation. Even when the slope becomes less steep and transport-limited denudation predominates, the

cross-section of the valley remains V-shaped as long as net vertical erosion takes place. Only when vertical erosion has ceased does the valley become flat floored.

If vertical erosion continues at a rate that is lower than the maximum possible rate of denudational retreat in all parts of the slope, then, because of feedbacks in the fluvial system, a dynamic equilibrium can develop between the rate of fluvial downcutting and the denudation rate on the entire slope, providing that the rate of downcutting remains almost constant for a long enough time. Should equilibrium be reached, all points on the slope are worn down at the same rate as the stream bed is lowered. The system is in a steady state and the cross-section of the

FIGURE 18.2 Saw cut gorge of the Hinterrhein, the Via Mala, in the Grisons, Switzerland.

FIGURE 18.3 V-shaped valley in the Cascade Mountains, State of Washington, USA.

valley has a constant form. The form of the valley slopes is then an expression of the dominating denudation process, concave for wash denudation and convex if slow mass movement predominates (section 9.4).

18.3.1 Two theoretical models of slope development

The slope development models in Figs. 18.4 and 18.5 were programmed in order to find out how a slope profile formed by wash denudation changed when the gross rate of downcutting at the foot of the slope was doubled, halved and then doubled again at regular time intervals. The gross rate of downcutting can be understood as a rate of uplift to which the stream reacts with a vertical erosion equal to the uplift.

The actual or net downcutting rate is equal to the gross rate of downcutting in a time unit, minus the thickness of slope debris transported by wash denudation from the slope to the stream bed in the same time unit. The time units and the length units are arbitrarily chosen. Compared to real denudation rates, one model time unit would correspond to about 3000–5000 years. The conversion of the rock to regolith takes place by weathering, according to equation (7.4). Only regolith material can be eroded by wash denudation and then transported, not bedrock.

The model shown in Fig. 18.4 alters the gross rate of downcutting (ZCUT1) every 500 time units between ZCUT1 = 4.0 m and ZCUT1 = 2.0 m per time unit. A succession of slope profiles is shown in the diagram. The relevant time units are shown at the end of the profile and the current gross vertical erosion rate ZCUT1 on the right of the profiles.

The initial profile is horizontal with fluvial erosion beginning on the left end of the profile at time unit = 1. Because only a short section is inclined at this beginning phase, the amount of debris arriving at the slope foot is small and the net vertical erosion rate is almost equal to the gross rate ZCUT1 = 4.0 m. This explains why the slope foot is cut down much more rapidly between $T = 1$ and $T = 26$ than between $T = 1201$ and $T = 1301$, although the gross rate of downcutting is the same in both cases. The high net rate in the initial phase runs ahead of the slope denudation, resulting in the strongly convex profile form at the beginning.

Between time units $T = 26$ and $T = 501$, the net rate becomes so small, because of the increased debris supply from the slope, that a dynamic equilibrium between fluvial erosion and slope denudation develops: the concave pro-file form at $T = 501$ shows the characteristic slope form for wash denudation used in the model programme. Between $T = 502$ and $T = 1001$ the amount of the gross downcutting rate is only ZCUT1 = 2.0 m. At the beginning of the phase some debris is still transported to the slope foot which is adjusted to the previous equilibrium. The debris can, however, no longer be removed completely by the stream and there is a net accumulation between $T = 501$ and $T = 526$, in spite of a gross downcutting rate of ZCUT = 2.0 m. Between $T = 526$ and $T = 601$ the slope adjusts its denudation rate to a new dynamic equilibrium. The slope profile is less steep than before. Consequently less debris

reaches the slope foot and its removal is accomplished by the gross vertical erosion rate without all its energy being used. The net downcutting rate begins again and keeps pace with the slope denudation. The visual expression of the equilibrium that now exists is the almost parallel retreat of the profile between $T = 601$ and $T = 1001$.

After the change to ZCUT1 = 4.0 m again at time $T = 1001$, the opposite takes place. Between $T = 1001$ and $T = 1026$ an excess of erosion energy is produced which leads to a steepening at the slope foot. This, in turn, accelerates the debris supply so that between $T = 1026$ and $T = 1051$ no net downcutting takes place despite the now higher gross downcutting rate. Only after $T = 1101$ is the slope denudation again equal to the changed erosion conditions. The regular vertical intervals and the approximately equal forms of the profile at $T = 1201$, $T = 1301$ and $T = 1401$ are an indication of the renewed development of a dynamic equilibrium between denudation rate and net downcutting. Because wash denudation, regolith thickness and weathering rate are linked by feedbacks to one another, weathering is also included in the equilibrium.

The material and process conditions of the slope development models in Fig. 18.5 are the same as those in Fig. 18.4 with the exception that the gross downcutting rates are changed after every 50 time units instead of every 500 time units. The short period during which the gross downcutting rate remains constant is not sufficient for an equilibrium to be established. As a result, at the slope foot, strong downcutting and steepening alternates with deposition and flattening of the lower slope in a way that is more complex than might be expected, with only a simple change in both gross rates. Very little of the changes are to be seen in the upper part of the profile. Here wash denudation wears down the slope at the divide at a rate that remains almost constant and is roughly the same as the mean net downcutting at the slope foot. In contrast to the profiles in Fig. 18.4, the changes from erosion and steepening to accumulation and flattening at the slope foot are, in this case, hardly transferred crestward because they succeed one another too rapidly. The slope acts as a kind of filter, absorbing and suppressing the impulses from the slope foot in its middle and upper

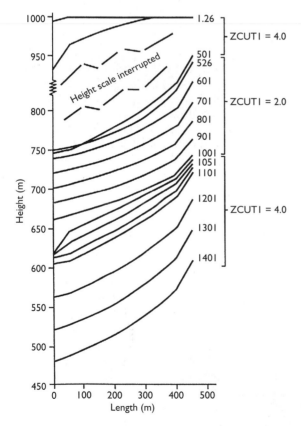

FIGURE 18.4 Theoretical model (program SLOP3D) of slope development with changes in the gross downcutting rate every 500 time units (after Ahnert, 1988b, p. 390).

slopes. These and other theoretical models dealing with the interaction of stream work and slope denudation are discussed in Ahnert (1988b).

18.3.2 Valley cross-sections after fluvial downcutting has ceased

Once downcutting has ceased, the formation of a flat-bottomed valley floor develops, either by lateral erosion or by deposition. Lateral erosion is possible if the stream continues the net transport of the bed load, that is when local accumulation and erosion by the bed load remain about equal. In this case the valley floor is formed of a thin layer of debris on the bedrock floor across which the stream swings. As the stream undercuts the slopes on either side of the valley at

various points, the bedrock floor widens and the slope denudation processes are accelerated initially on the lower slopes and later upslope as the denudation moves crestward. The slope retreats as a whole, similar to the undercut slope of a meander but without an accompanying slip-off slope on the opposite valley side.

In the valley, the course of the stream shifts so that undercutting takes place at different locations along the slopes and eventually almost the whole valley is affected on both sides. Those areas undercut earlier usually have a lower slope angle than those undercut more recently. The older locations also have colluvium deposits which are absent or only small where undercutting was later. The previously V-shaped valley becomes **box shaped**, a type of flat-bottomed valley with steep sides.

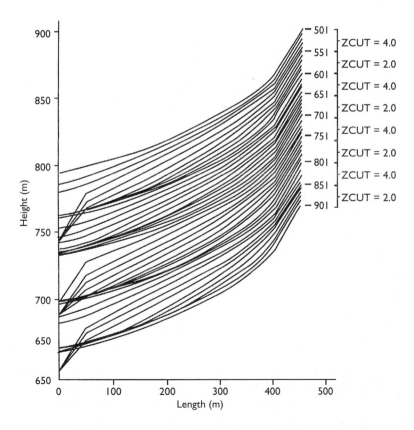

FIGURE 18.5 Theoretical model, similar to Fig. 18.4, but with a change in the gross downcutting rate every 50 time units (after Ahnert, 1988b, p. 391).

Fluvial deposition in a V-shaped valley buries the V-form in the valley bottom transforming it into an **accumulation valley floor**. In this case, the slopes above the valley floor are not usually undercut by lateral erosion; their form is determined by the prevailing denudation processes.

18.4 VALLEY HEAD TYPES

The **valley head** is the hollow in which the fluvially formed course of the valley begins (Kellersohn, 1952; Montgomery and Dietrich 1989). Lower down the valley has slopes on two sides, in the valley head there is a third side at the back of the hollow. The slopes and denudational transport lines converge from three sides. Out of the fourth, open side of the valley head the stream transports any material it receives from the slopes of the hollow down the valley or, if the valley is now dry, has done so in the past.

The valley head is neither a slope form nor a valley form and is best defined as a special component of the fluvial system. It is a special area because here the transition from denudational to erosive transport is characterized by the convergence of the slopes towards the upper end of the stream channel. The following classification was developed for valley heads in the Dahner Felsenland in Germany. The back and side slopes of valley heads in the humid temperate climates of the middle latitudes can be classified by the dells they contain. The number of dells converging in a valley head varies, as does their steepness. Using these two criteria, number and steepness of dells, valley heads can be divided into three groups of valley head types: **shallow-dell valley heads** (Fig. 18.6a–c), **steep-dell valley heads** (Fig. 18.6d–f) and **spring niche valley heads** (Fig.18g–i) (Ahnert, 1955).

Shallow-dell valley heads are subdivided into:

1. **single shallow-dell valley heads**, in which there is only one dell with an axis usually approximately aligned with the valley axis (Fig.18.6a);
2. **double shallow-dell valley heads**, in which two dells lead into the valley head, usually at an acute angle to the valley axis (Fig. 18.6b);

3. **spring hollow valley heads**, in which three or more dells converge in a semicircle (Fig. 18.6c).

Similarly, there are the three types of steep-dell valley heads: **single steep-dell valley head**, **double steep-dell valley head** (Fig. 18.6d and e) and **funnel-shaped valley head** (Fig. 18.6f).

The spring niche type of valley head develops with the headward erosion of contact springs. In the Dahner Felsenland, for example, the most important spring line is the boundary between the coarse-grained, well-jointed and pervious Trifels beds (main Bunter Sandstone) as the aquifer and the impervious clayey lower Bunter Sandstone, which has few joints, as the aquiclude. The groundwater flows mainly along the joints of the Trifels strata and the contact springs emerge where there are concentrations of joints. There is also unconcentrated seepage water from the pores and lesser joints of the aquifer at the spring line.

The headward spring erosion shifts the valley head in the direction of the joints from which the spring flows. The valleys that have developed as a result of this process are oriented in the dominant joint direction. Seepage outflow soaks the sides of the valley heads and accelerates the weathering and denudation at the slope foot. The seepage sapping steepens the head and side slopes of the valley heads above the spring line, causing them to retreat.

The spring niche valley head has a broad flat floor on the impervious rock, surrounded on three sides by the steep slopes of the aquifer. There are three types of spring niche based on the number of dells: a **single-dell spring niche valley head**, a **double-dell spring niche valley head** (Fig. 18.6g and h) and a **semicircular spring niche valley head** with three or more dells (Fig. 18.6i).

The number of dells in a spring niche has a morphogenetic significance. The more rapid the spring erosion in the valley head along the leading joint, the less time is available for lateral widening of the valley head by seepage sapping and for the development of dells joining from the side. The three types of spring niche probably

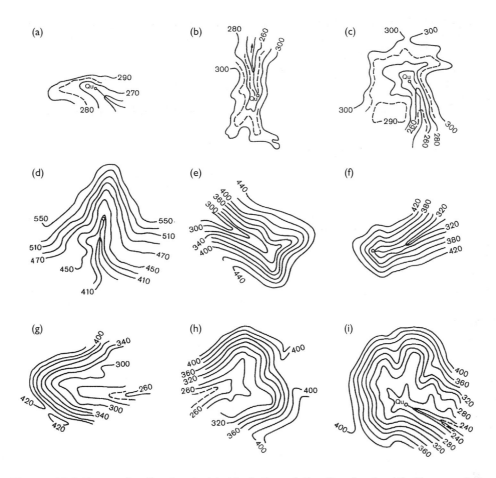

Figure 18.6 Types of valley head: (a)–(c) shallow-dell valley heads; (d)–(f) steep-dell valley heads; (g)–(i) spring niche valley heads (source: Ahnert, 1955, pp. 29, 36).

therefore indicate different rates of headward retreat relative to the rate of widening of the valley head (Ahnert, 1955). Possibly this also explains the differences in the number of dells in shallow-dell and steep-dell valley heads.

Where the climatic conditions favour wash denudation, converging gullies or wash hollows develop in the valley heads instead of dells. The classification system can also be extended to gully valley heads.

PENEPLAINS, PEDIMENTS AND INSELBERGS

On all continents, there are landscapes that have been almost completely levelled by denudative downwearing. Davis (1899) termed them **peneplains**. Their morphogenetic significance had, however, been understood by Powell (1876) even earlier. A peneplain is characterized by low relief, with surface forms that are not significantly influenced by either geological structure or differences in the rocks. Individual hills, known as **inselbergs**, rise above the otherwise subdued relief (Fig. 19.1). These are remnants left after denudation has worn down the surrounding land. The spatial association of peneplains and inselbergs indicates that their development is linked and for this reason they are discussed together in this chapter.

A number of possible explanations for peneplain development have been suggested, some of which are very different from one another. They include:

1. marine abrasion;
2. the end stage of the Davis cycle of fluvially determined landform development;
3. double planation;
4. pedimentation.

FIGURE 19.1 Peneplains with inselbergs near Cubal, Angola (after Jessen, 1936).

19.1 THE DEVELOPMENT OF SURFACES BY MARINE ABRASION

At many places on the cliff coasts of Great Britain, the surf at the foot of the cliffs has formed wide **abrasion platforms** that cut across the rocks and structure. They are produced by the abrading effect of the beach gravels as they are moved forwards and backwards beneath the cliffs (section 23.4.6). Ramsay (1863) postulated that similar forms in Great Britain now lying at higher elevations were originally formed by abrasion at sea level and subsequently uplifted.

Some peneplain surfaces were probably formed in this way, in Cornwall in England at the Lizard and near Land's End, for example (Fig. 19.2), but this explanation cannot be valid for the development of most peneplains for a number of reasons. The possible width of abrasion platforms is limited because the surf loses its erosive power as it runs out on the platform. Platforms that stretch as peneplains in all directions over many kilometres could only have been produced by surf if the sea level rose progressively at a rate that compensated for the energy loss of the surf waves which would otherwise have occurred as the platform width increased. If the rate of sea level rise had been too high, the surf would not have had enough time to abrade a platform at a particular level.

The minimum time period needed for the creation of a surface wide enough to be recognizable as a peneplain would depend also on the

Figure 19.2 Raised marine abrasion surface, Lizard Peninsula, southwest England.

strength of the surf and the resistance of the rock and would, therefore, vary from place to place. Also, peneplains rarely have the smoothness and absence of relief that characterizes the surfaces of abrasion platforms.

19.2 PENEPLAINS AS THE END STAGE OF THE DAVIS CYCLE

In an early qualitative model, the **cycle theory**, Davis (1899) describes the form development that results from the interaction of stream work and denudation on a rapidly uplifted crustal block without prior relief, such as a former sea floor. In Davis's model there is no further uplift as the forms develop. Initially, a simple drainage network follows the gradient and incises valleys between which are remnants of the original surface. Davis designates this stage as **youth**. As the valleys deepen and widen, their slopes intersect at the divides which are also lowered. The relief now no longer increases. The stream long profiles in this middle stage, termed **maturity** by Davis, are more or less adjusted. When valley deepening ceases, the divides continue to be lowered and are reduced to low ridges. According to Davis, this stage of **old age** changes gradually into the final landform of the development sequence, the peneplain.

Soon after its publication there were objections to the theory. One was that uplift in fact continues for a long time and stream erosion would take place already during uplift. Another was that after the initial uplift the crustal movement would not stop for a long enough period for the successive stages of landform development to be completed. It is more likely that in an area of uplift the crustal movement would take place in several phases.

However, it was not Davis's primary intent to describe the development of an actual landscape but to order the morphological developmental stages into a model. He supplemented his theory to take care of these and other objections; in particular, he allowed for successive phases of uplift and for the effects of climatic change such as increased aridity or the advent of a glacial climate (Davis, 1909).

One shortcoming of Davis's theory is that it requires tens of millions of years to lower a mountain to 10 per cent or less of its relief (section 3.4). Given the major tectonic and climatic changes that have been of lasting influence in the Tertiary and Quaternary, there has not been a long enough period, particularly in the mid-latitude areas in which Davis's cycle is primarily meant to be applicable, for the undisturbed development required in the cycle theory to take place. Possibly in earlier phases of earth history less affected by tectonic and climatic changes, peneplains could have developed in the way suggested by Davis. In the Mesozoic and earlier, there were few land plants, so that under similar climatic conditions, denudation processes were probably more effective than in the Tertiary or Quaternary.

19.3 THE DEVELOPMENT OF DENUDATION SURFACES AND INSELBERGS AS A RESULT OF DOUBLE PLANATION

The theory of **double planation** was developed by Büdel (1957) primarily to explain the existence of peneplains in the tropics (Fig. 19.3). He suggested that deep chemical weathering develops a thick regolith cover, at the subsurface of which the bedrock continues to disintegrate. On the surface, wash denudation is the dominating process and when its rate of removal of the

regolith material is less than the maximum possible rate of regolith production as a result of bedrock weathering, the two processes can be in a dynamic equilibrium (section 9.2). In effect, the denudation lowers the land surface at the same rate as the weathering lowers the surface of the underlying bedrock. Both levels are lowered simultaneously and parallel to one another, hence the term double planation. Credner (1931) and Wayland (1934) used the same approach. Thomas (1978; 1994, pp.28–310) has a comprehensive discussion.

The rate of chemical weathering of a homogeneous bedrock is mainly a function of the thickness and moisture content of the overlying regolith. Therefore, of two rocks, differently susceptible to weathering, the more resistant under a thinner regolith cover is weathered at the same rate as a less resistant rock under a thicker cover. For this reason, peneplains can extend over different rocks without the more resistant rock forming higher ground (Ahnert, 1987d, p.12f.).

Differences in the resistance of the underlying rock need not be petrographic. They can also be caused by differences in the spacing of joints. Narrowly spaced joints, for example, favour chemical weathering because they allow water to penetrate the rock mass more easily.

FIGURE 19.3 Peneplain in the Highlands of Kenya.

When the wash denudation rate exceeds the maximum possible weathering rate of the more resistant rock, this will eventually be exposed at the surface, initially as a low bedrock shield. The chemical weathering rate of the rock at the surface is considerably less than the rate under the regolith cover (section 7.2.3).

Beneath the regolith, the neighbouring less resistant rock continues to be lowered and the regolith cover is also lowered by wash denudation. The bedrock shield at the surface is, however, lowered more slowly and gradually forms a **shield inselberg** that rises above the surrounding area. Further lowering around the inselberg increases its relative elevation. This type of inselberg, which has a rounded summit and steep rocky flanks, is also known as a **bornhardt** after W. Bornhardt (1900), one of the first to publish on geomorphology in East Africa.

Tors develop in a similar way to bornhardts. They are usually formed of plutonic rocks such as granite in which there are perpendicular tectonic and horizontal pressure release joints. The tors were originally resistant remnants preserved in the regolith. When this was washed away, the free-standing tors remained (Linton, 1955). They are not limited to the tropics. In the granite areas of Dartmoor and Bodmin Moor in England, they are a conspicuous component of the landscape (Fig. 19.4).

19.4 PEDIMENTATION

A **pediment** in geomorphology is a gently inclined surface formed by either erosion or denudation and joined at its upper boundary to a steeper backslope by a sharp convex break of slope, the **pediment angle**. In general, the pediment has a slope of 7° or less at this boundary and the backslope an angle of 20° or more. The pediment becomes flatter downslope and usually merges into a sedimentary basin or a valley bottom.

Pediments occur very frequently in the semi-arid and subhumid tropics and subtropics but are also found in the arid tropics (Mensching, 1978), in western Argentina (Stingl et al., 1983), in Central Asia and in the Arctic periglacial climate (Schunke, 1988). They were first investigated in the fault block region of the Great Basin in the southwestern United States (Bryan, 1923). The numerous small mountain ranges in this area are tectonically uplifted crustal blocks, on whose margins pediments have developed. Between the

FIGURE 19.4 Hound Tor, Dartmoor, southwest England. Note the vertical tectonic joints and the horizontal pressure release joints.

uplifted blocks there are downfaulted basins filled with sediment which generally have no drainage outlets at the present time. The sedimentation area in this type of basin is known as a **playa** in North America; the transition area between pediment and playa is a **bajada**. The **kewir** in Iran is similar to the playa.

Because of the climate, the streams that reach the basin from the surrounding mountains flow only intermittently. They erode into the mountain spurs that lie between their valley exits and undercut the spur slopes by **lateral planation**. The spur slopes retreat and a continuous pediment forms along the foot of the mountain range which extends across marginal faults and causes the edge of the mountains to retreat further. Eventually the morphological edge of the mountains no longer lies along the original tectonic boundary between mountain front and basin.

The pediment does not extend itself at the same rate everywhere. If two pediments reach a divide from opposite sides, the gap they form in the mountain is a **pediment pass**. The development of several pediment passes may result in a range being separated into individual mountains or inselbergs. The development of this type of inselberg is quite different from shield inselbergs resulting from double planation. In the late stages of development, the pediments reach the

divide in a broad front so that only a few inselbergs remain (Fig. 19.5); the result is a **pediplain** sloping at a low angle from both sides of the divide to the neighbouring basins. Inselbergs occur on the pediments away from the divide if the lateral planation by the streams has left them as isolated remnants of former mountain spurs.

In some regions, pediments are in themselves complex features with several terrace like levels. These indicate base level changes which could be due to eksystemic changes that influence the pediment development such as tectonic movements or a series of climatic changes, or to ensystemic factors. The multilevel pediments of the Henry Mountains in Utah resulted from the ensystemic adjustment of the local drainage sytems to several resistant sandstone layers which acted as local temporary base levels (Schmidt, 1992).

In East Africa, the presence of rock shields and small shield inselbergs on pediments indicates that double planation has taken place there. It would seem, however, this was a secondary phase of development which occurred after the formation of the pediments (Ahnert, 1982, p.59).

The large inselbergs of the Machakos Highlands in Kenya are surrounded on all sides by

Figure 19.5 Two inselbergs in Rajasthan, India.

FIGURE 19.6 Inselberg Kyumbi near Machakos, Kenya with clear development of a pediment.

pediments and are being consumed by the pediments' retreat (Fig. 19.6). **Valley side pediments** have also been observed in Kenya which extend along the valleys and cause the slopes on both sides of the valley to retreat. The progressive development of valley side pediments in neighbouring valleys can lead to the eventual removal of the intervening ridge and to the formation of an integrated surface similar to a pediplain.

The difference in appearance between pediplains and peneplains formed by other processes is small, particularly in the late phases of pediplanation. L.C. King (1953) suggested that many peneplains result from the combining of pediments following pediplanation. Since pediment development is more rapid and more effective than other types of peneplain formation, this thesis may well apply in some areas. Research has to show yet how far it is valid.

19.5 PIEDMONT BENCHLANDS, ZONAL AND AZONAL INSELBERGS

In Kenya, the land surface rising in large steps from the Indian Ocean to the highland around Machakos and Nairobi consists of three planation levels at about 500 m, 1000 m and 1500 m.

Each level has been formed by denudation and is, therefore, a peneplain. Together with the **piedmont scarps** that separate them, they form a **piedmont benchland** (Fig. 19.7a).

W. Penck (1924, pp.165–186) identified and attempted to explain piedmont benchlands around the Fichtelgebirge in Germany. He termed them piedmont steps (**Piedmonttreppen**). There are several possible causes for their development. The most likely is a multiphase uplift interrupted by long pauses, during each of which a peneplain was formed that was adjusted to the new base level. The peneplain developed in the previous pause was raised during the intervening phase of uplift and incised on its outer margin by fluvial erosion.

The Kenyan piedmont benchland is an example of this type of development. A piedmont scarp does not form a sharp edge between the upper and lower peneplain, instead the slope rises for several kilometres. Because of its greater regional gradient, compared to the adjoining surfaces, the scarp is incised by streams whose local base level is the lower one of the two peneplains. Valley side pedimentation takes place in the valleys. At the valley exits onto the lower surface, the valley side pediments are older and wider than further up the valley and the ridges between the valleys narrower and lower because of the intersection of their slopes. Further up the valley where the valley side pediments are younger and narrower, remnants of the higher peneplain lie on the ridges between the valleys. Still further up the valley, valley side pediments do not yet exist, the valley cross-section is V-shaped and the gradient steep.

This form succession indicates that the key to the explanation of the piedmont benchlands is to be found in the piedmont scarps. At the exits of the piedmont scarp valleys, the most recently developed parts of the lower peneplain are created by the widening of the valley side pediments and the removal of intervalley spurs. The upper, steeper stretches of the piedmont scarp valleys shift by headward erosion into the upper peneplain and dissect its edge. This process causes the piedmont scarp to retreat as a whole. At the foot of the scarp, the lower peneplain increases in area and at the upper edge the higher peneplain loses terrain because of the headward erosion of the piedmont scarp valleys.

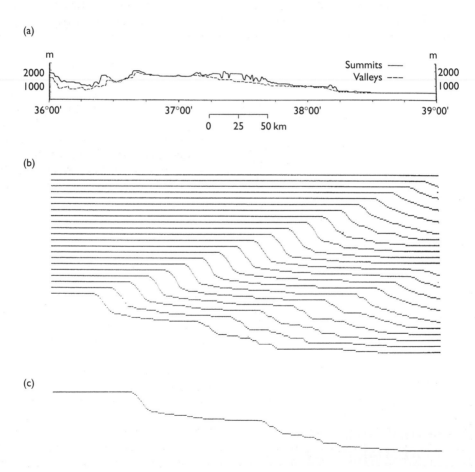

FIGURE 19.7 (a) Projected profile of a 20 km wide west-to-east strip of the piedmont benchland in southern Kenya, along 1°37.5'S (after Ahnert, 1982). (b) Model (program SLOP3D) of the profile development of a piedmont benchland with metachronous peneplains. The individual profiles have been separated vertically for clarity. (c) Individual profile from (b) for comparison with (a).

Of these three peneplain levels in Kenya, the lowest is extending at the expense of the middle level. The middle level is being reduced by the headward dissection of the piedmont scarp but is gaining in surface area on its upper, inner edge at the foot of the retreating next highest piedmont scarp. Only the uppermost peneplain, located above the upper piedmont scarp, is being consumed progressively.

The higher levels of piedmont benchlands have generally been thought to be older than those at lower levels. In fact, each level is older on its outer margins than in its interior segments. Also, areas on different levels are of the same age. Thornes and Brunsden (1977, p.30) have termed these levels **metachronous surfaces** because, at most, the beginning of the development of such a surface can be dated. Its different parts will have been formed at different times after this date (Fig. 19.7b and e).

The dissection of the piedmont scarp and the related valley side pedimentation takes place along each valley that lies in the direction of the regional gradient of the piedmont scarp and also along their side valleys. As a result parts of the intervalley divides become separated from the higher peneplain, leaving isolated hill summits surrounded by valleys or low saddles. As the valley side pediments of these valleys extend, the slopes of the isolated hills retreat and the hills become inselbergs. This separation process is active simultaneously at various places in the area of the piedmont scarp and is part of the retreat process of the scarp.

The inselbergs lying in front of the continuous edge of the piedmont scarp are termed **zonal inselbergs**. They are the separated remnants of the higher peneplain level. Their height and areal extent are gradually reduced by the extension of the surrounding pediment and the denudation of their summits and slopes.

A zonal arrangement of inselbergs can also develop after a long period of pedimentation if the inselbergs are the remnants of a former fault block mountain range. Another cause are zones of more resistant rock which have been worn down less than the weaker rocks in their vicinity. This type of inselberg occurs on the Appalachian Piedmont in the eastern United States and is sometimes termed a **monadnock** after Mt. Monadnock in New Hampshire, although Mt. Monadnock is in fact not made of a particularly resistant rock.

Azonal inselbergs occur over a wide area and have no apparent order. They are mainly present where there has been differentiated deep weathering and double planation. Thomas (1994, pp. 322–340) has summarized the various forms of inselberg development and their relation to planation surface formation processes.

19.6 CRITERIA TO DISTINGUISH PENEPLAINS AND PSEUDO-PENEPLAINS

Peneplains have been recognized in the uplands of Germany at least since Stickel's (1927) research and investigations. At the present time these peneplains lie above the regional base level and are older than the valleys incised into them. The incision of the valleys took place largely in the Quaternary and it is generally thought that the peneplains are Tertiary in age (Fig. 19.8). The climate, particularly of the earlier and middle Tertiary, was warmer than at the present time. Many authors consider the conditions to have been similar to the climates of the present day tropical and subtropical savanna and that the processes forming peneplains were similar to those prevailing today in these types of climates.

This generalization may not be entirely valid. A tropical or subtropical climate cannot be automatically equated with denudation surface development. Rather, valley formation takes place in all parts of the tropics where height differences and gradient are sufficient for the rivers to cut down. Bed load as a tool for erosion is less plentiful than in central Europe where glacial and periglacial conditions have left large quantities of coarse rock debris, but it is by no means absent in the tropics, especially where crystalline rocks with resistant quartz dykes are present (Ahnert, 1982, pp.41–4; 1983).

It is therefore likely that in central Europe areas that were considerably higher than their surroundings were incised by streams in the

FIGURE 19.8 Raised Tertiary peneplain near Vossenack in the Eifel, Germany, dissected by Pleistocene valley development.

Tertiary and the extensive formation of planation surfaces was limited mainly to those areas that lay close to their regional base level and were uplifted only after the peneplain had developed. This would explain the presence of large areas of Tertiary peneplains in the Harz, the Rhenish Slate Mountains and the Ore Mountains, and their almost total absence in the crystalline Odenwald (Zienert 1989) where there is evidence of high relief and valley incision during the Tertiary, as there is in the other marginal uplands of the Rhine Graben, such as the southern Palatinate Forest.

There are also ridges of resistant quartzite today that rise several hundred metres above the Tertiary peneplains of the upper Harz and the Rhenish Slate Mountains which were monadnocks when the peneplains were formed. The adjoining Tertiary surfaces at the foot of these ridges were probably large pediments.

In those areas in which Tertiary peneplains had been developed, the Quaternary incision of valleys altered the local gradient direction of their surfaces in many areas, making reconstruction of the prior morphography difficult. It is anyway not possible to differentiate peneplains on the basis of their elevation, particularly if they have been modified by stream incision. Additional criteria, such as a low surface gradient, a

minimum width of the surface at the divide, or the determination of a knick point against the adjoining slope are also not enough. In the German uplands the valley slopes are often convex so that the intervalley divides are mostly flat and often wide. Also, knick points in the gradient in the upper valley slopes may be the upper edge of old landslide scars within the convex slope profile and are not necessarily an indication of a boundary between an old peneplain and a young valley.

In order to prove that a surface is a peneplain and to determine its age, datable material is needed in addition to morphographic evidence but difficult to find because peneplains are denudational landforms, not landforms of accumulation, and in central Europe they lie high above the valleys. Fossil soils can be used to indicate age if they are identifiable as such.

Many of the areas that have been called peneplains in the central uplands of Germany include levels that are actually the **upper denudation levels** described in A. Penck's *Morphology of the Earth's Surface* (1894, I, p.365). He wrote 'for any area, the height of the divide above the streams is determined by the distance between the incising streams and the mean slope that results from the combined effect of runoff and mass movement'. In other words, when in an area of uniform rock resistance the streams of a given order are more or less equidistant from one another, their valley slopes have a similar profile and intersect at the divide, it follows that all divides of that stream order lie at about the same height. Consequently it cannot be assumed that if the divides lie at the same height a peneplain is present or, when from stream order to stream order the divide height is different, but within each stream order the height is the same, a piedmont benchland is present.

In areas where rock resistance varies, the divides on the weaker rock are lower than those on the more resistant rock. In this case, too, the different levels are not old remnants of a piedmont benchland but have developed as divides at their particular heights as a result of denudative adjustment of the relief to differing rock resistances. Hack (1960) in his classical study in the Appalachians in the eastern USA, demonstrated this type of form development for different divide levels that were formerly thought to be old peneplains. Römer (1993) has provided similar evidence in an area of different types of crystalline rocks in Sao Paulo State in Brazil.

In tableland and cuesta scarp areas with horizontal strata of differing resistance, the regional denudation is slowed down at the upper boundary of each resistant layer. This slowing encourages the development of a surface at this level, which can occur at any height above the regional base level, and is not the same as a peneplain but a **structurally controlled planation**.

There are still many aspects of peneplains to be explained. Perhaps some entirely new approaches with new methods are needed.

20

STRUCTURALLY CONTROLLED LANDFORMS

The **geological structure** includes the spatial distribution of the rocks, the dip and strike of the strata and also the unconformities, joints, faults and bedding planes. Structurally controlled landforms are those whose shape is clearly dependent on geological structure.

Structures are preserved as long as parts of the rock which makes up the structure are present. A fault, for example, remains as a boundary between an uplifted and a downthrown crustal block after the denudation has removed the fault scarp that was produced by the original crustal movement. Folded strata retain their inclined position after the land surface that was arched up by the folding has been flattened by denudation. The long-term geomorphological effect of many of the structures created by crustal movements lies in the fact that different rocks have been brought into proximity with one another and that variations in their resistance, perviousness and chemical composition determine the differences in the subsequent form development.

In relation to earth history, geological structures last much longer than landforms. For this reason geomorphologists have to be concerned with the past structural developments on the earth and, most importantly, the **orogenies** or phases of mountain building. Depending on its type and orientation, each orogeny left specific traces in the structure of the earth's crust that can influence geomorphological development long after the orogeny occurred.

20.1 FORMS DETERMINED BY JOINTS

20.1.1 Joint systems

Rocks in the earth's crust are subject to tension, compression and shear stress. When the threshold value of a rock's resistance to deformation is exceeded, it reacts either by plastic deformation or by fracture. The most frequent causes of stress in the crust are tectonic thrusting and folding of the rocks, pressure release when an overlying rock mass is removed by denudation, and, in igneous rocks, the contraction that follows cooling.

Only rocks such as clays that are plastic enough to compensate for stress alter the form of their volume without fracturing. Harder, more brittle rocks react by developing cracks and fissures known as **joints**. Joints are arranged in a **joint network** which varies according to the type of rock and the nature of the stress.

Contraction joints develop in lava as it cools and solidifies. The reduction in volume that accompanies cooling is simultaneous over the entire sheet of lava. The rock cannot, however, contract as a whole. Instead, numerous more or less evenly distributed contraction centres develop in the lava sheet. Each centre has its own area of contraction and on the boundaries between these areas the rock tears. If the distances between the contraction centres were equal, their affected areas would, theoretically, be circular but, because circles cannot cover an area without intersecting, ideally a hexagonal network of cracks develops. Of all the regular polygons that can completely cover a surface, hexagons have the most corners and are closest to a circle in form. In principle this also explains the existence, under ideal conditions, of other hexagonal patterns in nature such as stone nets, frost polygons and honeycombs. Many fine-grained thick sheets of basalt consist, therefore, of more or less **hexagonal basalt columns**, with each column having a width of tens of centimetres. Other igneous rocks, such as porphyries, also contract into columns but their structure is usually much less regular.

Pressure release joints are expansion fissures in the rock. The expansion takes place in the direction of the surface from which the rock mass which originally caused the pressure has been removed. The planes of the joint surfaces are, therefore, generally at right angles to the direction of the expansion. This type of joint is discussed in the chapter on weathering processes (section 7.3.6).

Tectonic joints are of far greater importance geomorphologically than pressure release joints or contraction joints. They are the result of compression, tension and shear stress brought about by crustal movements. Two joint directions tend to dominate, one in the direction of pressure and one more or less at right angles to it. If shear stress is present, joints can also develop at an oblique angle to the direction of pressure. There are three types of joint in folded structures: **longitudinal joints** that run parallel to the fold axis, **transversal joints** at right angles to the fold axis and **diagonal joints** at an oblique angle to it.

Tectonic joints lying parallel to one another in the rock are termed **joint sets**. Joint sets oriented in different directions but created by the same tectonic event constitute a **joint system**. Tectonic joints are generally at a steep angle or perpendicular in massive plutonic and metamorphic rocks and at right angles to the bedding planes in folded strata. In the field, the orientation of joints in a joint system is measured on natural outcrops or in quarries with a geological compass and plotted as a joint rose, similar to a wind rose. To show the joint orientation, a half circle is enough. Fig. 20.1a shows a joint rose for the Bunter Sandstone in the Dahner Felsenland in the southern Palatinate in Germany. Although the Dahner

Felsenland has joints in all directions, several maxima are apparent: north, northeast, east northeast, east, northwest and, less strongly, north northwest. Each maximum represents a joint set and the two joint sets lying at right angles can be regarded as a joint system. The joint system with the frequency maxima northeast and northwest corresponds to the northeast orientation of the main tectonic axis in the Dahner Land. Northeast is the direction of the longitudinal joints and northwest that of the transversal joints. Northeast is also the direction of several faults in the region. A second joint system corresponds to the Rhine or north–south orientation with transversal joints in an east–west direction. The north–south Rhine orientation relates to the general north–south direction of the Rhine Graben on the west flank of which the Dahner Land lies. There is a third system of east–northeast-oriented joints, although joints at right angles to it are weakly developed. Philipp (1931) suggested that the east–northeast

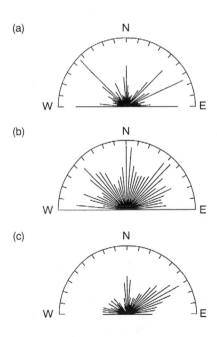

Figure 20.1 Orientation of linear structure and form elements in the Bunter Sandstone of the Dahner Felsenland (southern Palatinate Forest): (a) joints; (b) valleys; (c) compass direction of rock walls and bedrock spurs (after Ahnert, 1955).

direction was an indication of the Saxon tectonics which were active in the Mesozoic in southern Germany.

20.1.2 Joints as a factor in landform development

Joints can determine both the location of a landform and the shape of the form itself. The location of, for example, dolines is determined by joint crossings (section 21.2.3), the shape of bedrock outcrops are determined by joint surfaces and, in limestone caves, the location and orientation of cave tunnels are a function of the joint pattern (section 21.3).

The importance of joints in the orientation of valleys and of rock walls is apparent from Fig. 20.1b and c. In Fig. 20.1b the orientation of valleys in the Dahner Felsenland was determined for 500 m stretches on topographic maps and plotted, also in a semicircular rose because only the orientation of the valleys, not the flow direction of the streams, is of interest here. All possible orientations occur relatively frequently but the four that predominate are also north, northeast, east and northwest.

There are several reasons for the similarity in orientation. Narrowly spaced joint sets create zones of weakness in the rock along which incision can occur more easily than in other directions so that these zones determine the direction of valley downcutting. Because there are six different joint maxima in the Dahner Land, a large number of streams is adjusted to the orientation of the joint maxima. Also, many small valleys in the region have developed as a result of the headward erosion of niche valley heads on contact springs at the boundary between the impervious lower Bunter Sandstone and the acquifer, the main Bunter Sandstone (section 18.4). The contact springs are, in turn, related to narrowly spaced joints that help to determine their direction of flow. The springs erode back along the joints into the aquifer and create a valley oriented in the joint direction.

A comparison of Fig.20.1a and c shows the relationship between the orientation of joints and of rock walls and is another example of the form determining role of joints. Rock walls are formed primarily as the result of block disintegration and rock fall. The joints provide boundary surfaces in the rock from which the rock is removed on one side, leaving a bare joint surface on the other. In the Dahner Felsenland, the rock outcrops are mostly bedrock walls on ridges, bedrock spurs on protruding corners and spurs on the slopes. There are also small bedrock towers on cone-shaped bases. Because of the linearity of the rock walls, several joint maxima in the area, especially north, northeast and east or west, also recur in the orientation of the rock walls.

Other joint-determined forms include the high rock walls of the Alps and other high mountain areas, the dome-shaped granite peaks of the Sierra Nevada in California, which are shaped by pressure release joints, the walls of

FIGURE 20.2 Joints limiting a small outcrop in metamorphic rock in Maryland, USA.

columns of laterally eroded basalt sheets, such as the Palisades on the Hudson River in New York State, and the form of small, naturally occurring outcrops of massive rocks (Fig. 20.2). Even under a cover of soil, joints affect the location of weathering processes and the subsequent denudation. The tors of Dartmoor and Bodmin Moor in southwest England were formerly covered with soil (section 19.3).

20.2 FORMS DETERMINED BY FAULT STRUCTURES

20.2.1 Fault structures

Faults are single or narrowly spaced joints on which horizontal or vertical shear movements take place in the bedrock. They interrupt the former continuity of the bedrock. Fig. 20.3 shows the most important types of fault. In a **transcurrent fault** (c) the two blocks move mainly horizontally along the fault in opposite directions. One of the largest is the 500 km long San Andreas fault on the west coast of the USA. It runs almost parallel to the Californian coast from Tomales Bay northwest of San Francisco, through the city southwards, and ends in the vicinity of Los Angeles where it divides into several transcurrent faults that continue in a southeasterly direction to the Gulf of California. The mean rate of its fault block movement is 6–7 m per 1000 years. This mean value does not indicate much about the frequency and magnitude of the individual movements. The 1906 San Francisco earthquake alone involved a shift of approximately this magnitude. The San Andreas Fault forms the boundary between two large plates that move relative to one another: the Pacific plate in a northwesterly direction and the North American plate in a southeasterly direction.

Faults in which **vertical block movements** dominate are defined according to the angle of the fault plane. In a **normal fault**, the fault plane of the uplifted block is inclined toward the downthrown block. The movement includes a horizontal component, an expansion in which the downthrown block not only slides lower but

also away from the upthrown block (Fig. 20.3a). In a **thrust fault** the fault plane is inclined in the direction of the upthrown block, which means that besides the vertical movement there has also been a horizontal pushing together (Fig. 20.3b). If the fault planes are inclined at a low angle, the thrust fault becomes an **overthrust** and, at an even lower angle, a sheet overthrust or **nappe** (Fig. 20.3d). In the latter, older rocks are often pushed over younger rocks so that the age sequence of the strata is disturbed and the younger rocks are no longer on top. In the Alps, several large nappes have brought strata from a root zone of origin in the southern Alps several tens of kilometres to the northern margin.

A block that has sunk between two faults inclined towards one another forms, together with the flanks of the blocks on either side, a **graben** (Fig. 20.3f, section 4.3.8). The inverse

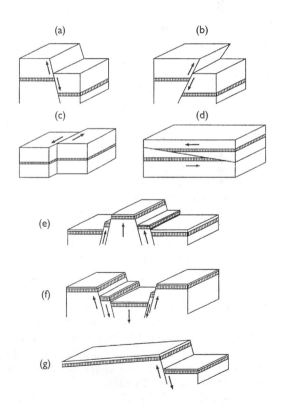

FIGURE 20.3 Types of faulted structures. (a) normal fault; (b) thrust fault; (c) transcurrent fault; (d) overthrust; (e) horst with step faults; (f) graben with step faults; (g) tilted block.

structure, a block uplifted between two faults facing away from one another is termed a **horst**. Graben and horst should be understood as tectonic structural terms that are used even when denudation and deposition have largely eliminated the height differences created by the original block movement. This also applies to a **tilted block**, a block uplifted and inclined along one side of a fault (Fig. 20.3g). **Step faults** are a succession of faults parallel to one another in which the total vertical amount of displacement is distributed over several fault steps (Fig. 20.3e and f).

The vertical movement between the two blocks is termed the **throw of the fault**. The rock on the fault plane is subject to an enormous strain because of the friction between the two blocks. Immediately at the fault plane the rock is, in many cases, broken up, forming a **shatterbelt**. When hardened, the coarse-grained debris of the shatterbelt is known as **fault breccia**, and the fine-grained debris as **mylonite**. The shatterbelt zones are often less resistant to weathering and denudation than the undisturbed adjacent rocks. On fault planes hard rocks develop **slickensides**, seemingly polished, smooth surfaces, usually traversed by micro rills which are, in part, the

direct result of the mechanical friction and, in part, due to recrystallization caused by the differential pressure of the fault movement (Barth *et al.* 1939, p.293).

20.2.2 Fault scarps, fault line scarps and fault block mountains

Both normal and thrust faults produce fault scarps along the fault line (Fig. 20.4). Maximally, their relative height can equal the throw of the fault, although it is usually lower, because during uplift there has been some denudation on the upthrown block and some deposition on the downthrown block. The throw of the main fault on the Rhine Graben near Heidelberg, for example, is about 1000 m. The fault itself is composed of narrowly spaced fault steps and is part of a broader series of step faults whose throw between the southern Odenwald and the lowest part of the graben totals 4500 m (Walter, 1992). Immediately south of Heidelberg the actual fault scarp is about 450 m high with the summit of the scarp at the top of the Königstuhl at 568 m and the foreland of the scarp in the Upper Rhine plain at 115 m. This total height difference includes the fault step of the Gaisberg (375 m

FIGURE 20.4 Young fault scarp with hanging valley on the northern edge of the Nyeboe Land peninsula in northern Greenland on the Arctic Sea. In the background is the continuation of the fault scarp on the Hall Land peninsula.

above sea level) which lies in between. The lowered fault steps below the main fault on the graben flank are covered by Tertiary and Quaternary sediments. From Heidelberg southward the height of the fault scarp declines gradually with the decline in the throw. Near Bruchsal in the transversal tectonic depression of the Kraichgau, the scarp is hardly visible in the landscape. Further south, the scarp rises again continually and reaches its greatest height on the edge of the southern Black Forest.

Fault scarp forms depend largely on the relationship between the rate of uplift and the intensity of the erosion and denudation processes. Downcutting by the streams that cross the scarp on the edge of the graben usually keep pace with the uplift rate and adjust to differences in their gradient on the fault scarp by depositing alluvial fans (section 16.1 and Fig.14.3). Hanging valleys can develop during uplift on small streams whose downcutting rate remains lower than the rate of uplift (Fig. 20.4).

Denudation also takes place on the slopes of the fault scarp between the valley exits but at a considerably slower rate than in the streams. If uplift is relatively rapid, parts of the fault surface lying between the valley exits initially form triangular facets (Fig. 20.5). These are triangular slopes with the apex upwards that have been

little altered by denudation and retain their almost plane slope surfaces. They are also an indication that the crustal movement is very young. Triangular facets are not present if the uplift is slower or has long ceased. In this case, the fault scarp consists of slope spurs between the valley exits which slope down to the line of the fault.

Fault line scarps differ fundamentally from fault scarps. Although they are escarpments located on faults, they are not directly the result of tectonic movement but have developed long after crustal movements have ceased, often after the original fault scarp has been denuded and levelled. Two conditions must be fulfilled for fault line scarps to be developed.

1. Rocks of differing resistance must lie on either side of the fault.
2. The local base level of erosion, usually the level of the main stream, must be lower than the general surface level of both the blocks displaced against one another.

Fig. 20.6 shows schematically the form development of the two types of fault line scarp. After the cessation of uplift, the upthrown block is first worn down to the level of the downthrown block. A peneplain extends across the fault. The local base level is then lowered below the level of

FIGURE 20.5 Triangular facets, Madison Range, Montana, USA.

the peneplain due to an event which occurs outside the region, for example, eustatic lowering of sea level, or isostatic uplift over an extensive area. At the fault block itself there is no renewal of block movement.

In Fig. 20.6a, the rock at the land surface of the upthrown block is more resistant than that of the downthrown block; the latter is, therefore, worn down more rapidly than the former. A new scarp develops along the line of the fault which can be deceptively similar to a fault scarp but is purely denudative and not tectonic in origin. Because, in this case, the new scarp slopes in the same direction as the original fault scarp, it is termed a **resequent fault line scarp**.

In the second case (Fig. 20.6b), the rock at the surface of the downthrown block is more resistant than that of the uplifted block. The more rapid wearing down of the weaker rock leads to the development of an **obsequent fault line scarp** which slopes in the opposite direction to the original scarp. The surface of the tectonically uplifted block is now topographically lower than

the surface of the tectonically lowered block and an **inversion of relief** has occurred.

Horsts and tilted blocks rising above the surrounding area are **fault block mountains**, as distinct from horst and tilted block structures that have been levelled. A graben that lies topographically lower between two uplifted flanks can be termed a graben depression as distinct from grabens that have been levelled or that are now inversions of relief limited by obsequent fault line scarps. The Rhine Graben, which extends 300 km from Frankfurt on the Main to Basel, is the largest graben in Europe.

In central Europe, all the upland blocks, which were originally part of the Palaeozoic Hercynian (Variscan) fold mountains that extended from the Central Massif in France to Silesia in Poland, now form fault block mountains. The original fold mountains were worn down and uplifted as separate blocks from the early Tertiary on. The Harz and the Ore Mountains are tilted blocks, as are the Odenwald, the Black Forest, the Vosges and the Palatinate Forest, which have been uplifted on the edge of the Rhine Graben. They slope away gradually on the sides that face away from the graben. The Thüringer Forest is a horst block mountain.

In the Great Basin in the southwest of the USA between the Colorado Plateau and the very large tilted block of the Sierra Nevada, there are a large number of fault block mountains. The climate is dry and most of the depressions between the individual mountain blocks have no outlet and act as sediment traps for material eroded from the mountains. Between the sediment-filled central areas of the basins, the playas, and the foot of the mountains are pediments which have developed across the boundary faults of the blocks (section 19.4).

20.3 FORMS DETERMINED BY BEDDING STRUCTURES

20.3.1 Bedding structures and form types

With the exception of volcanic lava layers, stratified rocks are of sedimentary origin. Their sediment was deposited more or less horizontally. If

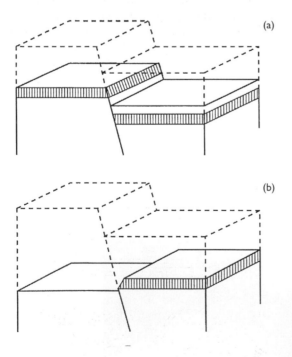

(a)

(b)

FIGURE 20.6 (a) Resequent fault line scarp; (b) obsequent fault line scarp. The original fault scarp is shown with a broken line.

the sedimentation conditions were similar over a large area, such as the continental shelves, the sediments and rocks that developed have the same **facies**, or properties. For this reason, the marine limestones of the upper Jurassic, for example, are very similar from northwest Switzerland, through the Swabian Alb to the Frankenalb. On continents, areas with uniform deposition conditions are usually smaller than those under the sea so that the composition of stratified rocks originating on continents, such as the terrestrial sediments of the Keuper in Europe, is more greatly varied and their facies change more often.

A second characteristic of stratified rocks is the vertical sequence of layers of different composition which result from changes in the deposition conditions. For example, if a sandstone with conglomeratic layers is overlain by a clay, above which is a marl and then a limestone, this would indicate that an initially terrestrial area of deposition had been covered first with shallow water and later with sea water of greater depth.

Differences in deposition over a large area are apparent from a comparison of the Triassic rocks in the Alps and the region to the north. The Trias began 230 million years ago and ended 195 million years ago. During this period, the area of the Alps was a geosyncline, Tethys (section 4.2), a large ocean basin between the Eurasian and the African plates in which environmental conditions were stable for a long time. Very thick deposits of pelagic limestone were laid down which became the rock of the Alpine Triassic. During the Alpine orogeny in the Tertiary, these sediments were folded and overthrust and today they form the mountain ranges of the northern and southern limestone Alps.

North of the limestone Alps, the Triassic is made up of the stratigraphically greatly varying deposits of Bunter Sandstone, Muschelkalk and Keuper – the threefold division that gives the Triassic its name. The Bunter Sandstone is composed almost entirely of fluvial sediments. The Muschelkalk was formed primarily in shallow marine conditions of varying depth and the alternating layers of shale, marl and gypsum, and sandstones of the Keuper were laid down partly in lagoons and tidal flats and partly on coastal plains and deltas. The marine deposits of the Jurassic and, in parts of western and central Europe, the Cretaceous, follow the Keuper.

Tectonic movements disrupted the horizontal, continuous layers of stratified rocks by faulting them into blocks that were uplifted, tilted or lowered, and by upwarping, downwarping and folding. Because folding is caused by the pushing together of the strata, it is often accompanied by fault tectonics and especially by the type of thrusting and overthrusting that occurred during the Alpine folding. The Palaeozoic folded structures that lie below the younger folds at the surface also have complex overthrusts (section 4.3.4).

The inclination of sedimentary rocks by crustal movements is described by their strike and dip. **Strike** is the compass direction in which an inclined layer maintains its altitude. The **dip** is the compass direction in which the layers are inclined and also the angle of inclination of the layer (Fig. 20.7). Strike is also used to indicate the direction of other structural units such as joints, faults, or entire folded mountain ranges.

The strike and dip directions always form a right angle with one another. Therefore, in a northwest striking **anticline**, the beds dip from the axis away to the southwest and northeast. In a **syncline**, the strata dip from both sides to the strike axis (Fig. 20.7b and c).

The position of the strata influences the development of the surface forms when different layers are weathered at different intensities and

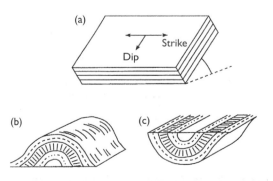

FIGURE 20.7 (a) Strike and dip on inclined beds; (b) anticline; (c) syncline.

removed by denudation, resulting in a rock specific spatial differentiation of form. If a resistant
rock, such as a sandstone or limestone, overlies a
less resistant rock such as clay or marl, and the
boundary between the strata lies at the land
surface, then, depending on the dip of the strata,
different types of landforms develop (Fig. 20.8):

1. with horizontal layers, a **tableland** or, if limited on all sides by steep slopes, a **mesa**;
2. with a continuously low angle of dip, maximally 5–6°, a **cuesta**;
3. with a greater angle of dip, a **hogback**.

Cuesta- and hogback-forming rocks are generally the more resistant rocks. The less resistant
bedrock lies below and outcrops on the lower
slopes.

Schmitthenner (e.g.1954), Blume (e.g.1971,
1976) and K.-H. Schmidt (e.g.1988) have investigated the morphology of cuestas and related
forms. The oldest and still largely valid description of a German cuesta and tableland landscape
is Hettner's monograph (1887) 'The structure
and surface forms in Saxon Switzerland'.

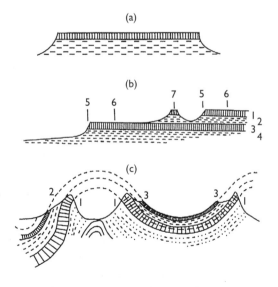

FIGURE 20.8 (a) Tableland or mesa; (b) cuesta scarp:
1 and 3, scarp-forming rock; 2 and 4, less resistant
rock; 5, front slope; 6, dip slope; and 7, outlier; (c)
hogbacks:1, front slope; 2, back slope; 3, back slope
with well developed ramp scarp. Broken lines show
folded structure already removed.

20.3.2 Tablelands

The land surface of a **table mountain** or mesa
forms a level plateau which falls away rapidly at
its margins. Table mountains are remnants of
tablelands that lie at a higher altitude than the
surrounding lowland. Tablelands and table
mountains are more or less symmetrical in cross-
section. The almost level surface of a tableland is
due to the high resistance of the scarp-forming
caprock. The less resistant rock that formerly
overlay the resistant bed has been removed and
the upper boundary of the scarp forming rock
becomes, with minor modifications due to denudation, the land surface of the tableland. The
altitude and form of the summit surface of tablelands are, therefore, largely controlled by structure and develop independently of the height of
the local base level. This distinguishes them from
peneplains.

20.3.3 Form elements of cuesta scarp profiles

The strata of cuesta scarps are inclined and their
form is asymmetric, with a steep **scarp slope**
against the dip (Fig. 20.8b). The upper scarp
slope, located in the scarp-forming rock, is
steeper than the lower part of the slope in the
less resistant rock.

The crest at the upper edge of the scarp slope
is the highest point in the scarp profile and runs
in the general direction of the strike of the strata,
with the exception of local indentations on the
scarp edge.

The **dip slope** of the scarp extends from the
crest at a low angle in the direction of the dip of
the strata. In arid and semi-arid regions with
strong surface wash denudation, the angle is
often the same as the dip angle so that the scarp-
forming rock forms a large part of the dip slope.
In the more humid climates of central Europe,
the angle of the dip slope surface is usually less
than the dip and younger, less resistant rocks
cover the scarp-forming rock at some distance
from the crest. The dip slope resembles an
inclined planation surface but, in contrast to a
planation surface developed on the structurally
independent pediment or a peneplain, the dip
slopes exist only because they are a structurally
integrated landform component of the cuesta.

20.3.4 Conditions for the development of cuestas

A prerequisite for the development of cuestas is that differently resistant strata lie in a slightly inclined position and that the boundary between the resistant and less resistant rocks is exposed at the land surface. They occur when:

1. the strata have been uplifted at a fault and the boundary is exposed on the slope of the fault scarp; or
2. the strata are incised as a result of fluvial erosion and the boundary is exposed on the valley slopes; or
3. the strata are truncated by an old peneplain and a new rock specific differentiation of weathering and denudation takes place on either side of the boundary between the resistant and the weak rock because, for example, of a lowering of the regional base level or a change in climate.

The further development of the scarps does not depend on which of these three causes was effective in the initial stages of formation. Important, however, is the dominance of erosion and denudation processes sensitive enough to react to the difference between the resistant scarp-forming rocks and the less resistant rocks below so that a differentiation of the landform occurs. Glacial processes are not sensitive enough because the ice removes nearly all rocks equally; aeolian processes are not suitable either because the power of the wind is insufficient to create large denudation forms. Only the combined processes of weathering, denudation and stream erosion in the fluvial process response system have the necessary energy and the necessary differentiating capacity to create cuesta scarp forms.

The degree of resistance of the scarp-forming rock is relative because it is significant only in comparison to the less resistant rock. Resistance can be due to:

1. mechanical rock hardness against the direct attack of weathering and denudation processes;
2. high porosity of the rock and the resulting areal perviousness;

3. the jointedness of the rock and the related linear perviousness.

These three conditions can be present either singly or together in any combination. A mechanically hard rock is also well jointed. Because of its low plasticity, it reacts to tectonic and other compression or tension by fissuring rather than by plastic deformation (Fig. 20.9). The attributes of the more resistant scarp-forming rock are most effective when the less resistant rock possesses contrasting characteristics such as mechanical weakness, easy disintegration and low perviousness. Most of the less resistant rocks in scarplands are clays or marls which have these characteristics. Their pores are too small, because of the clay content, for the rocks to be pervious, they are easily deformed plastically and possess few if any joints.

In most cases, perviousness is more important than hardness in determining the resistance of a scarp-forming rock. Water that infiltrates rapidly cannot wear down the surface of the land. A porous sandstone, for example, that abrades and disintegrates easily is not mechanically hard but can nevertheless be resistant to denudation (Fig. 20.10).

Scarp formations develop in homogeneous rock if crusts of calcrete, ferricrete or silcrete (section 7.5.5) create differences in resistance. The crusts then take on the role of the scarp-forming rock. Scarps of this kind occur primarily in the tropics and are known as **homolithic scarps**.

20.3.5 The formation of the scarp slope

Once the boundary between the scarp-forming resistant rock and the less resistant rock below is exposed, whether by tectonic uplift at a fault line or by fluvial downcutting, differentiation by the denudation processes begins and the slope acquires its characteristic scarp form.

On the land surface of the pervious scarp-forming rock, surface runoff is rare. The precipitation seeps down and is stored in the scarp-forming rock in a perched aquifer because the rock below is less pervious and acts as an aquiclude.

Figure 20.9 Rock walls determined by joint surfaces in the Permian deChelly Sandstone in Monument Valley, USA. On the right in the photograph are two small outliers whose form is also determined by joints. The underlying rock, Organ Rock Shale, is a shale with thin sandstone layers, one of which forms a rock surface in the foreground.

Figure 20.10 Cuesta scarp and outlier in the Mesa Verde Sandstone on the Colorado Plateau, USA.

Along the slope, at the boundary of the two rock types, the water reappears as contact springs and seepage water (section 10.3.2). The springs are usually where larger joints, in which groundwater concentrates, intersect with the slope and are exposed. The joints normally reach only to the boundary with the aquiclude which forms the spring horizon. Water also seeps out all along this boundary, moving through the pores and capillaries in the rock.

The area downslope from the boundary becomes wetted so that both the weathering intensity and the transport downslope of regolith produced by weathering increases. As a result, the slope above the spring line in the more resistant rock is undercut and steepened by **spring and seepage sapping** and a concave break in slope develops between the steep upper slope and the flatter lower slope. The cutting back and steepening of the upper slope is greater in the area of the contact springs than on the slopes between. The springs cut back into the scarp rock by means of **headward spring erosion** along the joints from which they flow. The headward retreat of the springs leads to the development of first-order valleys below the springs whose altitude is determined by the strata boundary and direction by the joint direction. These small valleys and the spring valley heads crenulate the scarp edge and define the outline of the cuesta scarp.

The form of the cuesta is therefore developed mainly on the scarp slope. It is here that resistance to denudation is least. Not only is spring sapping and seepage active on the slope but joints also provide surfaces on which weathering can take place. On the dip slope, resistance is much greater because most of the water infiltrates and is not available for denudation. The directional variability of resistance of the scarp-forming rock is of considerable significance for the formation of cuesta scarps. This has been confirmed in theoretical models of their development (Ahnert, 1976a).

20.3.6 Front scarps and back scarps

If a scarp-forming rock has a low angle of dip and the regional base level is lowered, the area at the back of the dip slope, facing away from the scarp crest, can be so deeply incised by fluvial erosion that the boundary between the resistant and weaker rocks is exposed here too. A **back scarp** can then develop that is oriented in the direction of the dip.

A back scarp (Mortensen, 1953) differs from a normal or front scarp (Fig. 20.11). The strongly indented outline of the back scarp indicates that headward erosion by the contact springs is more intensive. Their groundwater catchment is larger than that of the springs on the front scarp because the boundary plane of the aquiclude dips towards the back scarp and produces a groundwater gradient in that direction. The groundwater divide lies closer to the front scarp than to the back scarp (Fig. 10.3).

The back scarp is also more strongly indented than the front scarp in rocks that have a low perviousness. Infiltration is lower in this type of rock but the surface runoff also follows the inclination of the dip slope and, therefore, the dip of the strata. Because of the higher gradient on the back scarp, the surface runoff is largely concentrated linearly, especially along the joints that intersect the land surface and in which the runoff erodes incisions into the back scarp slope that are narrower than the broad valley heads of the contact springs.

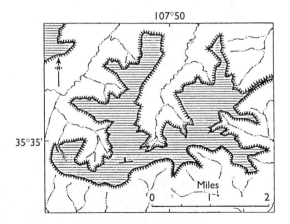

FIGURE 20.11 Front scarp, facing south, and back scarp, facing north, of the Point Lookout Sandstone near Crown Point, New Mexico, USA (after Ahnert, 1960b).

20.3.7 Retreat of cuesta scarps and the formation of residual outliers

Undercutting by springs and seepage steepens the slope of the scarp and, at the same time, causes it to retreat at a very slow rate. The rate is higher if rivers and streams and headward erosion by springs with a large flow contribute to the crenellation by erosion of the scarp edge.

If streams and their tributaries flow across the scarp edge and lower their valleys below the level of the boundary between the resistant and less resistant rock, parts of the scarp-forming rock become separated from their related cuesta scarp to form **residual outliers**. The outliers are tabular in form and an indication of the former extent of the scarp (Fig. 20.8b). Wherever a residual outlier becomes separated, the cuesta scarp retreats in one step in an amount equal to its width.

Once separated, a residual outlier is surrounded on all sides by scarp slopes which continue to retreat because of undercutting by springs and seepage and by denudation. The summit area is gradually reduced in size, it becomes a **butte** and eventually the rock that originally formed the scarp is removed entirely; the underlying less resistant rock remains as a low rounded hill.

Residual outliers are present only where the strata have a very low angle of dip. The steeper the dip, the higher an outlier's caprock lies compared to the cuesta scarp and, because of its greater height, the more rapidly it is worn down. If the angle of dip is 1°, a summit surface of an outlier 5 km from the scarp edge would be 87 m higher than the cuesta scarp's crest. With a dip of 2°, the elevation difference is 175 m and with 5°, 437 m. In the last case, no residual outlier would be preserved.

20.3.8 Cuesta landscapes in Europe, Great Britain and North America

Because of the structural and lithological requirements for their formation, cuesta landscapes occur in areas of sedimentary plateaus (*see* world map of morphostructural regions, Fig. 4.3). In Europe they are well developed in south Germany and in northern France in a cuesta region

that is divided from north to south by the Rhine Graben (Fig. 20.12). East of the graben the cuesta scarplands extend to the upper Danube River in the south, to a line from Regensburg on the Danube to Bayreuth in the east, and partly across the Main River into central Germany. West of the graben lie the cuesta scarplands of the Palatinate and the northern Vosges which form the eastern extremity of the large scarpland region of northern France centred on the Paris Basin.

The scarplands consist of essentially the same geological formations: sandstones, shales and limestones of mainly Mesozoic age. The only significant difference is that in Germany the youngest scarp-forming rocks are mainly the limestones of the Upper Jurassic, while in France Cretaceous and Tertiary strata are also present. The subsidence of the Rhine Graben in the Tertiary was accompanied by an upwarp of the graben flanks which exposed these sedimentary rocks on both sides of the graben, and initiated their progressive differential denudation. A sequence of cuestas developed. The oldest scarp-forming rock formation, the Bunter Sandstone (lower Triassic), was exposed last and lies, therefore, closest to the graben. The youngest in south Germany are the cuestas of the upper Jurassic limestone and in France, a cuesta of Tertiary beds in the Paris Basin. Since the early Tertiary, the scarp of the upper Jurassic limestone in Germany has retreated by up to 180 km from the edge of the graben, or about 3 km per million years.

In the area of Calais and Boulogne a gently warped anticline of Jurassic and Cretaceous rocks is cut off by the Channel coast, but continues in southeast England on the opposite shore as the Weald anticline. The central axis is made up of Jurassic beds flanked on either side by outward dipping Cretaceous beds, with the scarp of the Lower Greensand, prominent mainly on the north side of the anticline, and the chalk scarps of the North Downs and the South Downs on the outer margins of the anticline. Because of the anticlinal structure all these scarps face inward, towards the central axis of the Weald. To the southwest is the syncline of the Hampshire Basin and northward the synclinal London Basin, both in Tertiary beds. North of the London Basin, Cretaceous and Jurassic beds make up the cuestas of the Chilterns (chalk) and of the Cotswolds (limestone).

The lowland areas between the scarps, which include the lower dip slopes of the cuestas, are known as vales in England, such as the Vale of Sussex and Vale of Evesham. Because of their lithology and elevation the lowland areas of cuesta scarp landscapes in Europe have moister soil water regimes than the upland surfaces which has led to a differentiation of land use and influenced the location of settlements.

In North America, cuesta landscapes are in two main areas, the region of little deformed Palaeozoic sedimentary strata of the Interior Low Plateaus with the Cincinnati Arch, in which there are well-developed cuestas around the Nashville Basin and the Bluegrass Basin, and on the greatly varied Colorado Plateau in the western States of Colorado, Utah, Arizona and New Mexico. The sedimentary rocks of the Colorado Plateau are for the most part Mesozoic, although in some areas rivers have also incised their valleys into the underlying Palaeozoic strata. The rates of retreat of cuesta scarps on the Colorado Plateau have been estimated to range from 0.5 to 6.5 km per million years (K.H. Schmidt, 1988,

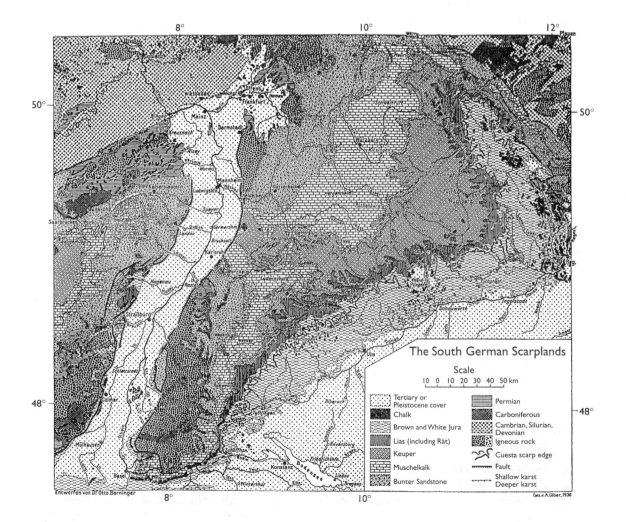

FIGURE 20.12 Geological and morphological map of the south German cuesta scarplands (after Gradmann, 1931, Vol. 1, Table 2).

p.110). The differences in the rates are due mainly to differences in thickness and resistance in the scarp-forming rock.

20.3.9 Denudation terraces

If incision by fluvial erosion takes place on horizontal strata of varying thicknesses, differentiated weathering and denudation related to rock resistance begins on the new valley slopes. The forms that develop are similar to those created in scarplands but are confined to the valley slopes.

The slope segments of the generally jointed and pervious resistant rocks are steepened at their boundaries with the underlying less resistant rocks. The latter are less pervious and are flattened by denudation processes over their entire surfaces. A stepped profile develops with steep segments in the resistant rocks and the levelled slope segments, **denudation terraces**, in the weaker rocks. They should not be confused with river terraces (Chapter 15).

The most important characteristic of a denudation terrace is its direct association with a particular rock stratum. If the strata dip across the valley, then the terraces are higher on one side of the valley than the other. A dip of the strata up or down a valley results in terraces that are inclined in the same direction and have the same angle as the dip.

The Grand Canyon of the Colorado River in Arizona (Fig. 20.13) is a well known example of a valley with numerous denudation terraces. The river has cut down 1800 m through the entire series of Palaeozoic strata, and locally also older sedimentary rocks, to the Precambrian crystalline basement. The resistance of the rocks varies greatly and denudation processes have formed terraces on the canyon sides. Fig. 20.14 shows the evolution of the terrace sequence on one side of the Grand Canyon in a theoretical model. The resistant strata are emphasized by hatching. The vertical erosion of the Colorado River is assumed to have taken place continuously and at a constant rate since the early Tertiary. The rate of

FIGURE 20.13 The Grand Canyon of the Colorado River, USA, showing the denudation terraces.

vertical erosion was probably variable but nothing is known of the variations. The estimates of the geological ages of the phases of development are therefore approximate. The process components of the model are differential weathering, slope steepening by spring and seepage sapping and general wash denudation. As the Colorado River cuts its channel downwards through the strata, resistant beds form steep, cliff-like slopes, while the weak intervening strata develop gentler gradients. It is clear from the sequent phases of development that the terraces are not produced by fluvial erosion but by differential slope denudation.

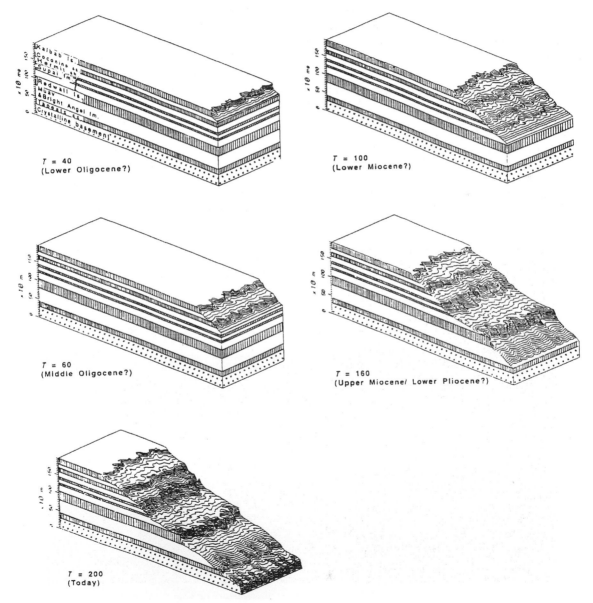

Figure 20.14 Three-dimensional theoretical model of slope development (SLOP3D) of the Grand Canyon of the Colorado River in Arizona, USA. Only one valley side is shown (after Ahnert, 1996).

20.3.10 Hogbacks

Hogbacks result from the differentiated denudation of steeply dipping, folded or tilted strata of varying resistances and are named after the Hog's Back, a ridge of nearly vertically dipping chalk in southern England. In areas of young fold mountains, such as the Swiss Jura, that have been little affected by denudation during the folding process, the structure and landforms are closely related, with the anticlines forming ridges and the synclines longitudinal valleys. A few narrow transversal valleys, **cluses**, have been cut through the ridges in the Jura, but denudation on the ridges has not progressed long enough for hogbacks to have developed.

In older fold mountains, the folded Appalachians in the eastern USA, for example, the crests of the anticlinal folds have been removed and the folds truncated. On their flanks, steeply dipping beds form long ridges oriented in the direction of the strike. Within the axes of the anticlines and synclines the strata are more or less horizontal. As denudation progresses, the less resistant rocks are lowered and become longitudinal valley zones between the more resistant rocks which remain as narrow **hogback** ridges (Fig. 20.15). This type of Appalachian terrain is referred to as ridge and valley topography. Hogbacks on perpendicular strata are rare. The vertical rock walls of the Dakota sandstone in the Garden of the Gods in Colorado Springs in the USA are one example. If the dip is more than 35°, the slopes on both flanks of the hogback have roughly the same inclination, but if it is less than 35°, the side of the hogback ridge that slopes in the direction of the dip, the **back slope**, tends to be less steep than the **front slope**. The mean gradient of the back slope is usually less than the angle of dip of the strata, so that the slope surface cuts across the strata. If some strata are relatively more resistant, small **ramp scarps** form with a scarp slope facing upslope and the dip slope following the gradient (Blume, 1971, pp.79–80; K.-H. Schmidt, 1988, p.78). Erosion channels following the back slope gradient divide the ramp scarps into individual triangular scarp segments, **flatirons**, with the point upwards (Fig. 20.15).

All hogbacks and intermediary forms between cuesta scarps and hogbacks whose dip slope or back slope is inclined more than about 5° are termed **homoclinal ridges**.

The difference in the angle of the back and front slopes can be used to identify geological strata on a topographic map. Figure 17.4 is part

FIGURE 20.15 Hogbacks with ramp scarps (flatirons) in the Canadian Rocky Mountains.

of a topographic map of Pennsylvania which shows two hogbacks, Evitts Mountain and Tussey Mountain separated by a longitudinal valley zone, the Snake Spring Valley. At the southern end of the map, near the gap made by the Raystown Branch, Evitts Mountain has an almost symmetrical cross-section. Further north this becomes increasingly asymmetric where the crest of the hogback bends to the northwest. The cause is a change in the angle of the dip. In the south the strata are very steep but where the ridge bends, the dip to the west and southwest, towards the neighbouring syncline, Dunning Cove, is less. Evitts Mountain and Tussey Mountain have a mirror image pattern of contours. Since Dunning Cove is a syncline, the Snake Spring Valley must, structurally, be the axis of an anticline.

In Germany there is an area of hogbacks, the Westphalian – Lower Saxon Bergland which lies between the Münster Basin and the Harz foreland. Here Tertiary tectonics, combined with deformation and flow movement (halokinesis) of Zechstein salts, have warped and tilted the overlying varyingly resistant Mesozoic rocks to form a number of anticlines and synclines. Hogbacks have developed on the flanks of these folds. The long narrow ridge of the Teutoburg Forest is a hogback of Cretaceous sandstone, the Wiehen, Weser and Ith hills are of Jurassic sandstone and limestone and the hogbacks of the Harz foreland are formed mainly in Triassic rocks. Where the strata dip less steeply, cuesta scarps have developed (Brunotte, 1978; Brunotte and Garleff, 1989; Spönemann, 1989).

20.3.11 Geometric and morphometric characteristics of cuesta scarps and hogbacks

The horizontal distance between two successive scarps of a cuesta scarp series or between two neighbouring hogbacks whose strata dip in the same direction and at the same angle depends on:

1. the angle of dip of the strata;
2. the total thickness M of the geologically younger scarp-forming rock and the less resistant rock at its foot;

3. the height difference H between the crests of the two scarps or hogbacks. The value H is positive when the crest of the younger, that is the upper scarp or hogback-forming rock lies higher than the older. In the reverse case, H is negative.

Fig. 20.16a shows the geometric relationship between the parameters L, α, M, and H. The following is valid:

$$L = M/\sin\alpha - H/\tan\alpha \qquad (20.1)$$

From this follows that if the crests of both scarps or hogbacks are the same height, i.e. $H = 0$, the distance grows proportionally to the thickness M and decreases with increasing angle of dip. With perpendicular strata ($\alpha = 90°$), $L = M$. With an angle of dip of $\alpha < 90°$, the distance L between the two crests is smaller when $H > 0$ and larger when $H < 0$ than when the crests are of equal height.

Cuesta scarps in particularly resistant rocks are usually higher than the neighbouring scarp that lies in front of them, that is $H > 0$. The distance from the neighbouring scarp is, therefore, smaller than if the crests were of equal

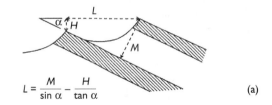

$$L = \frac{M}{\sin\alpha} - \frac{H}{\tan\alpha} \qquad \text{(a)}$$

$H = 0$ (b)

$H > 0$ (c)

$H < 0$ (d)

FIGURE 20.16 The geometry of cuesta scarps and hogbacks.

height (Fig. 20.16b and c). The smaller distance is also an indication of the lower rate of retreat of the higher scarp. Conversely, a cuesta scarp whose crest is lower than the scarp in front retreats more rapidly and lies at a greater distance from it (Fig. 20.16d), an indication also of the relatively lower resistance of its scarp-forming rock. Differences in the resistance of rocks that form scarps are caused by either the properties of the rock or the thickness of the strata.

More than 100 cuesta scarps on the Colorado Plateau in the USA have been investigated using a large number of parameters related to their form and evaluated statistically by K.H. Schmidt (1988, pp.118–44). Among the functional relationships of general interest is the dependence on the properties of the rock and its structure on the frequency and size of the crenulations along the scarp. A crenulation index, similar to the sinuosity index (section 13.3.1) was developed which is the quotient resulting from the measured distance between two points along the indented edge of the scarp and the straight line distance between the same two points. The index shows the difference in the crenulation of the back and front scarps (section 20.3.6) and also that the dip and thickness of the cuesta-forming rock influence the crenulation index. The indices of individual scarps are widely scattered but the size of the crenulation index and the scatter decrease noticeably with increasing angle of dip and increasing thickness of the strata; the lower is the angle of dip of the strata and the less thick is the scarp-forming rock, the greater is the crenulation of the scarp edge. Other factors that may play a role include the mechanical and chemical resistance of the rock, the joint density available for groundwater flow and contact springs as well as regionally variable climatic factors such as the amount and intensity of rainfall.

20.3.12 The development of cuesta scarps in a theoretical model

The development of cuesta scarps takes a long time, usually tens of millions of years, and can be inferred only indirectly by comparing a number of cuesta scarps that are assumed to be at different stages of development. Using this assumption, quantitative theoretical process response models can reconstruct the development process spatially and temporally and be used to test a hypothesis of scarp development based on observation. The model is based on weathering and denudation process equations that have been developed from empirical observation (Chapters 7 and 8; Ahnert, 1996, pp.100–101).

In the model shown in Fig. 20.17, the height and length dimensions and the time unit T are arbitrary. One time unit equals approximately half a million years. In the block diagram of the model, the rocks dip from the front right to the back left at about 5°; the cross-hatched layer is more resistant than those above and below and potentially, therefore, a scarp-forming rock. The rock also contains some lines of vertical joint sets that run in various directions and reduce the resistance of the rock along their paths.

At the right front and back left of the model field, two streams incise at constant and equal rates. Each forms a valley whose deepening determines the relief of the model and, by means of the slope angles that result, advance the denudative processes that develop the forms.

The model includes a weathering process of the bedrock (equation (7.4)) and wash denudation of the regolith (equation (8.16)). It also contains a mechanism for undercutting by springs and seepage at the contact between the scarp-forming rock as the aquifer and the underlying aquiclude.

The land surface is shown by contours. In the beginning phase ($T = 10$) the resistant rock has not yet been reached by erosion. The area in the centre of the model field is almost horizontal.

At time $T = 30$, the two valleys already determine the size and the direction of the slope angle over wide areas of the surface; only the narrow area at the watershed is still flat. Side valleys have developed by headward erosion from the two main valleys on the margin of the model. Their courses depend, in part, on the zones of weakness in the areas of the joint sets. Side valleys in the resistant rock are much narrower and steeper than the valley stretches in the less resistant rock.

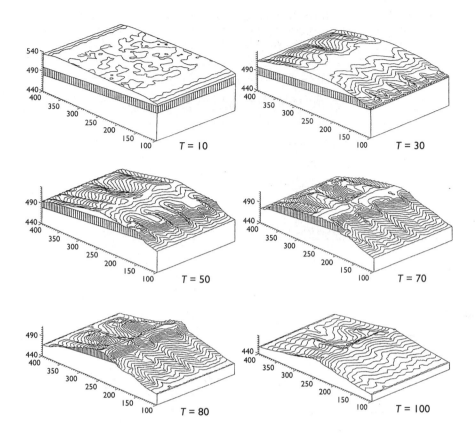

FIGURE 20.17 Theoretical model (program SLOP3D) of the development of a cuesta scarp. The time and length units have been chosen arbitrarily. A model time unit *T* equals about half a million years of earth history.

At time *T* = 50, the front edge of the resistant layer has clearly retreated and developed a scarp that is subdivided by several small steep valleys. At the watershed and the back slope which follows the dip, parts of the upper less resistant strata remain. The valleys on this dip slope are relatively wide and have gentle gradients compared to the valleys on the front scarp side.

At *T* = 70, the stream at the rear of the model field has also cut through the resistant rock. The development of a back scarp begins at this stage.

It is increasingly cut into by the side valleys that flow in the direction of the dip. The front scarp is, however, relatively straight, similar to the scarp in the Point Lookout Sandstone in Fig. 20.11.

At *T* = 80, the front and back scarps form a narrow ridge with several spurs on the back scarp side. At *T* = 100, the scarp-forming rock has been removed completely and a hilly landscape with low relief remains on the less resistant rocks.

21

KARST FORMS

21.1 INTRODUCTION

Karst (Slovenian *kras*, Italian *carso*) is the landscape name for the limestone uplands in western Slovenia. In geomorphology, karst forms are those that develop when weathering and removal by solution dominate. The first comprehensive study of the characteristics and the development of karst was made by Cvijic (1893) in the karst region of Slovenia. Discussions of karst morphology have been published by Jennings (1985) and Ford and Williams (1989). Dreybrodt (1988) has specialized in the physics, chemistry and geology of karst.

The two most important prerequisites for the development of karst are the solubility of the rock and the presence of sufficient water in a fluid state. Karst forms are, therefore, dependent on particular rock types and most developed in climates that are humid all year round. Long dry periods and long periods of frost impede their development.

Karst forms are most frequently developed on limestone, which is chemically a calcium carbonate ($CaCO_3$). The rock is attacked by carbonate weathering, a chemical weathering process (section 7.4.2.4), and is easily soluble. The less soluble **dolomite**, a calcium magnesium carbonate, and **gypsum** (hydrate of calcium sulphate) are other rock types in which karst can develop. The solubility of gypsum increases with increasing temperature to a maximum at 37°C (Jennings, 1985, p.26).

The most soluble rocks are rock salt (NaCl, **halite**) and potash (KCl, **sylvite**). Because of their high solubility, these rocks are not present at the land surface in humid areas and even in arid areas only rarely appear as outcrops of any size. An exception is Mount Sdom, a large salt stock at the southern end of the Dead Sea where the infrequent and very low precipitation has been sufficient for karst to develop.

Another prerequisite for the formation of karst is the perviousness of the rock. There is little or no surface drainage in karst areas. Precipitation infiltrates rapidly through the pores or joints in the rock which it dissolves as it moves down to the groundwater. The water flows out again at karst springs (section 10.3.2) and carries away the rock material in solution. The spatial variability in the intensity of solution at the surface produces the surface karst forms and the solution weathering within the rocks leads to the formation of karst caves.

In order to maintain the perviousness necessary for karst development, the rock must be relatively pure and have a low proportion of insoluble components. These are usually clay or silt particles left on the surface after the soluble components have been transported away. They may form impermeable deposits of residual clay or loam at the surface which fill the underlying joints and limit further karst development.

21.2 KARST SURFACE FORMS

21.2.1 Dry valleys

Dry valleys are linear hollows, eroded originally by a stream or river which no longer has a channel bed. They can develop in several ways. One mode of formation is probably due to the progressive development of the karst itself. If, in the beginning phases, the rock, usually limestone, was relatively dense and not very pervious and the joints in it were so narrow that only small amounts of water could infiltrate, then most of the precipitation would have flowed on the surface and incised the valleys. The water that did infiltrate the rock would, nevertheless, have gradually widened the joints and pores by solution weathering so that the proportion of precipitation seeping directly into the groundwater rather than flowing on the surface increased. Eventually the stream courses fell dry, their channel beds filled as a result of local denudation processes and the valleys became dry valleys (Williams, 1983).

Dry valleys can also develop if the rock is pervious but the streams flow directly on the groundwater surface. In this case, the water table is lowered as the valley is deepened. If the main stream downcuts more rapidly than its tributaries, the water table is lowered at a similar rate, so that when it sinks below the level of the tributary streams these become dry. This type of dry valley development was investigated by Warwick (1964) in the limestone region of the Pennines in England where the perviousness of the tributary stream beds was initially reduced by a layer of claystone through which the tributaries incised before deepening their valleys into the underlying pervious limestone.

In areas that have been dominated by a periglacial climate and permafrost during the Pleistocene cold periods, the frozen ground created a general imperviousness which caused all meltwater and precipitation to flow over the surface. Streams incised their valleys where the gradient and discharge were sufficient for erosion to take place. At the end of the ice age the permafrost thawed and in areas where the material on the surface and the underlying rock were pervious and the water table also lay deep enough, the valleys became dry and their channels disappeared. The dry valleys in the chalk of the South Downs in southern England have probably developed in this way.

Dry valleys are, therefore, almost always an indication that the underlying rock was less pervious in earlier phases of their development. They are not limited to areas of karst. Lowering of the water table following incision by a main stream often occurs in other pervious rocks such as sandstone, in which dry valleys can also form. Dry valleys not associated with karst can also develop when a contact spring (sections 18.4 and 20.3.5) which originally supplied a valley with water erodes headward and thereby progressively reduces its groundwater catchment area and source of supply until it falls dry.

21.2.2 Lapies

Lapies are small forms a few centimetres to a metre in size, caused by solution on the surface of the rock. Depending on the inclination of the surface and the rock's structure, various types of lapies can develop.

Hole lapies are rounded, oval or elongated hollows on the surface, with a width of a few centimeters; their depth can be greater (Fig. 21.1). Their development on bare karst surfaces is initiated in small depressions in the limestone in which rainwater can be temporarily stored and solution take place. Further rainfall washes out the dissolved material. They develop more intensively in rock lying under a soil cover, sometimes forming interconnected series of hollows (Jennings, 1985, p.72).

On bare, inclined, joint-free surfaces, solution by flowing rainwater forms **rill lapies** (Fig. 21.2). The rills are usually a few centimetres wide and

FIGURE 21.1 Hole lapies in the Gran Sasso Massif, Italy.

run more or less parallel to each other in the direction of flow. Some rills come together and combine, similar to a parallel valley network. The rill depth is, in general, only a few centimetres but can be deeper if the size and frequency of flow events and high solubility of the rock favour their development.

Joints on the surface of the rock become guiding lines for flowing water and are, at the same time, infiltration pathways for seepage. Initially,

therefore, a type of rill lapies can form on a jointed surface, although, in this case, the rills follow the line of the joints and not the surface gradient of the rock. Solution by infiltrating water widens the rills until they form broad deep **joint lapies** which direct the rainwater downwards so that all surface runoff ceases.

Joint lapies under a regolith of residual loam are filled with the loam but can continue to develop if it is pervious enough. In a cross-

FIGURE 21.2 Rill lapies on the Stockhorn near Thun, Switzerland.

section of a quarry wall, for example, filled joint lapies are known as **karst chimneys**. True karst chimneys in the form of vertical pipes occur only where two joint lapies intersect and where, therefore, the solution has been particulary intensive. If there is a high density of joints on bare karst, the ridges between the widened joints form sharp crests. In covered karst, the effects of solution are more evenly distributed and the crests are more rounded. Rounded joint crests exposed later by denudation are known as **grikes**.

Hole, rill and joint lapies are the most important types of lapies. Larger rill lapies are also known as channel lapies or meander lapies if the channel meanders. Various exceptional forms exist. If conditions are suitable, lapies can develop in other rock types, particularly in crystalline silicate rocks such as granite. In this case, hydrolysis is the chemical weathering process involved. The forms are known as **silicate karst** (Louis and Fischer, 1979, p.386) and are well developed on the almost horizontal summit surfaces of the granite tors in Dartmoor and Bodmin Moor in southwest England and other granite areas, where they occur as more or less round hollows, several tens of centimetres in diameter and 10–20 cm deep, termed **opferkessel** (Hedges, 1969) or solution pits (Fig. 21.3). They develop similarly to hole lapies on bare rock, where rainwater collects long enough in puddles for intensive chemical weathering to take place, including hydrolysis of the feldspar and oxydation of the iron in the biotite. The dissolved substances, particularly the potassium carbonate from the feldspar and the iron oxide as well as the residual solids, clay and quartz sand, are washed out of the developing hollow by subsequent rains, or blown out by the wind. The organic acids from algae in the water in the hollows intensify the weathering process. Channel lapies have also been found on granite and related crystalline rocks in a number of areas, particularly in the tropics and subtropics (Louis and Fischer, 1979, pp.385, 386). Lapies forms in silicate rocks occur much less often and are usually much less developed than in carbonate rocks, sulphates and salts, although in the Niger Republic karst forms, including lapies, dolines and caves, have been reported in silicate rocks in

FIGURE 21.3 Opferkessel in granite, Dartmoor, England.

very dry areas of the Sahara (Busche and Sponholz, 1988; Sponholz, 1989, 1994).

21.2.3 Dolines and uvalas

Cvijic (1893) used the word **doline** (Slovenian *dolina* = valley) to describe a hollow with a more or less circular form measuring a few to several hundreds of metres across. They are also known as sinks or sink holes of which there are two types: **solution sinks** and **collapse sinks** (Fig.21.4). Solution sinks develop where denudation by solution is particularly strong and the dissolved materials are transported into the rock and down to the groundwater. This occurs mainly at the intersection of joints or joint sets at karst chimneys where there are more internal rock surfaces available for attack by chemical weathering. A slight depression forms on the surface at these points and once the hollow has been created, rainfall from the surrounding area flows into it either as surface runoff or as subcutaneous flow (Williams, 1983), a type of karst interflow, along the boundary between the soil and rock. Solution denudation at the centre of the doline and in the karst chimney below increases and the doline is enlarged and deepened.

Solution dolines range from shallow depressions to deep funnels (Fig. 21.5), depending on the relationship between the rates at which the centre of the doline and the surrounding area are deepened by solution. Centripetal runoff also brings insoluble weathering residue into the doline and deposits it. If this residue material is mostly clay and impervious, a small doline lake may develop. Its depth depends on the height to which the doline sides are sealed by clay. Above the lake surface, the water continues to seep into

the rock. Differences in the height of the water level in neighbouring dolines indicate that the lakes formed independently, unrelated to a common groundwater source. Equal water level heights, for example in karst lakes in Florida, USA, are an indication of their connection with groundwater.

Collapse dolines are created following the collapse of a cave roof. If the original cave was large enough, the doline is, at first, an open hole and only later as the walls fall in, does it acquire its characteristic funnel shape. The sides of collapse dolines tend to be much steeper than solution dolines although as denudation continues they become more rounded and the form differences lessen.

Two or more dolines whose sides intersect to create an elongated depression, often with several deeper areas within it, are termed an **uvala**. Large uvalas are also described as elongated karst depressions.

Dolines, uvalas and elongated karst depressions are located on limestone plateaus and in karst dry valleys whose fluvial development took place much earlier. Once the water table sinks below the valley bottom, depressions begin to form in the valley floor. The gradient of the original valley floor becomes divided into sections, each with its own local base level in which surface runoff flows only a short distance before it disappears in a shallow hole, a **ponor**, in the floor of the depression. The valley becomes a **blind valley** because its drainage no longer has an exit on the surface but continues in a cave system underground (Jennings, 1985, pp.95–7; Pfeffer, 1978, p.33; Leser, 1993, pp.189–90).

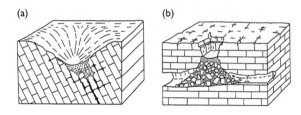

FIGURE 21.4 Doline types: (a) solution doline; (b) collapse doline (after Jennings, 1985, p. 107).

FIGURE 21.5 Two small dolines on the North Island of New Zealand.

21.2.4 Poljes

Poljes are closed depressions with a length and breadth of several kilometres and are the largest of the karst forms (Fig. 21.6). Many poljes in the Dinaric Alps are several tens of kilometres in length. Their floors are generally flat and covered with sediment removed from the surrounding slopes by wash denudation and deposited. The polje slopes in the Dinaric Alps are bare and rocky and only their floors can be used for agriculture. Polje is the word for field in Serbocroat.

Figure 21.6 Small polje (centre of photograph) in the Gran Sasso Massif, Italy.

If a polje is crossed by a stream, it appears on the surface in one or more karst springs (section 10.3.2) and disappears into a **ponor** or swallow hole at a low point on the polje floor. True poljes have no outlet on the surface although the word polje is also used locally in the Dinaric Alps to describe depressions that do have a surface outlet. Cvijic (cited by Jennings, 1985, pp.124) has called these open poljes.

The long axis of the polje generally follows the strike of the geological structure. In young structurally uncomplicated fold mountains such as the Dinaric Alps or the Swiss Jura, poljes are usually synclinal basins. In these structural depressions the solubility of the limestone has encouraged the development of underground drainage. Other poljes are associated with fault structures, such as the high tectonic basin of the Campo Imperatore in the Gran Sasso Massif in the Italian Abruzzi. **Corrosion poljes** have developed following the collapse of solution forms but without significant structural influence (Gerstenhauer, 1977). They are usually smaller than the poljes dependent on structure and may form if several uvalas become interconnected, or as a result of deepening by solution in a large former dry valley.

21.2.5 Polygonal karst, cockpits, cone and tower karst

The distances separating the dolines in a field of dolines depend on several factors. One of them is the location and frequency of joint crossings into which dissolved load can be transported to the groundwater, although a doline does not develop at every joint crossing. A sufficient flow of water must be present for effective solution weathering and transport to take place. The higher the precipitation, the greater is the amount of water per surface unit available and the more structually determined weak points in the rock are potential dolines.

In the karst region of New Zealand, precipitation is high and the dolines larger and deeper than those in central Europe. They also lie so close together that only narrow ridges remain between them. The lack of space prevents them increasing in size and they are, in effect, in each other's way. Instead of the normal circular form of individual dolines, the dolines are polygonal in outline. Their ideal form is hexagonal. Williams (1972) has defined this type of karst as **polygonal karst**.

Polygonal karst actively developing at the present time, and probably during the Quaternary in general, is confined to areas between 45°N and 45°S with more than 1000 mm precipitation. The limitation within these latitudes may be because, towards the poles, processes during the ice age interfered with the karst development (Williams, 1993).

Polygonal karst has also developed in New Guinea and China. The density of polygons varies regionally. The mean diameter in the Sandu karst region in Guanxi Province, China is about 400 m and in the Waitomo area of the North Island of New Zealand, about 80 m (Fig. 21.7). Geological structure is the major determining factor for the differences. The limestones in Guanxi are Palaeozoic, 1000 m thick, have few bedding planes and widely spaced joints. The Waitomo limestones are Oligocene, 100 m thick and have a high joint density (Williams, 1993).

The diameters of **cockpits** are similar to polygonal karst forms but their floors are usually flat, either covered by sediment or formed in the outcropping rock. Unlike dolines, cockpits have been widened and the slopes of the hollows are not funnel-shaped and inwardly concave but composed of several segments curving convexly upwards. Cockpits are more or less star-shaped in outline.

FIGURE 21.7 Polygonal karst on the North Island of New Zealand.

FIGURE 21.8 Cone karst on the island of Bohol, Philippines (photograph by W. Knörzer).

Cockpits develop from dolines in a number of ways. One possible cause is that the floor of the doline reaches the water table of the karst and does not become any deeper but does widen in all directions. The foot of the limestone slopes inside the doline are undercut by solution denudation and steepened. Should the water table, which is anyway very variable in karst areas, sink and the cockpit floor become permanently dry, new small dolines may develop within the cockpit floor.

Sediments deposited in the cockpit can also cause its floor to widen. If, as often occurs in limestone areas, the sediments have a high clay content, the solution is concentrated at the margins of the cockpit floor. The limestone at the foot of the surrounding slopes is then dissolved and the slopes retreat and become steeper. Similarly, cockpits develop if more or less impervious and insoluble rocks, such as claystone and siltstone, lie beneath the limestone layers in which the doline has originally formed are reached.

The widening of the cockpit floors can be regarded as the first step in the development of a **karst plain**. All three of the mechanisms described could contribute to its formation. As the floors of neighbouring cockpits interconnect at their extremities, the level areas join and gradually form a surface. This is similar to the growing together of dolines to form uvalas but differs in so far as they result from a denudative lowering of the crests between the dolines rather than a widening of their floors.

Isolated masses of the former side slopes remain after the cockpit floors are interconnected. Levelling continues around these high forms and causes their slopes to retreat further. The remnant forms are termed **cone karst** or **tower karst** (Fig. 21.8). There is no sharp distinction between cones and towers. Cones usually have a greater surface area in relation to their height so that their slopes are often less steep. A number of regional names are applied to these forms. In Cuba they are known as mogotes (haystacks), in North America as haystacks and in Puerto Rico as pepinos (cucumbers). The Croat word hum is applied to all karst hills on karst plains, not only cones.

The tower karst landscape in Guilin in southern China is the most well-known (Fig. 21.9). The towers rise several hundred metres above the karst plain. The plain itself continues to develop, forming a local denudation level which is slightly above the karst water table and the level of the Lijiang River. The height of the towers is due to the long period of their development, probably reaching into the Tertiary, the great thickness of the limestones, a geological structure containing numerous faults which have favoured vertical movement of water in the rock, and a long period of very slow tectonic uplift (Sweeting, 1990).

Essential to the development of towers is the slower rate of lowering of the divides and summit area of the karst compared to the surrounding lower ground. At the summits only rain

FIGURE 21.9 Tower karst, Guilin, South China (photograph by W. Symader).

falling on these points is available for solution, whereas the lower slopes receive both the local rainfall and the flow from above. In addition, water diverges rapidly from the divides and summits in all directions so that the soils dry out, solution is interrupted and the vegetation is less dense, which also increases the erodibility of the soils in the summit area (Gerstenhauer, 1966; Pfeffer, 1978, pp.95–6).

Cockpits, cones and tower karst develop only if the limestone is pure and its strata are thick. They form in warm and humid climates with at least nine months of rainfall annually. In the middle latitudes they are present only rarely as fossil forms. The Cretaceous cone karst in the Frankenalb in Germany is an example (Pfeffer, 1989). It was thought that these tropical karst forms were the result of very intensive solution weathering in subtropical climates and were, therefore, to be regarded as forms associated with particular climatic morphological conditions (Lehmann, 1954). Not all karst morphologists agree with this interpretation (Sweeting, 1993). For example, the saturation concentration for carbon dioxide (CO_2) which is important for the dissolving of limestone in water, is only half as large at 20°C as at 0°C. On the other hand, precipitation is higher in the humid tropics than in the middle latitudes. Also, the supply of biogenic carbon dioxide from organic decay is higher in the humid tropics than elsewhere, but this additional carbon dioxide has first to be absorbed by the water.

Available solution rates of limestone in comparison to other denudation rates, expressed as a lowering by denudation of the land surface, vary as a function of precipitation, but do not indicate any significant dependence on temperature conditions (Miotke, 1975; Gerstenhauer, 1977; Jennings, 1987, pp.192–94). Cockpit, cone and tower karst are all limited for the most part to regions of warmer climates but the present day warm climates cannot be considered as the decisive factor in their formation.

It is possible that the absence of these forms at higher latitudes may be due to the lack of a long enough period of undisturbed chemical weathering (Corbel, 1959b). The tower karst in China has had an undisturbed period of development of a million years or more (Sweeting, 1990). The glacial and periglacial conditions during the Pleistocene ice age were dominated by mechanical weathering and denudation processes that were destructive of karst development, which perhaps explains why, with the exception of the structurally determined poljes, karst forms are mainly small, lapies and dolines for example, and little developed on the surface in the areas affected by the ice age.

The development of caves and subterranean karst forms did, however, continue during the warm periods of the Pleistocene. Cave-forming processes were interrupted during the cold periods but the caves and their forms remained largely intact.

The single climatic prerequisite for the development of karst forms is, therefore, the availability of sufficient water in liquid form. In the case of cockpit, cone and tower karst, there is the additional factor of past climatic conditions in that the Quaternary climate must have been free of cold phases that would have disturbed their development. Grund (1914) suggested that the formation of dolines, polygon karst and karst plains with cone karst was a continuous development series which formed a **karst cycle**. Quantitative models confirm that this is the most probable development series of karst forms and that it occurs without the need to invoke any particular climatic conditions or change in the climate (Fig.21.10).

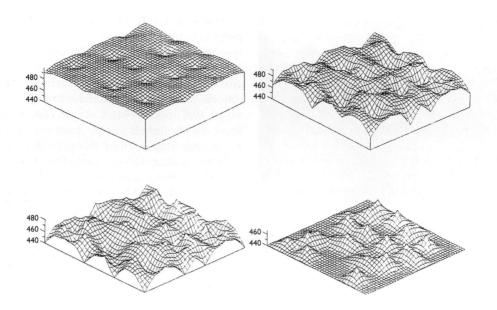

FIGURE 21.10 Dolines, polygonal karst, cone and tower karst as stages of continuous development in a theoretical model (program SLOP3D, after Ahnert, 1994a).

21.3 KARST CAVES

A large part of the chemical weathering in limestone takes place below the earth's surface. Joints and bedding plane contacts provide paths for the water to flow into the rock and are widened by solution to form caves.

The joints in most limestones form a network of fissures more or less perpendicular to the bedding planes and at right angles to one another. Their enlargement produces a right angled network of cave passages. Where large joints cross along which large amounts of water can flow, the intensity of solution is relatively higher and large cave chambers develop (Fig. 21.11). The systems of passages in large karst caves are more than 100km long. Mammoth Cave in Kentucky in the USA, one of the largest cave systems in the world, has a known total length of over 250 km.

The widening of joints to cave passages usually takes place at the level of the karst water table, below which all hollow spaces are filled with water. Above the water level is the **vadose** zone through which the water moves downwards. Below sink holes or other areas where water flows downwards in large quantities, solution activity in the vadose zone can lead to the formation of pipes and shafts (Jennings, 1987, pp.140–1).

With each sufficiently heavy precipitation event, water flows from the surface into the joint

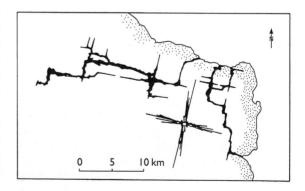

FIGURE 21.11 Joints widened to passages and caves, Fuchsloch cave, Ostalb, Germany. Cross indicates predominant joint directions (after Dreybrodt, 1988, p. 197).

system until it reaches the water table, which rises rapidly. The CO_2-rich rainwater then follows the piezometric gradient downwards via the open joint system and reappears at the surface as a karst spring. Because it is continually supplied with fresh groundwater, the area underground immediately above the long term groundwater level is where most solution takes place. The joint passages gradually become tunnels. Above this level the water does not remain long enough for much solution to occur. Below the permanent dry-weather water table, the groundwater is older and more stagnant, its disposable content of carbon dioxide and, therefore, its activity as a solvent, is lower than in the zone of fluctuating groundwater. As a result cave passage systems tend to be at particular levels underground.

The karst hydraulic surface of the groundwater does not lie at the same level throughout a particular underground karst system. The groundwater in karst is made up of a complex system of communicating pipes with differing pressures that cause spatial variations in the water level. Because of these height variations it is termed the **karst water table** to distinguish it from the normal water table found in other rock types. Differences in height in the karst water table can lead to variations in the level of the cave system.

Apart from local variations, the overall height of the karst water level in the rock depends primarily on the altitude of the streams into which the karst springs of the system drain. Frequently this is a stream that itself flows at the height of the karst water level. The longer the stream maintains its altitude and does not erode downwards, the longer the karst water level remains stable; also, the larger and more complicated the cave system can become that develops at this particular level (Dreybrodt, 1988). Such cave systems are termed **epiphreatic caves** because they develop at the upper boundary of the phreatic groundwater zone.

If the stream cuts down, the karst water level sinks also, together with the zone of groundwater fluctuation. The development of passages and caves at this level then ceases. A pause in the stream's downcutting, when long enough, allows the development of a new system of caves at the new lower position of the karst water level.

Large karst caves are often made up of several stories of caves and passages that are related to earlier pauses in the stream's downcutting. The stories are usually connected by only narrow pipes or shafts at joint intersections because the karst water level did not remain at this level for any length of time. Sometimes the various stories in the cave correspond to different terrace levels in the stream valley and indicate pauses in downcutting. Mammoth Cave has five corresponding cave and terrace stories (Lobeck, 1939, pp.140–1) and the Hölloch cave in the Muota Valley in Switzerland, three (Bögli, 1970). Progressive valley deepening can expose the higher stories of cave systems on the valley sides, providing a relatively easy access to the caves. The Beatus cave on the Lake of Thun in Switzerland is an example.

Karst caves can develop below the permanent karst water level, but at a much slower rate than in the area of groundwater fluctuation. These **phreatic caves** not only show solution forms on their walls and roofs but also traces of erosion as a result of the high piezometric pressure of the water flowing through some parts of the cave. Purely phreatic caves have no stories. If, however, the karst water level sinks because the local stream lowers its bed, a phreatic cave may develop further afterwards in the area of groundwater fluctuation and acquire the characteristics of an epiphreatic cave.

The development of phreatic caves is also encouraged by the process of **corrosion by mixing** (Bögli, 1964; also a discussion by Dreybrodt, 1988, pp.7, 242–3). This takes place mainly in the upper level of the karst water zone where old and new groundwater mix. The temperatures and carbon dioxide contents of the two water bodies are different and therefore also the calcium carbonate content when they are saturated. If both are saturated with calcium carbonate and are then mixed, the mixed water is not saturated and has further carbon dioxide available for solution activity because the functional relationship between the carbon dioxide content and the calcium saturation is not linear; the mixture, however, expresses the linear proportionality of both water bodies (Fig. 21.12).

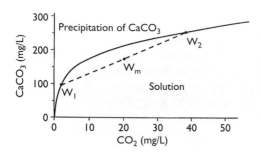

Figure 21.12 Saturation curve and corrosion by mixing (after Bögli, 1964). The mixing of the two saturated carbonate solutions W_1 and W_2 produces the unsaturated carbonate solution W_m so that further solution of calcium carbonate is possible.

In the chambers and passages of karst cave systems there is also deposition. Unstable blocks fall from the roofs of the caves. In extreme cases collapse dolines are formed at the surface following rock falls in the caves below. The Goufre de Padirac in the Causses in southern France is a doline of this type with a round open hole at the surface from which a lift now descends to the cave below. **Cave loam**, a fine-grained deposit of the insoluble remains of the limestone, usually lies on the floors of caves. This also includes sediment brought in by water from karst chimneys and openings at the surface.

The most impressive deposits in caves are the various forms developed as **speleothems**, or **dripstone**, the precipitates of **calcite**, crystalline calcium carbonate ($CaCO_3$). The calcium remains in solution as long as the equilibrium in the water between CO_2 and $CaCO_3$ is present. When the water, saturated with calcium carbonate, seeps through the rock and out on to the roof of the cave in drops, it gives off CO_2 into the air in the cave. As a result, the water cannot hold as much calcium carbonate in solution as before and the surplus precipitates out mineral calcite where the water seeps out on the cave roof. Continual deposition in the same location in the cave produces **stalactites**; these are cone-shaped, like icicles, which thicken and lengthen as the water continues to drip down their sides.

Water also drops on to the floor from the ends of the stalactites. As the drop impacts on the cave floor it is shaken and releases more CO_2 so that calcium carbonate is precipitated out on the cave floor as well. Long-term repetition of this process leads to the development of a steep-sided calcite cone, a **stalagmite**. Stalactites and stalagmites can grow together to form dripstone columns. Curtain-like deposits of dripstone cover the walls of many caves.

The decisive factor in the development of calcite deposits is the transfer of CO_2 from the water to the air in the cave, not any evaporation of the water. Evaporation rates are extremely low because of the high humidity in caves. Dripstone deposits only exist in epiphreatic caves or in cave stories that now lie above the level of active cave development, not in caves filled permanently with water.

22

THE GLACIAL SYSTEM

Glacial landforms (Latin *glacies* = ice) are landforms produced by the work of glaciers and are, therefore, limited to areas that are or have been glaciated. Glacial erosion forms in areas of denudation are related to glacial accumulation forms in areas of deposition. The meltwater streams from the glaciers transport waste material produced by glacial erosion beyond the limits of the ice margin where it is deposited to form **glaciofluvial landforms**.

The glacial geomorphological process response system describes the functional relationships between the glacial and glaciofluvial processes, the material that is eroded, transported and deposited by them, and the landforms that result. **Glacial geomorphology** examines the formation and development of landforms created by glaciers. **Glaciology** is concerned with the study of the glacier itself.

22.1 THE FORMATION AND CHARACTERISTICS OF GLACIER ICE

Glaciers develop in the polar regions and high mountain areas of the earth where, over a period of many years, more snow falls than melts. This continual production of an annual surplus of snow is the only necessary prerequisite for the formation of a glacier. Very cold winters are not essential. On the contrary, during a mild frost the humidity of the air may be higher and result in heavier snowfalls than would occur at lower temperatures. The absolute amount of snowfall is, in effect, less important than the absence of

sufficient heat energy to melt the snow entirely. It could be said that a glacier develops when, each year, in one and the same place, at least one additional snowflake lies unmelted throughout the summer until the onset of the next winter. The qualitative presence of the surplus is more important than the quantity. The latter only determines whether the development of the glacier takes place rapidly or slowly.

The metamorphosis of snow to glacier ice takes place in several stages which last tens of years for Alpine glaciers and about 150–200 years for the Greenland ice (Embleton and King, 1975, pp. 78–9). Freshly fallen snow has a specific density of 0.02–0.08. The first stage in the change from snow to ice is the melting of the tips of the star-shaped snow crystals. The snow becomes granular and the snow mass denser and firmer. The weight of additional snowfall contributes to this change. Granular snow that has accumulated over several years compacts and becomes **firn** with a mean density of 0.5–0.7. The pressure created by subsequent snowfalls causes the firn grains to crystallize into glacier ice with a density of about 0.8–0.9, depending on the proportion of air bubbles enclosed in the ice mass.

The glacier begins to move when its thickness exceeds a specific threshold which is dependent on the quality of the ice and the gradient of the underlying surface. The direction of flow is generally the same as the gradient on the surface of the glacier but may not be the same as the gradient of the land surface below the ice. The type of movement depends largely on whether the temperature of the glacier ice lies near the melting point or considerably below it. The melting point is, in turn, a function of the pressure that exists within the ice and which increases from the surface of the glacier downwards commensurate with the increase in the load from above. The melting point of the ice decreases as the pressure increases at a rate of about 0.007°C per atmosphere (1 atm = 1.033 kgf/cm^2). With a specific ice density of 0.9, the mean pressure increase per 100 m of depth is around 9 kgf/cm^2 and the melting point is lowered by about 0.06°C per 100 m.

Local tension caused by variable movement of the ice produces small spatially and temporally delimited deviations from the mean values for pressure and melting point. In glaciers whose ice temperature lies close to the depth related melting point, local melting and refreezing occurs in areas within the ice largely independently of the air temperature above the glacier. This melting and refreezing, which may cover areas of several millimetres or centimetres of ice, causes changes in the crystal structure and form of the moving ice. In mountain glaciers with a pronounced gradient, the ice pressure is also varied by the downslope directed shear tension that affects all slope processes (section 8.2.1).

In glaciers with a very thick ice mass, the temperature in the deepest parts of the glacier increases because heat from **geothermal energy** is added from the earth's interior. Measurements by Hansen and Langway (cited by Embleton and King, 1975, p. 88) in a bore hole in the Greenland ice cap near Camp Century about 160 km from the edge of the inland ice, indicated an ice temperature at 10 m depth of − 24°C which is similar to the mean annual temperature at the ice surface. The minimum temperature in the bore hole was − 24.6°C. at a depth of 154 m. At 1100 m below the surface it was about − 18°C and at the base of the ice at 1387 m depth, − 13°C. Between 1000 m and 1387 m the rate of temperature increase was approximately constant at about 1.7°C per 100 m. The temperatures at the base of the ice in central Greenland at a depth of 3000 m range between − 9°C and − 10°C (Gundestrup *et al.*, 1993). The pressure melting point at this depth is about − 1.8°C

which means that the ice is too cold to be generally deformed by melting unless lateral pressure from shear movement is also present. In the middle latitudes, the ice temperature at the base of the glacier is often close to the pressure melting point so that the glaciers in these areas are more likely to be deformed.

The deformation and, therefore, the **flow movement** of the glacier is oriented in the direction of the gradient. Displacement within the glacier can also develop along **shear surfaces**. Very cold glaciers, with temperatures well below the pressure melting point, are deformed less easily and tend instead to slide in coherent blocks ('Blockschollen', *see* Embleton and King, 1975, p. 113) along shear surfaces above the underlying rock. The ice near the surface of the glacier also generally moves in blocks. **Crevasses** form on the surface where lateral shearing tension produces variations in the speed of flow within the ice mass. The crevasses do not extend down into the zone of deformation flow in the glacier.

22.2 THE MASS BUDGET OF GLACIERS

The purpose of any budget is to determine, for a given period of time, the changes in the balance that result from additions in income and subtractions due to expenditure. The mass budget of a glacier as a whole is determined by the addition of snowfall and the subtraction by **ablation**, the melting of snow and ice and the flow of meltwater off the ice. Part of the meltwater is refrozen and remains within the glacier and is not part of the external budget but constitutes a component of the internal mass budget of the glacier.

To express the local budget, the thickness C is used to represent the mass of the glacier. The supply, or income, is composed of two components: S, the depth of the local snowfall compacted to glacier ice and A, the ice supplied by the movement of the glacier. The latter can contain meltwater which has refrozen. On the loss side of the balance is M, the melted ice, which also includes losses due to sublimation, and R, the ice removed as a result of the glacier's flow.

The ice thickness C is a state parameter whereas S, M, A and R are rates: quantities added or removed per time unit. The most suitable time unit is a year. In this case, C' is the local ice thickness at the beginning of the year and C the ice thickness at the end of the year. The budget equation is therefore

$$C = C' + S + A - M - R \qquad (22.1)$$

Equation (22.1) has the same basic structure as equation (9.1), the denudative mass budget at a slope point. The substances of ice and regolith and the processes involved are very different, but process-controlled changes in the respective systems can be understood using the same type of budget equation.

Figure. 22.1, showing a mountain glacier flowing from the summit area to the valley, illustrates equation (22.1). At the upper end of the glacier at the mountain crest $A = 0$ and the budget equation here is

$$C = C' + S - M - R \qquad (22.2)$$

At the lower end of the glacier, the ice no longer flows downslope so that here $R = 0$ and

$$C = C' + S + A - M \qquad (22.3)$$

Between these extreme positions all the process components in equation (22.1) normally have a value greater than zero.

In the mountains the temperature decreases with height at a mean rate of about 0.7°C per 100 m. The temperature in a mountain glacier such as the Glacier de Bossons in the Mont Blanc region of the Alps near Chamonix, which extends from a height of 3000 m down to 1200 m,

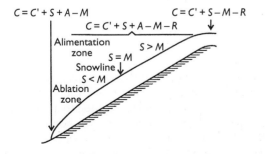

$C = C' + S + A - M$
$C = C' + S - M - R$
$C = C' + S + A - M - R$
Alimentation zone
$S > M$
$S = M$
Snowline
$S < M$
Ablation zone

FIGURE 22.1 The budget zones of a mountain glacier (schematic).

is therefore, considerably lower in its upper section than at its lower end and a much larger proportion of precipitation falls as snow higher up than lower down. At the same time, the amount of heat available to melt the snow is much less in the upper part of the glacier. In the lower section the reverse is true in that less snow falls here than the available quantity of heat could melt.

The upper part of the glacier is an area of surplus snowfall, the **alimentation zone** of the glacier which is covered throughout the year by snow. In the lower part of the glacier there is a deficit, because the entire snowfall is melted and the ice that arrives as a result of the glacier's movement has also been reduced by melting. The area is a **zone of ablation**, recognizable by the snow free surface of the glacier in late summer before the onset of the autumn snowfall. The boundary between the white snow above and the usually grey appearance of the snow-free glacier is the **snowline**. This separates the zones of alimentation and ablation and is the line below which the available heat was sufficient to remove the snow completely. At the snowline, there is a **dynamic equilibrium** between snowfall and potential melting; dynamic in so far as a shifting of the snowline produces a strengthening of those factors that would limit any further shifting. If the snowline moves downslope, it moves into an area of warmer climate and less snow. If it shifts upslope, it comes into an area in which a larger proportion of the precipitation falls as snow but also where less heat energy is available. Figure 22.1 shows the alimentation zone ($S > M$), snowline ($S = M$) and ablation zone ($S < M$).

The **temporary snowline** shifts its position seasonally or even with single snowfall events and sunny days. Its highest position in summer is described as the actual or **orographic snowline**, which varies every year depending on the quantity of snow that falls and the available heat energy. It is higher on slopes exposed to the sun than on those in the shadow, and its location can be affected by differences in the vegetation cover and exposure to prevailing winds. The orographic snowline on glaciers is also known as the **firn line**.

The **climatic snowline** is the mean height of the firn line in a mountain region. It is lowest in the polar regions and rises towards the middle latitudes. In the Alps it lies at about 2500 m to 3500 m above sea level and in the Atlas Mountains of North Africa at about 4200 m. Its maximum height is not at the equator but in the tropical dry belt where solar radiation is particularly high because of the absence of clouds and low precipitation. In the Andes the snowline is at 6750 m on the Tropic of Capricorn and at 5000 m on the equator (Hermes, 1964). Within mountain ranges, the snowline is lower on the windward side with higher precipitation than on the lee side where precipitation is less. On the humid northwestern slopes of the western Alps it lies at 2500 m, compared to 3400 m in the drier central Alps of the Valais (Wilhelm, 1975, p. 104).

The components of the glacier's process response system expressed in equations (22.1)–(22.3) are connected by feedbacks. The more snow falls in the alimentation zone, the thicker is the glacier, and the thicker the glacier, the more mobile it is. The flow of ice from the zone of alimentation to the zone of ablation counteracts further increases in the thickness of the ice in the alimentation zone. The rate of flow of the glacier determines, together with the ice thickness at the snowline and the heat energy available in the ablation area, how far below the snowline the glacier extends. The further down the glacier reaches, the higher are the air temperatures on the surface and the greater is the intensity of the ice melt. The snout of the glacier is located where all the ice that arrives is melted.

If the supply of ice in the ablation area is constant and the temperature relationships remain unchanged, the position of the glacier terminal is also constant. Such a constant position is in fact a **steady state** of the glacial process response system (section 2.2.4) whereby all processes, including the movement of the ice, are still active but in a state of dynamic equilibrium with one another in which neither the mass nor the length, breadth or thickness of the glacier change.

Expressed in equation (22.1), the steady state means that the ice thickness C remains constant, that is, $C = C'$. At the upper end of the glacier, the equilibrium is then identified by

$$S = M + R \qquad (22.4a)$$

and on the glacier as a whole by

$$S + A = M + R \qquad (22.4b)$$

whereby in the alimentation zone $S > M$ and $R > A$ and in the ablation zone $S < M$ and $R < A$. On the snowline, $S = M$ and therefore also $R = A$. At the glacier terminal, where $R = 0$,

$$S + A = M \qquad (22.4c)$$

with $S =$ the addition through snowfall, $M =$ the loss by melting, $A =$ ice supply and $R =$ ice removal as a result of the glacier's movement.

If there is a change in either of the eksystemic factors, snowfall or available heat energy, then the process response system of the glacier reacts by changes in its process rates M, A and R, which result in a change in the size of the glacier. Increased snowfall and/or a lowering of temperature during the summer months cause an increase in the ice thickness, an acceleration in ice flow and relatively less melting. At the glacier's terminal, more ice arrives than can be melted and the glacier moves foward; a **glacial advance** takes place. During the advance the glacier front moves down on to the lower slopes of the mountains, into an area of higher temperature and more intensive melting, until supply and melting are equal and there is a new steady state.

A reduction in the snowfall or an increased melting intensity in the summer months leads to a reduction in the thickness of the glacier ice. The glacier's movement becomes slower and the terminal melts more rapidly than additional ice can be supplied. The end of the glacier moves upslope. This **glacier retreat** is, nevertheless, generally still associated with an ice flow in the direction of the glacier's terminal. In an extreme case, the ice movement can cease if the supply ceases. Such stagnant ice masses are termed **dead ice**.

The retreat of a glacier also leads to the establishment of a new dynamic equilibrium between supply and loss or removal for the glacier. The end of the glacier retreats upslope to areas where the winters are longer and the summer temperatures lower so that the supply of snow increases and melting is reduced. Eventually the position of the glacier terminal stabilizes and the process response system is adapted to the altered process conditions, and there is a new steady state.

Only where the supply conditions of snow and heat energy change progressively in the same direction does the establishment of the steady state of the process components become impossible.

22.3 GLACIER TYPES

Depending on their form, size, position in the terrain and mass budget, glaciers can be subdivided into several types. Only the more important are discussed here. A more detailed typology of glaciers appears in Embleton and King (1975, pp. 84–107).

The **inland ice** of the Antarctic is the largest glacier in the world with an area of more than 12.5×10^6 km². The Greenland inland ice covers 1.7×10^6 km² (Fig. 22.2). Together they account for 96 per cent of the ice covered land surface of the earth. The maximum thickness of the Antarctic inland ice is about 4 km and the Greenland ice about 3 km. The mean thickness of ice in the Antarctic is about 1.8 km and in Greenland about 1.5 km (Flint, 1971).

Because of the low temperatures, the annual precipitation on the inland ice is also low with snowfalls usually about 50 cm per year, the equivalent of 30–50 mm precipitation. The local annual turnover in the mass budget is, therefore, very small, but this is compensated for by the enormous surface area involved. The alimentation area of the ice caps is huge and the speed of the ice movement very low, usually a few metres annually, although at the ice edge, where the ice flow becomes increasingly linear and concentrated along lines of low lying terrain, **outlet glaciers** form (Fig. 22.3) which flow much more rapidly than the ice further inland. Outlet glaciers are similar in form to the **valley glaciers** found in high mountains but have a considerably larger alimentation zone and generally more rapid movement. The Jacobshavn Glacier in western Greenland has the highest recorded speed at more that 24 m per day.

The contrast between the very large reservoir of the slow-moving inland ice and the relatively small, rapidly flowing outlet glaciers is similar to the apparent calm of a lake and the turbulent flow of the stream that flows from it.

On the uplands of northern Greenland, beyond the northern edge of the ice cap, there are small areas of ice known as **plateau glaciers**.

FIGURE 22.2 The edge of the Greenland ice near Thule. The dark stripes on the edge are ablation moraines on shear planes of the ice which has been pushed up in this area.

FIGURE 22.3 The Moltke Glacier, an outlet glacier on the edge of the Greenland ice, calving icebergs in Wolstenholme Fjord, near Thule in northern Greeenland.

They hardly move and their size is determined both by their mass budget and the size of the upland on which they have developed. The Hans Tausen Ice cap is one of the largest with a diameter of about 50 km. Small glaciers flow from some of the plateau glaciers to the neighbouring valley floors.

The Vatnajökull, the largest plateau glacier in Iceland, and the Austfonna Ice Cap in Spitsbergen are both more than 100 km long and 50 km wide. The Jostedalsbreen in Jötunheim in Norway is also of this type.

In high mountains, large areas at high altitudes with little relief on which plateau glaciers

FIGURE 22.4 A cirque glacier in the Ötztal, Austria, in summer. The grey ice surface of the ablation zone can be clearly distinguished from the alimentation zone.

could form are rare. The high precipitation means, however, that only a relatively small alimentation area is necessary for a large glacier to develop.The turnover of the mass budget in high mountains is correspondingly large.

The **cirque glacier** is the simplest and most common type of glacier in high mountains. It begins as a perennial snow accumulation in an existing slope hollow. As the snow accumulates it changes into firn and eventually into glacier ice. The ice moves downslope and has an erosive effect that enlarges and deepens the slope hollow which is known as a **cirque (cwm, corrie)**. As the ice moves away from the back of the cirque, it 'tears' and a characteristic crevasse develops, a **bergschrund**, between the moving ice and the ice frozen to the back slope of the cirque. Cirque glaciers are small. Their alimentation area lies entirely within the cirque and their ablation zone, a small glacier tongue, often extends hardly beyond the cirque itself (Fig. 22.4). Local climatic differences affect the height at which cirques develop. On northerly and northeasterly slopes with limited solar radiation they lie at

lower altitudes than on the sunnier southern and southwestern slopes.

Cirques that have a large surplus of snow produce longer glacier tongues which become **valley glaciers** if they follow the line of an existing valley. Most valley glaciers have formed by the flowing together of ice streams from several cirques. The largest glacier in the Alps, the Aletsch Glacier in the Bernese Oberland (Fig. 22.5), is about 24 km long and has a maximum thickness of 500 m. Other large Alpine glaciers are the Mer de Glace and the Argentière Glacier in the Mont Blanc region, the Gorner Glacier near Zermatt and the Pasterze in the High Tauern in Austria. In general, they move less than 1 m per day.

Some valley glaciers flow across watersheds at low points, **transfluence passes**, into neighbouring valleys and combine with other glaciers to form **ice stream nets**. During the Pleistocene ice age, many Alpine glaciers developed into ice stream nets. The Seefeld Saddle in Tirol in Austria and the St Gotthard pass in Switzerland are transfluence passes that were formed at this

FIGURE 22.5 The Aletsch Glacier in the Bernese Oberland. In the background, from left to right, the Jungfrau, Mönch and Eiger. The dark bands on the glacier are medial moraines.

time. In Alaska, the Juneau Ice Field is one of the largest existing ice stream nets.

Crevasses are formed on the steep stretches of the glacier and along the margins of the ice. Where the ice tears deeply, a jumble of fissures and ice towers, known as seracs, develops (Fig. 22.6). If the glacier flows over a steep rock wall with a height greater than the thickness of the ice, the ice detaches from the upper part of the wall and falls in the form of ice blocks to its foot, where the blocks freeze together again. The regenerated glacier continues to move downslope but with another structure.

Because of the similarity of their elongated form and linear flow and, consequently, the nature of their erosive action, both valley glaciers and outlet glaciers are termed **ice stream glaciers**. A glacier that flows into the sea and whose

FIGURE 22.6 The Tschierva Glacier on the Piz Bernina in the Grisons showing a bergschrund on the upper edge of the glacier and crevasses on the glacier. The ablation zone of the glacier begins at the lower edge of the photograph.

tongue floats is known as a **fjord glacier**. Inland ice moving along a broad front into the sea is described as an **ice shelf**. The Ross Ice Shelf and the Weddell Ice Shelf on the coast of Antarctica are two of the largest.

Ice shelves also develop where there has been continued snow accumulation on the sea ice. They are not related to any form of glaciation on land. An accumulation of this type lies on the north coast of Ellesmere Island in the Canadian Arctic. Occasionally large table-shaped, **ice islands** break off and are carried by ocean currents into the North Polar Sea. They have a longer life than normal sea ice floes and are thicker. Some have been used as a drifting base for research stations.

The mass budget on the edge of the glacial ice shelf and at the end of a fjord glacier is not determined, as it is on land, by the intensity of the melting but by the rate at which the ice breaks off the floating end of the glacier and drifts away as an **iceberg**. The rate of **calving** of the glacier is a function of the ice supply; the size of the iceberg depends on the crevasse density present in the glacier. Antarctic glaciers have few crevasses and produce large table icebergs. The outlet glaciers in Greenland, by contrast, have numerous crevasses and generally produce small icebergs greatly varying in form (Fig.22.3). An exception is the Steensby Glacier, an outlet glacier in north Greenland between Warming Land and Nyboe Land from whose broad disintegrating floating tongue, icebergs break off that are up to 1 km across (Ahnert, 1963b).

In a sufficiently cold climate, valley glaciers may flow out of the valleys onto the flat foreland where, no longer confined, they spread laterally and often combine with neighbouring glaciers as **piedmont glaciers**. In Alaska, ice from the Mt. Elias Range spreads over the Pacific Coastal Plain and covers an area of more than 2640 km^2 to form the Malaspina Glacier. The debris on its outer edges is so thick that firs grow on it.

Piedmont glaciers also existed on the northern and southern borders of the Alps during the Pleistocene. Today, lakes and landforms of glacial deposition on the alpine foreland of Switzerland, southern Germany and in northern Italy indicate the former presence of these glaciers.

22.4 GLACIAL EROSION: PROCESS AND FORM

22.4.1 Detersion and detraction

The pressure of the moving ice erodes the land surface. Loose material is taken up as ground moraine at the base of the glacier and this contributes to the abrading or **scouring** of the bedrock beneath. In this way the rock is smoothed by glacial polishing or **detersion**. Rocks embedded on the underside of the ice erode grooves and small scratches, **glacial striations**, whose orientation on outcrops indicates the direction of ice movement after the ice has disappeared. The rocks in the ground moraine and the rocks beneath the glacier are worn down by friction that produces a fine-grained **glacial rock flour**; transported in streams as suspended load, it is the cause of the greenish white clouding of the water known as **glacier milk**.

Detersion is particularly effective on the **stoss side** or upvalley side of hummocks in the rock floor, where the slope faces the direction of flow. The pressure is less on the **lee side** directed away from the ice movement. If the ice moves fairly rapidly across a bedrock hummock, it may become detached from the floor for a short distance forming a hollow, similar to water flowing over a ledge or at waterfalls.

The typical erosion process on the lee side of hummocks on the rock floor under the ice is the **detraction** of blocky debris that has been produced by local pressure release weathering. Under glaciers whose ice is at or near the pressure melting point, the freezing and remelting of the melt water in joints in the rocks can also produce debris, a process termed **plucking**. The glacier takes up the debris into its basal ice as ground moraine and transports it away.

2.4.2 Rôches moutonnées and bedrock basins

Detersion and detraction together transform elevations in the bedrock under the ice into **rôches moutonnées**. These are asymmetrical rocky humps with a streamlined, rounded and flattened stoss side and an angular lee side limited by joint surfaces (Fig.22.7).

FIGURE 22.7 Rôches moutonnées in Linköping, Sweden. The ice moved from the right to the left side of the area shown in the photograph.

Large numbers of rôches moutonnées occur in the glacially eroded core areas of the Pleistocene inland ice in central and northern Sweden, Finland and Labrador. From their orientation, it is possible to reconstruct the former local direction of ice flow. Very probably many exist under the inland ice of Antarctica and Greenland. In areas of mountain glaciation, rôches moutonnées have developed mostly where movement was moderate, particularly on the transfluence passes of the ice stream nets.

Rock basins are hollowed out by glacial erosion in the bedrock under the inland ice. After the ice melts these basins become **rock basin lakes**. Many were formed in the core areas of the Pleistocene inland ice. In Finland and on the Laurentian Shield in Canada the density is particularly high. Compared to the relatively short life of lakes in regions of glacial deposition, rock basin lakes persist for a long time because very little loose material is available which might be brought in to fill the basins.

22.4.3 Cirques

The typical glacial erosion forms in mountains are products of the various types of glacier. A **cirque** is the result of the spatially differentiated erosion of a cirque glacier. At the bergschrund on the back slope of a cirque glacier, frost weathering is the dominant process. This produces debris which is taken up and removed by the

FIGURE 22.8 Cirques in the Canadian Rocky Mountains.

glacier. Weathering and denudation are more intensive in the area of the bergschrund than on the slope above, which is steepened at its base and retreats. With the progressive deepening of the cirque the same process becomes effective on the sides of the cirque so that it takes on the form of an armchair open on the downslope side (Fig. 22.8). The ice is thickest in the centre of the glacier where it moves most rapidly and erodes most intensively. If the cirque glacier is short, the amount of vertical erosion declines sharply from the centre to the terminal of its tongue; the cirque basin then becomes over deepened and is closed off by a **cirque threshold**, a rock bar in the area of the glacier tongue where there has been much less erosion. Once the ice has disappeared, the over-deepened cirque basin is filled by a **cirque lake**. Most small lakes at high altitudes in the Alps are cirque lakes. Glaciers that produced an over-deepened cirque basin reached only a short distance below the actual snowline. The altitude of such cirques is a useful indication of the location of the snowline at the time of their formation. During the Pleistocene ice age in Europe, cirques were also formed in, for example, the Black Forest, the Vosges, the Bohemian Forest, the Riesengebirge, upland Britain and the Massif Central of France.

The headward expansion of cirques on either side of a mountain ridge narrows it to form an **arête** into which the back walls of the cirque erode. Similarly, spurs can be narrowed by erosion on the lateral slopes of neighbouring cirques. If several cirques lie on the slopes around a summit, the headward erosion of steepened cirque slopes may result in a pyramid-shaped summit or **horn**. The Matterhorn in the Alps is an extreme example, but there are many alpine peaks with this shape which is characteristic for glaciated mountain regions. The angular form of the summits indicates that they stood above the ice during the glaciation period. On the edges of the Greenland and Antarctic inland ice rocky outcrops known as **nunataks** rise above the ice surface. Their flanks are eroded by the movement of the ice. Where the ice moved over the summits during the ice age, as is the case on many of the peaks in the Scottish Highlands, they have a more rounded form.

22.4.4 Glacier troughs

Valley and fjord glaciers flow along valleys that existed in the preglacial landscape. The glacier eroded the V-shaped or flat bottomed valley into a **glacial trough** with a U-shaped cross-section and a floor that lies lower than the river valley floor. The sides of the former valleys have also been widened and steepened by glaciation and the spurs between the entrances of side valleys eroded into triangular facets similar to the triangular facets of young fault scarps (section 21.2.2).

A glacial trough is in fact a glacier bed, not a valley. At the upper edge of the trough is the **trough shoulder**, the area below which the valley was, or is, filled with ice and above which processes of slope denudation have prevailed. The trough shoulder usually appears as a distinct zone in the slope profile lying above the steep slopes of the trough itself. In the Alps, these gentler slopes above the trough shoulder are often used as summer pastures. The high mountain summits lie on the watersheds further back from the troughs of the main valleys and, because of the trough shoulders, are often not visible from the valley floor.

After the ice melts, the glacier bed, or **trough valley** remains (Fig. 22.9). If the floor has been over-deepened in some stretches, rock basins may have formed that are oriented in the direction of the valley and elongated in shape. Filled with water they become **trough basin lakes** but the period of their existence is limited compared

FIGURE 22.9 Trough valley on the Trollstig near Andalsnes, Norway.

to the rock basin lakes in the areas of continental glaciation. Debris transport by denudation from the trough slopes and sedimentation by streams has caused most of the smaller trough basin lakes in the Alps to disappear already. The larger lakes of this type, such as Lake Thun and Lake Brienz in the Bernese Oberland, not only fill out the eroded trough basins but are often additionally dammed by walls of end moraines or by alluvial fans.

Another consequence of the variable deepening of glacial troughs is the development of rocky barriers and valley steps. The barriers lie at right angles to the valley direction and often separate trough basins that have been eroded out upvalley and downvalley from the barrier. A zone of more resistant rock may be the cause for such a barrier to have been less intensively eroded. Today streams cross the barriers in narrow chasms that were usually first eroded subglacially by meltwater streams. Rock basins are often used as locations for reservoir dams. The barrage at Tignes in the Val d'Isére in the French Alps is such a dam.

Valley steps are large breaks in slope in the long profile of trough valleys. They develop in a variety of ways. **Confluence steps** occur when two glaciers of similar volume flowed together resulting in increased vertical erosion. Other valley steps are caused by a change in rock resistance where a weaker rock outcrops downstream. If a confluence and change in resistance occur in the same place, the impact is greater. Many large valley steps were formed at preglacial knick points in the valley which have then been strongly accentuated by glacial erosion (Sölch,1935). Some large trough valleys in the Alps, such as the Inn valley and the Rhône valley in the Valais, have been filled with sediment to the extent that they have become flat-floored glacial trough valleys.

Before glaciation, the side valleys entered the main valley at the same altitude. Glaciation deepened the main valleys more than the side valleys with the result that they have become **hanging valleys** whose streams now fall or flow through narrow gorges into the main valley (Fig. 22.10, section 14.5.3.). On the outer margins of the glaciated area, rock basins are deepened by both ice and subglacial meltwater. After the ice retreated, the basins formed linear glacial lakes. Typical are the Finger Lakes in New York State, west of Syracuse, although some aspects of their development have not been explained completely (Coates, 1968). The foerden, linear embayments on the Baltic coast of Germany, can also be explained by a combination of glacial erosion, subglacial meltwater erosion and moraine deposition.

FIGURE 22.10 U-shaped hanging valley of the Ova dal Vallun near Silvaplana in the Upper Engadine, Switzerland. The alluvial fan on which the village of Silvaplana lies also separates, as a delta, the Silvaplana lake from the Champfer Lake, leaving only a small channel between them.

22.5 GLACIAL DEPOSITS: MATERIALS, PROCESSES AND FORMS

22.5.1 The term moraine

The term moraine refers to the rock debris carried or deposited by the glacier and is applied in at least three different ways:

1. debris moved by the glacier that lies either within the glacier or on its surface;
2. debris deposited by the glacier, which still covers the land surface long after the end of glaciation;
3. the landforms made up of these deposits.

22.5.2 Moraine in and on the glacier

The debris transported by the glacier is produced either by glacial erosion of the rock beneath the glacier or by denudation processes on the slopes that rise above the glacier surface. Material eroded by the glacier is carried primarily at the base of the glacier as **ground moraine**. The debris from the slopes above the glacier is transported as **lateral moraines** along the outer margins of the glacier. Where two valley glaciers of similar size flow together, the lateral moraines on the inner margins of the glaciers come together below the junction to form a debris strip in the direction of flow in the middle of the glacier, a **medial moraine** (Fig. 22.5). Many large valley glaciers have several medial moraines, an indication that several glaciers have come together. The lateral moraines of small glaciers that join large glaciers are often pushed to the side of the glacier bed of the larger glacier and disappear in the latter's lateral moraine rather than forming a medial moraine.

Some of the material from lateral and medial moraines falls into crevasses and becomes enclosed in the ice. Parts of the ground moraine can also move up into the body of the ice as a result of the flow of the ice on shear surfaces. Regardless of its origin, debris transported within the ice is termed **englacial moraine**.

The ice melts continually on the glacier tongue in the ablation zone so that the debris on the ice in this area increases relatively and may eventually cover the glacier surface as an **ablation moraine** (Fig. 22.11). The debris cover at the end of the glacier can become so thick that the underlying glacier ice is no longer visible. Very

FIGURE 22.11 The end of the Lang Glacier in the Lötschental, Switzerland with an ablation moraine on the ice and a meltwater tunnel exit at the glacier snout.

thick ablation moraines are known as **ice core moraines**.

22.5.3 Deposited moraine material

While streams sort transported bed load material according to grain size, a glacier transports its material like a conveyor belt, moving the finest dust and large blocks next to each other at the same rate of movement over the same distance. Moraine material remains, therefore, unsorted both during its transport and after it has been deposited. Stones and blocks in the moraine often show striations that have been caused by friction between them and also with the bedrock beneath the ice.

The general term for unsorted, deposited glacial material is **till**. If it contains large quantities of finer-grained material, including silt and clay, it is termed **till loam**, and **till marl** if there is a calcareous content. A mixture of all grain sizes from boulders to clay particles is known as **boulder clay**. A more general term than till is **drift**; it denotes all glacial deposits whether they are sorted, that is, stratified, such as ice contact deposits, or whether they are unsorted till.

Some moraines are composed only of coarse material. The ground moraine under the Mer de Glace, for example, is made up of large blocks over long stretches. **Erratics** are rocks in the moraine that differ from the bedrock of the deposition area. They are important as source indicators for the moraines of the Pleistocene glaciation. Fragments of crystalline rock from Scandinavia are found in north Germany, for example. If a rock outcrops in only a small part of a source area, it can show the path of glaciation fairly accurately. The largest erratics are several metres across. Some have been used as building material for Megalithic graves in northern Germany.

22.5.4 Moraines as landforms

Once the ice has retreated, the moraine deposits are exposed at the land surface. Their various landforms indicate their position relative to the glacier during deposition and provide information about the shape and ice flow of the now vanished glacier. The investigation and mapping of moraines is an important component of field work in glacial morphology.

The lateral moraines of glaciers lie as debris ridges along both sides of the valley. Young lateral moraines have a sharp crest line from which the moraine slopes outward to the valley side and inward to the valley centre. The inner slope is often steeper because it was formerly supported laterally by the glacier ice and acquired the natural angle of repose of the moraine material only after the ice melted. The outer slope of the moraine which was generally not supported by the ice and had been exposed to denudative processes when the ice was still present, has a lower angle. Differences between the inner and outer slopes of older moraines become less noticeable as denudation obliterates them.

The height of the lateral moraine is a function of the rate of debris accumulation per unit of length on the side of the moving glacier which in turn depends on the rate at which material has been delivered to the glacier from the slope above and the rate of movement at the side of the glacier. The more slowly the glacier moves the more time is available for debris to accumulate on the glacier from a given length of slope above its surface.

The lateral moraine develops as a distinct landform when melting dominates and before the glacier has retreated completely. Melting lowers the level of the ice so that the inner slope of the lateral moraine also rises above the glacier surface. This can be seen in the Alps where all valley glacier tongues have been reduced in size and thickness since the middle of the nineteenth century because of a general warming trend in central Europe.

All material brought by the glacier and deposited in the area of its tongue is the **end moraine**, also termed terminal moraine. Its volume depends on the quantity of debris supplied by the glacier and the length of time the glacier tongue remains stationary. The more debris arrives at the end of the glacier and the longer the glacier remains stationary, the higher is the end moraine.

End moraines also have a ridge-like form, although rarely with a single crest. During a long stationary period, the glacier terminal hardly ever remains in the same position; it advances and retreats, usually over distances of only tens of metres but sufficient to widen the area of the end moraine and give the crest an irregular form.

Because of the oscillations at the end of the glacier, end moraines and ablation moraines are difficult to distinguish from one another. The ice under the debris of a thick ablation moraine is relatively sheltered from the sun's rays so that its melting rate is low and may also vary over short distances, depending on the debris thickness. Parts of the glacier melt away completely while other areas become isolated from the main mass of ice and, because they no longer move, become **dead ice**. When the dead ice melts, the overlying ablation moraine subsides and a **kettle hole** is formed. Small kettle holes are generally round and funnel shaped. Larger ice masses leave more irregularly shaped basins.

At the end of a glacier with a large number of crevasses, blocks of dead ice become separated even with small oscillations of the glacier's terminal, so that the accumulation of the end moraine and the formation of kettles is more or less simultaneous and in proximity. The resulting **knob and kettle** landscape is typical of young end moraine.

The valley and outlet glaciers receive debris from the valley side slopes that rise above them and, since the ice also moves relatively rapidly, they can accumulate high end moraines if the glaciers reach into the foreland. The Pleistocene end moraine of the Inn piedmont glacier forms a semicircular hilly area with a radius of 25 km on the north side of the Rosenheim Basin in Bavaria, the former tongue basin of the glacier (Fig. 22.14b).

The comparatively slow movement of the inland ice cover and plateau glaciers and the almost total absence of debris slopes above their surfaces means that their end moraines have a much lower rate of debris accumulation. The margins of large masses of the inland ice oscillate much less than valley glaciers. Their stationary phases are, however, much longer and locally, high end moraines can accumulate; also contributing to their height is debris that was pushed up in a frozen state from the underlying preglacial sedimentary deposits by the advancing glacier. These are known as **push moraines**. The 164 m high push moraine near Oldenburg in north Germany formed by the inland ice during the last Pleistocene glaciation is a good example.

A series of end moraines, **retreat moraines**, lying one behind the other, reflect several stationary phases of the glacier during a period of overall retreat. The end moraine lying farthest inward is the youngest in age, since a renewed advance of the glacier would destroy the most recently deposited material.

Moraine in the area of the glacier's retreat that forms a more or less irregular pattern of hillocks and hollows is usually described as **ground moraine**, although the material was not necessarily transported at the glacier's base. If a glacier has retreated slowly without lengthy stationary periods, the end moraine, or, more precisely, the ablation moraine on the retreating glacier, accumulates in a similarly irregular form. Once the glacier has retreated or no longer exists, it is not possible to determine from the material or the form whether it originated as ground moraine or ablation moraine. Moraines are an important component in the spatial distribution of glacial landforms in the glacial system (section 22.7).

22.5.5 Drumlins

Drumlins (Irish *druim* = narrow ridge) are streamlined hills, occurring usually in groups, in the deposition areas of glaciers. They are composed of unconsolidated material, often moraine material but can also include fluvial deposits from meltwater. Their outline is oval with the long axis in the direction of ice flow. The long profile is asymmetrical, the steeper side being on the side towards the direction of ice flow. They are generally up to several tens of metres high and up to a few 100 m in length.

Drumlins indicate that previously deposited glacial material, in some cases also fluvioglacial material, has been overrun by a readvance of the glacier. Their streamlined shape is a form of interface between two mobile media, the glacier ice and loose material. Drumlins are generally not evenly distributed over the entire area of the glacial tongue but seem to be limited to those areas where the rate of ice flow is suitable and there is sufficient unconsolidated material available for their formation. In the glacial tongue basin of the Pleistocene Rhine Glacier between Lindau on Lake Constance and Wangen in the Allgäu and in the basin of the Pleistocene Inn-Chiemsee Glacier there are large numbers of drumlins. They are also found in northern Wales,

the Eden Valley and other low-lying areas near the northern Pennines and the Lake District in England, in the Scottish Lowlands and in northern Ireland. In North America they occur in central-western New York, New England and adjacent areas of Canada, Michigan and Wisconsin, as well as in Manitoba and British Columbia (Embleton and King 1975, pp. 416-18).

22.6 GLACIOFLUVIAL PROCESSES, DEPOSITS AND LANDFORMS

22.6.1 Glacial meltwater

Glacial meltwater discharge is a function of temperature. On the glacier surface, the commencement, intensity and duration of melting depends on the air temperature so that there are both annual variations and daily fluctuations in the meltwater flow on the ice. When the meltwater flows into crevasses and perhaps reaches the base of the glacier, the volume, the amount of stored heat in the meltwater and the temperature of the surrounding ice determine whether, along the way, the discharge is refrozen or forms a subglacial meltwater stream under the ice until it flows out of the glacier snout.

On the surface of the glacier, meltwater follows channels which have been melted into the ice by the warmth of the water. Turbulence can lead to small meanders in the channels. Depending on their grain size and the velocity of flow, fragments from the lateral, medial and albation moraines are transported by the current and deposited lower down the glacier, often in crevasses. Isolated occurrences of englacial moraine probably originate as fluvioglacial deposits in crevasses that have been transported further after movement of the glacier closed the crevasse. The temperature at the base of the glacier is similar to the mean annual temperature of the air, or may be slightly higher if influenced by geothermal heat from the earth's interior (section 22.1). The ice insulates the base of the glacier almost completely from the daily fluctuations at the surface. Any influence they may have is indirect, for example, when warmed meltwater from the surface reaches the base through a large crevasse, losing only a small amount of its heat energy on the way.

In the middle and lower latitudes, the mean annual temperature of the air in the ablation zone of glaciers is above freezing and meltwater often flows during the entire year in these areas. Pressure-determined lowering of the melting temperature can also generate flow in some parts of the glacier which, because of spatially changing pressure of the ice, may refreeze again in other areas. Continuous meltwater flow under cold glaciers in summer is possible if the meltwater brings enough heat from the ice surface down into the glacier.

The geomorphological effect of meltwater is, like all flowing water, controlled by the laws of fluvial hydraulics (section 10.5). There is a difference, though, because both on and in the ice, changes in channel beds come about in other ways and other gradient conditions prevail than in a stream bed on rock or gravel. Also, under a glacier, pressure flow can occur which runs against the gradient of the glacier bed.

Rapid movement of the glacier causes subglacial flow of the meltwater to change its position constantly. In ice that moves only slowly or is stationary, the course of the meltwater flow under the ice remains stable and there is time for the meltwater to change the subsurface by erosion or accumulation and also the underside of the glacier itself. Repeated phases of high discharge lead to the melting out of tunnels which end at the glacier snout (Fig. 22.11). A well-developed tunnel exit is an indication of limited ice movement.

Dust particles blown on to the glacier surface are picked up and carried away as suspended load by the meltwater. The glacier flour produced by detersion under the ice also contributes to the suspended load. The bed load in the meltwater consists of coarse moraine particles which initially are entirely angular but become rounded like stream gravels after a longer period of transport under the ice in contact with the bedrock. The dissolved load in the meltwater contains components from various sources including substances that fall from the atmosphere with precipitation, others that were dissolved in the ice, particularly at the base of the glacier, and those resulting from the chemical weathering of the transported rock particles and of the bedrock under the ice (Souchez and Lorrain, 1991, pp. 148–55).

22.6.2 Glaciofluvial landforms: kames, kame terraces and eskers

The Gaelic word **kame**, meaning a steep-sided hill made of unconsolidated material, is used as a term to describe isolated debris deposits under stagnant glacier ice that, after melting, remain in the landscape as small hills. Their formation takes place in various ways. Kames of unsorted material in mountain valleys can originally have been crevasse fillings derived from medial moraines. The material of many kames is sorted, indicating that they are local accumulations of meltwater loads, although in this case the material could also have been deposited in a hollow or crevasse in stagnant ice and have subsided on to the land surface only after melting. Alternatively, the meltwater could have accumulated the debris in a locally limited space under the ice.

Kame terraces do not have much in common with kames. They are not developed under the ice but are glaciofluvial deposits on the sides of valley glaciers in the groove between the glacier and the bordering valley slopes. Streams from the side of the valley flow along the glacier in this groove and the glacier itself contributes meltwater and load from the lateral moraine to the streams. Their existence depends on the glacier having no crevasses near the edge of the glacier into which the water could otherwise disappear.

The sand and gravel in the kame terraces is layered and sorted and their surfaces are smooth. After the glacier melts the deposit remains on the valley side. Because of their composition and form, they can be confused with fluvial accumulation terraces (section 15.2).

Eskers are related to kame terraces. They are long, often sinuous, embankments of sorted layers of sand and gravel. Their form is obviously the result of flowing water but paradoxically they lie higher than the surrounding area. They originate as gravel filling in glacier tunnels where their sides were confined by the tunnel and remain as elevated landforms after the ice melts. The ice in which they accumulated must have been dead ice because movement of the glacier would have destroyed the form. Esker development belongs therefore to the end phase of an ice age. Eskers are often discontinuous. The breaks probably are due to locally increased flow velocity as a result of pressure flow in the subglacial meltwater which prevented deposition.

There are large numbers of eskers in the areas of continental Pleistocene glaciation, some of considerable length. Many Finnish lakes are separated by the sinuous gravel deposits of eskers. The Uppsala esker in Sweden extends 100 km from near Stockholm, northward through Uppsala, whose cathedral is built on it, to Gävle on the Gulf of Bothnia (Fig. 22.12).

22.6.3 Outwash plains and varves

Glacial meltwater which initially flows under the glacier and has sufficient transport power to cross the bordering moraines, carries large quantities of bed load which is deposited as alluvial fans in the glacial foreland. The alluvial fans of neighbouring glacial streams coalesce laterally to form **outwash plains**.

Outwash plains begin in front of their respective end moraines and slope from there away outward. Their gradients and also their grain sizes decline with increasing distance from the end moraine. Pauses in the retreat of a glacier create new outwash plains which partially cover, and in part dissect, the glacial and glaciofluvial deposits of the previous period that lie in front of the new end moraine. Since the younger outwash plain begins glacierward from the earlier one, the older deposits become dissected by the development of the younger. Remnants of the older outwash plain are left as an accumulation terrace in which the accumulation surface of the younger outwash plain is a flat bottomed narrow valley which broadens outwards to a **trumpet-shaped valley** (Troll, 1926, p.170).

The Pleistocene terraces of the Swabian–Bavarian Alpine foreland, for example, were originally outwash plains related to the different positions of the glaciers in the various Pleistocene ice ages. In north Germany a 10–20 km wide outwash plain extends from Flensburg to Schleswig in front of the series of end moraines from the last Pleistocene ice age. To the west, the

FIGURE 22.12 A quarry showing the gravels and sands of the Uppsala esker, in Sweden.

outwash partly covers the ground moraine of the previous ice age. Owing to pauses in the last retreat of the ice, this outwash also became terraced, although the height difference between the different levels is only a few metres (Liedtke, 1989, p.17).

Outwash fans are also deposited by local meltwater streams. They cover small areas on the outer margins of end or lateral moraines and have steeper surface gradients than the large outwash plains (Louis and Fischer 1979, p. 443; Kuhle, 1984).

The rock basins and hollows exposed as the ice retreats in the area of ground moraine become lakes when meltwater flows into them. The bed load of clay and silt in the meltwater is deposited in layers in a cyclical pattern that is the result of a seasonal change in flow. In winter the low discharge deposits a thin clayey dark-coloured layer containing organic material. In summer the discharge is greater and a thicker, coarser grained light-coloured layer is deposited. Together the two layers form an annual deposit a few centimetres thick; their annual repetition appears in cross-section as a series of alternating light and dark layers known as **varve clays** (Fig. 22.13). By comparing the layers in proglacial lakes in Scandinavia formed during the retreat phase of the last Pleistocene ice age, De Geer (1912) estimated an age for this retreat which has

FIGURE 22.13 Varves near Uppsala, Sweden. The pen is about 13 cm long.

since been confirmed using other dating methods.

22.7 THE GLACIAL SEQUENCE

In their important work *The Alps during the ice age*, A. Penck and E. Brückner (1901–1909, p. 16) have shown in a diagram the typical spatial distribution of the landforms of unconsolidated material produced by the ice and its meltwater. It was termed the **glacial sequence** (Fig. 22.14a) and remains a useful graphic model for the spatial pattern of landforms in former piedmont glacier regions. With modification the scheme can also be used to describe the arrangement of forms related to the glacial terminals of the Pleistocene inland ice and of other types of glacier.

The basic pattern is that of a large valley glacier flowing from the mountains, its tongue spreading radially to form a piedmont glacier which lies in an extensive **glacier tongue basin** eroded by the glacier. Ground moraine is present together with a more or less semicircular series of end moraines. Between the centre of the glacial tongue basin and the inner edge of the end moraines lies an area of older end moraines formed into drumlins after being overridden by the ice. Beyond the end moraines are gently sloping outwash plains.

The glacial tongue basin of the Inn Glacier in Bavaria which was mapped by Penck and Brückner (1901–1909) and later by Troll (1924, p. 32), shows this general pattern well (Fig. 22.14b). It is repeated elsewhere in the Swabian–Bavarian Alpine foreland where piedmont glaciers were present during the Pleistocene.

Glacial tongue basins were also eroded in the marginal zone of the inland ice in north Germany. They tend to be connected laterally, in contrast to the more widely spaced basins in the Alpine foreland and the end moraines curve only slightly outwards. The outwash plains also cover a much larger area than those in the south.

A unique form in the glaciated area of north Germany are the **glacial spillways (urstromtäler)**, valley-like forms running along the outer edge of the outwash plains which carried the

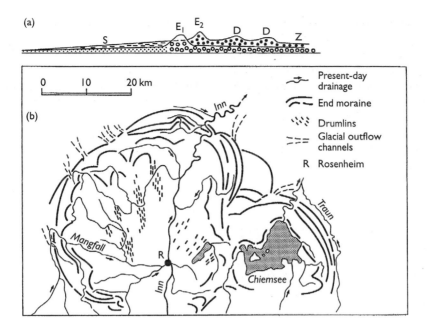

FIGURE 22.14 (a) Schematic long profile of the glacial sequence modified from Penck and Brückner (1901–1909, p. 16). Z: glacier tongue basin; D: drumlins; E_1: end moraine of an older glacier advance (small circles); E_2: end moraine of a younger glacier advance (small dots); S: outwash. (b) The landforms of the Inn-Chiemsee Glacier (based on Troll, 1924).

meltwater streams from the north European inland ice in a westerly direction to the basin of the present-day North Sea. Their geomorphology is discussed in the next section. The valley of the Danube lying along the northern edge of the Alpine foreland was a glacial spillway for the Alpine piedmont glaciers.

22.8 THE PLEISTOCENE ICE AGE AND ITS MORPHOLOGICAL SIGNIFICANCE

22.8.1 The glacial periods and possible causes of the ice ages

The Pleistocene lasted about two and a half million years during which there were at least six major **cold periods** with intervening **warmer periods. Glacials** are cold periods when the inland ice spread extensively into areas that are without glaciers today. The **interglacials** are the warmer periods between the glacials.

Table 22.1 shows the main glacials and interglacials with the names commonly used in north Germany, the Alps, the United Kingdom and North America. Quaternary research has taken place locally in many areas and other names are often used for these periods in, for example, the Netherlands, Poland or Russia. The dates in Table 22.1 are given for orientation. They vary greatly for different parts of the earth and for different authors. Penck and Brückner recognized from work on river terraces in the Alpine foreland (section 15.4) that there had been four major ice ages with interglacials. They named them after four small streams in the area, Günz (the oldest), Mindel, Riss and Würm. Research in other areas substantiated this division although a pre-Günz glacial has since been confirmed and some of the glacials in which there has been more than one major advance have been subdivided into **stages**.

Since the middle of the nineteenth century various explanations have been suggested for the cause of the ice ages. There is still no general

TABLE 22.1 Quaternary time units

Years before present	North Germany	Alps		United Kingdom	North America
−10 000		**Beginning of the Holocene**			
	Weichsel	Würm		Devensian	Wisconsin
−70 000			(Glacial)		
	Eem-	Riss/Würm		Ipswichian	Sangamon
−120 000			(Interglacial)		
	Saale	Riss		Wolstonian	Illinoian
−180 000			(Glacial)		
	Holstein	Mindel/Riss		Hoxnian	Yarmouth
−260 000			(Interglacial)		
	Elster	Mindel		Anglian	Kansan
−420 000			(Glacial)		
	Cromer	Günz/Mindel		Cromerian	Aftonian
−820 000			(Interglacial)		
			(Glacial)	Beestonian	
			(Interglacial)	Pastonian	
	Menap	Günz		Baventian	Nebraskan
−1.2 million			(Glacial)		
	Waal	Donau/Günz		Antian	Unknown
−1.4 million			(Warm period)		
	Eburon	Donau		Waltonian	Unknown
−1.7 million			(Cold period)		
	Tegelen	Biber/Donau			Unknown
−2.2 million			(Warm period)		
	Brüggen	Biber			Unknown
−2.4 million			(Cold period)		

Length of time units from Walter (1992, p. 131).

agreement today. Some of the hyotheses are described below.

1. Long-term changes in received solar radiation:
   (a) over the entire earth because the sun moved through a cloud of cosmic material or because the atmosphere was polluted by a series of major volcanic eruptions;
   (b) over the whole earth or regionally as a result of provable periodic changes in the earth's orbit (the angle of the earth's axis, eccentricity of the orbit, procession of the equinox).

2. Changes in the storage components of the heat budget:
   (a) as a result of a changes in the albedo because of changes in the cloud cover or

of the extent of the snow and ice surfaces;

(b) changes in the humidity and/or carbon dioxide content in the atmosphere (the greenhouse effect).

3. Regional cooling with simultaneous changes in atmospheric circulation and distribution of precipitation due to the uplift of large mountain ranges.

4. Changes in the position of the poles in relation to the distribution of the continents and oceans, as well as changes in the ocean currents as a result of tectonic plate movements.

Flint (1971, pp. 788–809) and Embleton and King (1975, pp. 27–35) have described and criticized these and other hypotheses in some detail. Most of the possible causes are not mutually exclusive and could in combination have generated greater climatic fluctuations. Further research is needed by palaeoclimatologists.

22.8.2 The distribution and spatial order of Pleistocene glacial forms

The present-day glaciation of the Antarctic (12.5×10^6 km^2) and Greenland (1.7×10^6 km^2) gives an indication of what a large continental area covered by ice might have looked like during the ice ages. The Pleistocene glaciations covered 45×10^6 km^2, three times the area of the present-day ice sheets. The largest glacial was the one before the last (Riss, Saale, Wolstonian, Illinoian). In Europe it reached from Scandinavia northwards to the Arctic Ocean as far as Spitsbergen, Franz Josef Land and Novaya Zemlya. Westwards it covered most of the British Isles except for southern and southwestern England, to the south it extended to the German and Polish central uplands and to the east into the centre of European Russia and beyond the northern Urals as far as the Ob River. Inland ice also covered the areas of the central Siberian upland and the east Siberian mountains east of the Lena.

In the mountains of Central Asia and the Himalayas mountain glaciation dominated with glaciers extending to the foreland in some regions, similar to the Alps. In North America the inland ice covered the entire Canadian Shield and was met in the west by the piedmont glaciers of the Rocky Mountains. Further south, the

ice covered the Middle West as far as about 40° latitude near St Louis, and reached even a little further south in a few areas.

Near the equator, the Andes, the Ruwenzori Mountains and the volcanoes Mt Kenya and Kilimanjaro were more strongly glaciated than at present; the mountains of New Guinea were also glaciated (Löffler, 1977, pp. 63–7).

In the southern Hemisphere there are, besides the inland ice sheet of the Antarctic, much smaller ice fields in the mountains of the South Island of New Zealand and in the Patagonian Andes. The largest, in Patagonia, has a surface area of 13 500 km^2, and where there seem to have been at least seven ice ages during the Pleistocene and piedmont glaciers extended into the foreland (Wenzens *et al.*, 1994; Clapperton, 1993).

The glaciers of the last Pleistocene ice age (Würm, Weichsel, Devensian, Wisconsin) covered a smaller area than the previous glacial and their end moraines accumulated on the ground moraines of the Riss (Saale, Wolstonian, Illinoian). There is as a result a spatial sequence of glacial landforms and materials that is typical for all regions affected by several glaciations in the Pleistocene which can be termed the **multiglacial sequence** of continental glaciation.

In its centre lies the area of **continental glacial erosion** with a discontinuous soil cover, rock surfaces, rôches moutonnées and rock basins. Insofar as these forms have a long axis, it is oriented in the direction of the ice movement. In North America the central area lies around Hudson Bay and in Labrador and in Europe in north and central Sweden and Finland. In Norway, because of its greater height and proximity to the sea, the outlet glaciers that flowed down the steep slope to the Atlantic have eroded fjords. The same is true in the mountainous west of Scotland where the fjords are called **sea lochs**, to distinguish them from trough basin lakes which are also known as lochs. Numerous Scottish sea lochs have their continuation as glacial troughs on the land. The valley of Glencoe in the western Highland is a particularly good example of such a trough. Where glacial or fluvioglacial sediments occur in the central area of continental glaciation, they have usually been accumulated during a retreat or melting phase of the ice.

Examples are the Salpausselkä end moraine series in Finland and the Uppsala esker in Sweden.

In Europe and North America, primarily to the south, the area of erosion is linked to the **zone of young moraines** of the last glaciation. These moraines are relatively unmodified by subsequent denudation processes and still possess their original relief. Knobby ground moraine lies on their inner margins and is followed by the belt of young end moraines. Beyond the young moraines is the **zone of old moraines and young outwash plains**. The old moraine from the previous glaciation has been altered and flattened by periglacial denudation processes and is covered by young outwash deposits from the last glacial, although higher areas of the old moraines do lie above the outwash. In Schleswig Holstein in north Germany, the end moraines of the Weichsel glaciation are densely clustered along the inner end of the foerden and in the Bay of Lübeck. The young and the old moraine landscapes of the area are clearly distinguishable, not only by their relief but also their soil, so that there is an almost model spatial distribution of the multiglacial sequence. The young moraines have mostly developed fertile brown earths and parabrown earths while podsols and moors dominate on the old moraines. The infertile old moraine landscape has long been known as **geest**.

The hydrography and drainage of the old and young moraine areas also differs. Many enclosed basins still exist in the young moraine landscape because of the irregular pattern of moraine accumulation and hollows left by the melting of dead ice. Today the hollows form lakes in Holstein, Mecklenburg and Brandenburg in Germany and Pomerania and Masuria in Poland.

A large number of the lakes that existed in the old moraine area between the last two ice ages have since been filled by sediments or have become moor. The lakes of the young moraine will also eventually be filled from the surrounding areas of easily transported glacial and unconsolidated glaciofluvial material.

The drainage network in the young moraine area is largely chaotic (section 17.4). On the old moraines, by contrast, local erosion, capture and sedimentation have resulted in the development of a dendritic drainage pattern. The landform development in the young and old moraine areas of the Lüneburg Heath has been described by Hagedorn (1964, 1989).

In Mecklenburg-Vorpommern and Brandenburg, the young end moraines lie farther apart than farther west and extend in a southeasterly direction. The Weichsel ice age had several stationary phases, including the Brandenburg stage south of Berlin, the Frankfurt stage and the Pomeranian stage. The old moraine landscape lies to the south of the Brandenburg end moraines. This broad zone of young moraines gives a complexity to the multiglacial sequence in this area which is further increased by the fact that each stage developed its own spillway.

In central Europe the **glacial spillways** are more or less parallel to the edges of their respective ice sheets. They are continuous low-lying strips of land running in a more or less east–west direction. Present-day major rivers flow along them for part of their course, leaving again downstream to flow seawards. The central European spillways are known as **urstromtäler** (singular: urstromtal). Between the stretches occupied by major rivers, the urstromtäler form linear hollows with very low gradients. Many are used by canals which connect the navigable rivers.

The three stages of the Weichsel produced three urstromtäler. The Brandenburg Stage created the Glogau–Baruther urstromtal which extends from the Oder River near Glogau to the Elbe south of Tangermünde, the Warsaw–Berlin urstromtal, associated with the Frankfurt stage, running from Warsaw to Eisenhüttenstadt and onward to the confluence of the Spree with the Havel in Berlin, and the Thorn–Eberswalder urstromtal, belonging to the Pomeranian stage. It begins in the Weichsel valley, between Warsaw and Thorn, then follows the Netze and Warthe valleys to the Oder River and the Oder bog to Eberswalde. All three urstromtäler combine west of the lower Havel into a single urstromtal which follows a northwesterly direction and is occupied by the lower Elbe along its entire distance between Havelberg and the North Sea. In this part of the northwest German lowland the ice margins of the three Weichsel stages were too close together for separate urstromtäler to develop.

The outermost spillway in central Europe is the Breslau–Magdeburg urstromtal, which was formed in the Saale ice age. It runs northwestwards from Magdeburg along the Mittelland Canal and the Aller River to the lower Weser River.

The east–west orientation of the central European spillways is a function of the general northward slope of the land surface and the southward slope of the ice sheets and their meltwater deposits. The streams that flowed northwards from the central uplands had to turn west to avoid the ice. Meltwater from the ice could not flow southwards because of the slopes of the central uplands. Both the central upland streams and the meltwater therefore followed the low-lying areas in a westerly direction and transformed them into urstromtäler. The gradient of these spillways was very low, an indication that the seasonal peak discharges must have been very large in order to transport the load that accrued.

After the ice sheets retreated, the Baltic Sea and the North Sea provided new base levels. Capture and course changes have altered the flow direction of the north German and Polish streams and they now follow the urstromtäler for only part of their courses.

In North America the gradient patterns were different. The centre of the continent slopes south from the zone of glacial accumulation and deposition around the Great Lakes, that is, in the same direction as the inclination of the inland ice sheet. Glacial spillways were, therefore, not parallel to the edge of the ice but mostly led away from the ice carrying their load to the largest river on the continent, the Mississippi.

In high mountain areas that were, or are, glaciated, the vertical zonation of glacial forms at different heights is more important than their horizontal distribution. The piedmont glaciers are an exception (section 22.6).

The advance of the mountain glaciers on to the foreland is possible if the snowfall increases and if the altitude of the orographic snowline, or firn line, is lowered. Both causes, which can combine in their effect and are anyway often functionally related, increase the supply of ice from the alimentation area so that the terminal can advance, despite decreasing altitude and increasing ablation.

The **depression of the snowline** is the difference in height between the Pleistocene snowline of the last ice age and the present-day snowline. It is obtained by comparing the height of snow-free cirques with the present orographic snowline (section 22.2). The depression of the snowline varies considerably, depending primarily on precipitation but also influenced by other factors. The smallest values are in the very dry high plateau of Tibet. Wissmann (1959) estimated that in the Pleistocene the snowline lay only 250 m lower than now. In the low-lying area west of the Kun Lun mountains, bordering the Takla Maklan Basin, Hövermann and Hövermann (1991) estimated a snowline depression of about 1000 m. Similar values have been given for the mountains of Central Asia (Rost, 1993; Lehmkuhl and Rost 1993). In the Alps the snowline depression varied between 1700 m in the humid outer ranges of the western Alps and 800 m in the relatively dry Gurktaler Alps in Carinthia, Austria (Wilhelm, 1975, p. 105). In the uplands of central Europe the snowline sank during the Pleistocene below the highest upland summits. The Black Forest, the Vosges and Bohemian Forest all have cirques and other glacial landforms. (Woldstedt and Duphorn, 1974, pp. 251–2.

The other climatic altitude zones in the mountains were also lower than at present but not necessarily by the same amount. For example, in the Alps, the height difference between the terminals of the Pleistocene piedmont glaciers and the present-day glacier terminals is greater than the depression of the snowline, with a few exceptions, such as the Bossons Glacier on Mont Blanc.

22.8.3 The geomorphological effects of the ice ages outside the glaciated areas

The ice ages were climatic events that affected the entire globe. Glacio-eustatic changes of sea level caused changes in the exogenic process response systems and the base level of the exorheic stream systems. The high sea levels in the interglacials left behind eustatic coastal terraces. Glacio-isostatic crustal movements mainly affected glaciated areas, but they continued as the ice retreated and still take place today in regions that are now ice free (section 4.1). The glacio-isostatic terraces on the coasts of these areas also indicate where pauses in the post-glacial uplift relative to sea level occurred.

As the ice spread outwards from the core areas during each glacial, the climatic zones changed position. Instead of the temperate climate of today, a periglacial climate prevailed in the middle latitudes of Europe and North America, although with some differences when compared to the present day sub-polar climate in the tundra around the Arctic circle. Because of the very long summer days and very short winter days, radiation conditions in the tundra at the present time differ from those in the periglacial climates that dominated the middle latitudes in the Pleistocene when the length of day was the same as it is now. The duration of winter frost, the depth of thaw in summer and the freeze–thaw frequency, together with the permafrost regime that resulted, created conditions different from those in the high-latitude tundra of today. Periglacial conditions similar to those in the ice ages in the middle latitudes can be observed in the high mountains in Europe, although the areas affected are only small and because of local conditions such as a high slope angle, there is a large contrast between received radiation on sunny and on shaded slopes and great differences in the debris cover, depending upon whether the slopes lie beneath a summit of exposed bedrock or not.

A reconstruction of periglacial processes in the middle latitudes during the ice ages can therefore be only approximated. Numerous traces of processes do exist in the form of fossil ice wedges, gelifluction regolith and cryoturbated soil. Büdel (1937) was first to recognize the significance of Pleistocene gelifluction in the development of central European landforms. He explained the presence of allochthonous debris, 2 km from its area of origin, which lay on a surface with a very low gradient in the Ore Mountains in Germany. The debris could not have moved as a result of any ·process active at the present time because it lay embedded below a Holocene moor. A map of periglacial process traces in the Würm ice age has been made by Poser (1948).

During the ice ages, the entire area between the Alps and the edge of the inland ice in north Germany was a periglacial tundra. Tree vegetation existed only in the Mediterranean area farther south. Once the ice melted only a few tree species could spread across the Alpine passes and through the Burgundy gap into central Europe. In North America, however, the plains of the Middle West posed no such barrier, and the tree species that had moved south during the ice ages, moved north again in the Holocene with an undiminished range of species. For this reason there are a great many more tree species in North America than in central Europe.

Stream processes in the Pleistocene periglacial climates of the two continents also differed from those of the present day. The importance of mechanical weathering in a climate dominated by freeze–thaw meant that the streams carried large quantities of bed load. This, combined with the variable flow which peaked during snow melt, resulted in strongly braided channel beds. Today the bed load consists largely of debris inherited from the ice age. The streams continue to redistribute this ice age material and adjust their channels to the environmental conditions of the Holocene.

An ice age deposit of importance in the ice-free regions of Europe and North America is the windblown **loess**. The interglacial soil horizons that lie between the layers of loess of different glacials enable its stratigraphy and age to be determined. The loess material, which consists mostly of silt, was blown from the valley bottoms and branching channels of the Pleistocene streams and deposited as a layer of dust on the land surface. A large part of the loess cover on slopes, particularly in the central uplands of Germany, has been removed by wash denudation or mixed with slope debris during mass movement. Extensive deposits have been preserved in the lowlands, especially bordering the central uplands and on the dip slopes of some south German cuesta scarps. On the edge of the upper Rhine Valley and in the Kaiserstuhl, the loess deposits are up to several tens of metres thick.

Weathering has changed the loess into **loess loam**. A large proportion of the calcium carbonate content of loess was dissolved in the soil water and redeposited in the form of concretions in a lower horizon. Fertile black and brown earths have usually developed on the loess.

Loess is also found in other areas of Europe that have not been covered by Pleistocene ice sheets, particularly in those areas where the sedimentary deposits of major river systems supplied large quantities of silt. In Great Britain occurrences of loess, locally known as brickearth, are almost entirely limited to the Thames valley. Larger loess areas exist in France, on the Danubian plains in Hungary and, especially, in the Ukraine and adjacent parts of southern Russia. In the USA, the river system of the Mississippi with its major tributaries the Missouri and Ohio is the source area for the extensive region of loess in the Middle West.

Unlike the European and North American loess, the deposition of loess in China was not confined to the ice ages but took place over a much longer period and attained much greater thicknesses.

In the arid areas bordering and also in the subtropics, there are indications that the climate of the lower mid-latitudes was more humid than now. These phases, which coincided with the glacials, are known as **pluvials**. Evidence for their existence is found in lake deposits and lakeshore terraces. There were numerous lakes during the pluvials in the Great Basin in the western USA, including Lake Bonneville which had a surface area of more than 50 000 km^2 and a depth of 300m. Today only a very shallow lake, about 5000 km^2, the Great Salt Lake, remains. Traces of pluvial lakes are present in other arid regions.

Not all present day deserts were more humid during the pluvials. Schwarzbach (1974, pp. 224–6) suggests that there was a pluvial climate only on the polewards margins of the desert belts and that other areas of this zone were more arid than today. Research is continuing on the subject (for example, Buch and Zöller, 1992; Heine, 1992; Besler et al., 1994).

The composition of gravel deposits in dry river beds and wadis in present day arid regions is very varied and can be an indicator of earlier pluvial climates. Stream flow during the pluvials was more continuous than today. Gravels were transported greater distances and tend, therefore, to be more much homogenous than those

deposited by the rare runoff events that occur at the present time. These recent gravel deposits are usually little sorted and have moved only short distances.

The morphological effects of the ice age were probably least in the humid and subhumid tropics and on the tropical east coasts and tropical islands then, as now, in the trade wind belt.

The effects of past climates is the main theme of climato-genetic research (Büdel, 1963, 1977) in geomorphology and is closely connected to the study of climatic history (Schwarzbach, 1974). In order to be effective qualitatively, the climatic change must lead to a qualitative change in the processes, for example, from fluvial to glacial erosion or from predominantly chemical to predominantly mechanical weathering. To these qualitative conditions, the quantitative has to be added because the intensity of the altered geomorphological processes must exceed certain thresholds in order to bring about the changes in form. A change in the climate must also last long enough for a particular form to develop under a particular set of conditions. The fact that today many landforms are inherited from the Pleistocene ice age indicates that the development and the alteration of many landforms lasts longer than the time span of a period of constant climatic conditions.

Many climatic changes, particularly in the tropics, have led to an acceleration or slowing down of existing processes rather than to a qualitative change in the nature of the processes so that the process response system of the form development remains unchanged. The development takes place more rapidly or more slowly but is still directed qualitatively to reach the same type of dynamic equilibrium as before. The development of inselbergs in the Kenyan upland, for example, has remained qualitatively unchanged since the middle Tertiary, that is for several tens of millions of years, despite the climatic fluctuations that occurred during this period (Ahnert, 1982). Changes in climate occur more often than climatically determined significant changes in the development of landforms. It cannot be assumed that every known climatic change has generated a geomorphologically significant change in the processes.

23

THE LITTORAL SYSTEM

23.1 INTRODUCTION

The **littoral system** (Latin, *litus* = coast) is made up of the geomorphological processes that influence the formation of coasts, the rock materials that these processes remove, transport and deposit, including related organisms and organic substances, and the coastal landforms that result from the interaction of the materials and processes.

Coasts are by far the most important of all the natural boundaries on the earth. They are the dividing line between the continents and the sea and lie, therefore, between very different physical, chemical and biological process response systems.

Locally, these systems are not always sharply separated. On the North Sea coast of Germany and the Netherlands, for example, there are several kilometres of mudflats seawards of dyke-protected marshes that are invaded twice a day at high tide. During storm floods, the land behind the dykes and the lower courses of the rivers are also sometimes invaded by the sea.

Coasts include the zone of surf in which the waves influence the adjacent land area, the dune belt composed of redeposited sand that has been removed from the beach by the wind and the zone in which the vegetation is influenced by salty groundwater, salt dust in the air and storms that affect plant growth.

Neither seawards nor landwards does the coast have a well-defined boundary. It is a broad area which reaches landwards as far as the influence of the sea reaches and seawards as far as it is influenced by the land. How far this influence reaches depends in each case on the purpose of the definition. Coastal climates or the distribution of a coastal population extend much further inland than the beach and dune belt and a coastal fishing industry extends further seaward than the outer boundary of the mudflats.

The **shore** can be more precisely defined. It is the area of direct, reciprocal interaction of land and sea. Geomorphologically this is primarily the surf zone and the strip of land formed by the surf. On tidal coasts, the shore zone changes its position in relation to the rhythm of the high and low tides. Lakes and rivers also have shores. The **beach** is a shore zone composed of sand or pebbles.

Peculiar to the development of coastal landforms is their immediate linkage to sea level. Every change in sea level relative to land brings with it a vertical and usually also a horizontal change in the position of the shore, so that after each change the development of a coastal landform begins again at a new location.

The postglacial eustatic rise of sea level, known as the **Flandrian Transgression**, which ended about 5–6000 years ago, has had a worldwide influence on the development of present-day coastal forms. A **regression** is a fall in the sea level relative to the land, together with the retreat of the sea that accompanies it.

Sea level has changed frequently throughout earth history both eustatically and in response to vertical movement of the land. During any given period of sea level only a relatively short time is available for the formation of coasts although the limited time span is, to some extent, offset by the high spatial concentration of energy in the waves, tides and currents.

The transformation of this energy into geomorphological processes takes place mainly in the narrow zone of the shoreline. Coastal forms can change rapidly, occasionally within a few minutes. The spatial concentration of energy in fluvial, glacial or aeolian processes and the accompanying slope development are much smaller but influence much larger areas and generally take place over much longer periods of time.

23.1.1 The length of coasts

The measured length of a coast depends on the degree of generalization at different map scales. The larger the scale the more detailed it is and the seemingly greater is the length of the coast. Using maps with scales ranging from about 1:200 000 to 1:2 000 000, the total length of the coasts of all continents, excluding islands, has been estimated to be 286 300 km (Kossack, 1953, p. 239).

The length of a coast at different scales can be estimated using fractal geometry (Mandelbrot, 1982). A stretch of coast is measured in constant straight line steps L. If the total number of steps is n, the length of the coast G, then $G = nL$. Other step lengths for the same stretch of coast result in different total lengths. The shorter the steps the more closely the measurements follow the shape of the coast. On a straight coast the total lengths of the coast would, of course, be the same, whatever the step length used. On an indented coastline the measured total length is a nonlinear function of the step length used.

Figure 23.1 shows the functional relationship for the strongly indented west coast of Scotland from the Solway Firth to Cape Wrath. Both scales of the diagram are logarithmic. Step lengths of 10, 20, 35, 40, 50, 100 and 200 km were used. The measured total lengths range from $G = 500$ km for $L = 200$ km to $G = 1610$ km for $L = 10$ km. A straight regression line expresses the distribution of the measured values:

$$G = 3945L^{-0.42} \tag{23.1}$$

Between $L = 35$ and $L = 100$ km, the position of the data points deviates noticeably from the regression line because of the polygenetic origin

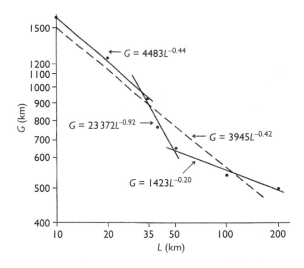

FIGURE 23.1 Determination using fractal geometry of the length of coast G as a function of the length L of measuring steps, on the west coast of Scotland between the Solway Firth and Cape Wrath.

of the coastline. Measurements with $L > 50$ km contain, primarily, the more or less straight, structurally influenced outer coastline, without the numerous bays and lochs (fjords). Calculated separately, the measurements with $L > 50$ km result in a much flatter regression line:

$$G = 1423L^{-0.20} \tag{23.2}$$

Between $L = 35$ km and 50 km, the change in length is more strongly expressed with

$$G = 23\,372L^{-0.92} \tag{23.3}$$

With these shorter step lengths, the line of measurement swings into the bays and fjords ignored by the longer steps. With even shorter lengths ($L < 35$ km) the coastline length changes accordingly:

$$G = 4483L^{-0.44} \tag{23.4}$$

which is similar to equation (23.1).

The method can be used diagnostically. In this example, the form of the coast is controlled by three morphologically explainable measured lengths. Those over 50 km reflect the major structural trends of the coastline, between 35 km and

50 km, the regional glacial coastal form and under 35 km, the local details.

23.1.2 Form components of coastal cross-profiles: the littoral sequence

The **littoral sequence** describes the typical spatial pattern of landforms created by the work of the sea. It is represented by two schematic profiles, one for coasts formed of unconsolidated material (Fig. 23.2a) and one for cliff coasts formed in resistant outcrops of rock (Fig. 23.2b). In a natural profile the number of components may be greater or smaller depending on the local conditions.

The littoral sequence of coasts of **unconsolidated material** begins in the continuously flooded shallow water areas with a **sand bar** or **gravel bar**. The bars indicate that in this area the wave movement transports and redeposits the material on the sea floor. Several bars can occur one behind the other. Their long axes are parallel to the beach and they are separated by channels, also parallel to the beach.

The shore is the surface that rises more or less continuously to the beach on which the waves

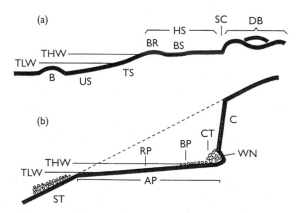

FIGURE 23.2 Littoral sequence (schematic). THW: mean high water; TLW: mean low water. (a) Unconsolidated material coast. B: bar; US: underwater shore; TS: tidal shore; BR: beach ridge; BS: beach swale; HS: high shore; SC: sand cliff; DB: dune belt. (b) Rock cliff coast. ST: submarine talus; AP: abrasion platform; RP: rock platform; BP: beach pebbles; CT: cliff foot talus; WN: wave cut notch; C: cliff.

break and spread out. Below the average tidal low water line is the **underwater shore** and between the average low water and the average high water, the **tidal shore**. Above the line of the average high tide is the less frequently flooded **high shore**. The highest part of the latter is reached only during storms and is also known as the storm shore. On the **high shore** and running parallel to it, there is often a long, low sand or gravel ridge, the **beach ridge**. Landwards of this ridge there is also a shallow **beach swale** which, however, is generally developed along only some sections of the beach ridge.

The high shore, also termed **beach platform**, has a much lower gradient than the tidal shore. Inland from the high shore lies the belt of **dunes**. A **sand cliff** usually forms on their seaward side, an indication that storm waves sometimes reach and undercut the foot of the dunes.

The most important components of the littoral sequence of **cliff coasts** are the **cliff** made of bedrock and the **abrasion platform** at its base. If the rock is resistant, a **wave cut notch** can develop in the foot of the cliff. A **cliff foot talus** of rock fragments that have fallen from the cliff and not been removed by the waves may cover some parts of the lower cliff. The edge of the abrasion platform towards the cliff is usually covered with **beach pebbles**, towards the sea it is generally bare rock. The upper part of the abrasion platform is comparable to the tidal shore of the sand or gravel beach. Instead of an underwater shore, the outer margin of the abrasion platform is either an underwater bedrock slope or a **submarine talus** of transported pebbles. The upper edge of the cliff forms a well-defined knick point in the profile against the upper slope where terrestial denudation processes dominate.

23.2 COASTAL CLASSIFICATION

The only classification of coastal types which can be applied to all coasts of the world and which describes both the nature of the coast as well as the processes that formed it in a geomorphologically satisfactory way, is that of Valentin (1952). The classification is hierarchical. Figure

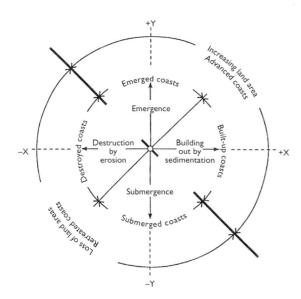

FIGURE 23.3 Coastal development shown schematically (after Valentin, 1952).

23.3 shows the basic approach. Essentially two process systems affect the formation of coasts:

1. vertical movement of sea level or of the land which causes the coast to emerge or submerge;
2. the work of the tides, waves and ocean currents that either denude and cause the coast to retreat, or deposit sediments and extend it seawards.

Vertical movements include eustatic changes in sea level and isostatic or tectonic movements of the land. Because the land surface near the coast generally rises from the sea floor landwards and continues to rise inland, the whole forms a sloping surface which causes a landward or seaward horizontal shift in the position of the coast with every change in sea level relative to the height of the land, providing no other processes counteract this tendency.

When the coast retreats inland because of denudation by the sea, the altitude of the surface is also changed, since the sea now covers the coastal area that has been lowered by the denudation. When the coastline is extended seawards by sedimentation, with the development of a delta (section 16.2), for example, deposition raises the

surface of the sediment above sea level. Valentin (Fig. 23.3) integrated these changes in position into a conceptually Euclidean coordinate system in which the vertical y axis represents the relative changes in sea level, and the horizontal x axis, the extent of the destruction by removal of or building out through sedimentation. The classification was developed as a qualitative system but can easily be expressed quantitatively. The net change in the position of the coast can be expressed using the x and y identified in the coordinate system. When both are positive, the coast advances, when negative it retreats. When one value is positive and the other negative, either advance or retreat takes place, depending on which coordinate has the greater absolute value.

In all cases the following are valid:
the advance of the coast:
$$x + y > 0 \qquad (23.5)$$
the retreat of the coast:
$$x + y < 0 \qquad (23.6)$$
a stationary position of the coastline:
$$x + y = 0 \qquad (23.7)$$

The diagonal $y = -x$ in Fig. 23.3 separates the data points (x, y) of all coasts that advance or have advanced from those that are retreating or have retreated. The stationary coasts lie on the diagonal.

In the few thousand years since the last Pleistocene ice age, coasts have been submerged as the sea level rose world-wide by more than 100 m as a result of glacio-eustatic change (section 5.2). Exceptions are coasts that have been strongly uplifted tectonically, as in California, or glacio-isostatically, as has occurred on the coast of Norway. The eustatic rise in sea level has also been compensated for where rivers with large loads have deposited sediment on the coast and formed deltas. Apart from these exceptions, most coasts in the world have receded. If sea level remains stable for more than ten thousand years, exogenic processes, especially sedimentation, will nullify the effects of postglacial drowning on many coasts. The present is an exceptional period in the geomorphological development of coasts.

Valentin's classification (Table 23.1) contains up to five hierarchical levels. The first two are identical with Figure 23.3 and the other three

TABLE 23.1 Coastal classification

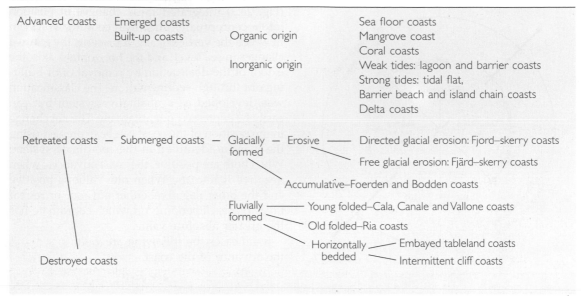

Source: After Valentin (1952).

relate to process or structural systems. The terms in the table are self-explanatory. The classification was accompanied by a world map of present-day coastal forms.

King (1972) compares 13 classification systems. Some are based on structural and geometric criteria (e.g. Guilcher, 1958), others on the energy budget (Price, 1955) or climatic change (Davies, 1964). Kelletat (1989, p. 164) expanded Valentin's classification to include drumlin coasts or doline and cone karst coasts which have not been formed by the action of the sea but are landforms reached by the postglacial transgression which now lie on the coast.

23.3 THE TIDES AND THEIR GEOMORPHOLOGICAL EFFECT

23.3.1 The physical basis

The tides are the rise and fall of sea level that occurs with a periodicity of approximately 12 h 25 min or 24 h 50 min. The latter corresponds to the moon day, the time between two successive transits of the moon through the meridian. The moon's daily retardation of about 50 min compared to the sun day of 24 h results from the

moon's own movement from west to east around the earth.

The relation between the moon's position in the sky and the occurrence of the tides was first understood in the seventeenth century after Newton had developed the theory of gravity. The mechanics of the tides are, in detail, very complex (*see*, e.g. G. Dietrich, 1963) but the basic tidal phenomena can be explained using a few models. The simplest model is that of an earth entirely covered by water. The gravitational pull of the moon on the solid earth body acts upon the centre of the earth, but the pull on the water cover on the surface is spatially differentiated. Newton's law of gravity states that the strength of the gravitational pull decreases with the square of the distance. The water cover on the side of the earth facing the moon is about 6300 km, or 1.6 per cent of the earth–moon distance, closer to the moon than the centre of the earth and is consequently more strongly pulled by the moon than the solid earth. The result is a high-water bulge on the side of the earth facing the moon. A second high-water bulge occurs on the side of the earth facing away from the moon. Here, the gravitational pull of the moon is less than at the earth's centre because of the greater distance and the water 'stays behind' (Fig. 23.4).

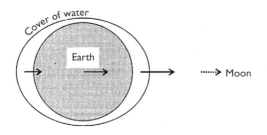

FIGURE 23.4 Simple schematic model of the development of flood bulges as a result of the effect of the moon on tides. The arrows in the centre of the earth and on the water bulges show the direction and the varying strength (indicated approximately by the arrow's length) of the moon's gravity.

The moon's position in relation to the sun also influences the height of the tides. When the moon, the earth and the much more distant sun lie in a straight line, the gravity effect of the sun is added to that of the moon. At full moon and new moon, therefore, the rise and fall of the tides are particularly high and they are known as **spring tides**. In the first and last quarter of the moon's phase, the moon and sun are at right angles to each other and the moon's gravitational pull is diminished by the sun's. The **neap tides** are the result.

In a simple model, the earth rotates beneath both high-water bulges. Because the moon in its own orbit moves further in the direction of the earth's rotation, thereby pulling the high-water bulges with it, the rotating earth needs 24 h 50 min until the bulges lie in the same place on the earth as the day before. From a point on the earth that rotates beneath the two bulges and the two low water levels in between, the change in the water level appears as a wave with a periodicity of 12 h 25 min from one wave crest to the next, a **semi-diurnal tide**. In reality the tides are an undulation of the water relative to the solid earth.

The tides are diurnal with a periodicity of 24 h 50 min on some coasts, most frequently in the tropics, due in part to the interference of the tidal waves. There can also be a variation depending on whether the moon is vertically above the equator or another latitude.

In the simple model the transmission velocity of the tidal waves corresponds to the ratio of the earth's circumference (40 000 km) to the tidal period of 24 h 50 min, or approximately 1600 km/h (447 m/s). In water that is relatively shallow compared to the wavelength, the transmission velocity C (m/s) of a wave is

$$C = (gt)^{0.5} \qquad (23.8)$$

where g is gravity acceleration (9.81 m/s^2) and t is the water depth (m), from which follows

$$t = C^2/g \qquad (23.9)$$

In order that the tidal waves move around the entire solid earth at the required velocity $C = 447$ m/s, the depth of the water cover must be

$$t = 447^2/9.81 = 20\ 368\ \text{m} \qquad (23.10)$$

The mean depth of the seas on the earth is, however, only 3790 m so that rotation of the tidal waves as postulated in the model in which the entire earth is covered with water, is not possible. With the actual mean ocean depth the tidal waves would have a velocity of about 193 m/s and lag behind the earth's rotation. Their movement is also hindered by the distribution of the seas in ocean basins separated by continents.

Because of the lower transmission velocities of the tidal waves and the barriers formed by the continents, independent tidal waves develop within the ocean basins. Instead of travelling around the entire earth, the tidal waves describe a more or less circular movement within their particular area, termed an **amphidromy**, in which the wave moves through a full circle during each tidal period. The **Coriolis effect** of the earth's rotation causes the waves to move in an anticlockwise direction in the northern hemisphere and clockwise in the southern hemisphere.

The area of the ocean occupied by an amphidromy depends on the velocity of the tidal wave and therefore on the mean water depth of the area. The critical circumference U_k of an indivdual amphidromy is the distance that a wave can pass through during a tidal period ($P = 12$ h 25 min = 44 700 s):

$$U_k = P(gt)^{0.5} = 44\ 700 \times 3.13^{0.5} \qquad (23.11)$$

The product $Pg^{0.5}$ is constant and has a value of almost exactly 140 000. It corresponds to the distance covered in meters of the tidal wave at a mean depth of $t = 1\,m$ during a semi-diurnal tidal period. The only independent variable in equation (23.11) is the water depth $t(m)$. The equation can, therefore, be simplified to

$$U_k = 140\,000t^{0.5} \qquad (23.12)$$

whereby U_k and t are expressed in metres. This is also correct dimensionally because the constant 140 000 has the dimension $m^{0.5}$. If the amphidromy, with a circumference U_k and a water depth t, were a perfect circle, its diameter would be

$$2r = U_k/\pi \qquad (23.13)$$

Because the circular form seldom occurs and the mean water depth on which U_k is dependent is only an estimated value, a critical length L is usually used, instead of the diameter $2r$, as a measure of the size of the area covered by an independent amphidromy. This measure is based on the assumption of the path distance covered by a tidal wave moving in both directions in a tidal period and is therefore approximately

$$L = U_k/2 = 70\,000t^{0.5} \qquad (23.14)$$

The relatively shallow North Sea, for example, contains three amphidromies (Fig. 23.5) of which the central amphidromy between northeast England and the Danish coast has a diameter of about 600 km. If this value is expressed in metres and used as L in equation (23.14), the resulting water depth is 73.5 m which is the approximate mean depth of the North Sea in this area.

The North Atlantic between northwest Africa, Western Europe, Iceland, southern Greenland and Canada forms one single amphidromy. The mean depth of this very large area of ocean would be similar to the mean depth of all the oceans, 3790 m. If this depth is used in equation (23.14), the critical length of the amphidromy should be 4300 km. In fact the great circle distance from Grand Canary off the northwest coast of Africa to the southern tip of Greenland is 4220 km and therefore very close to the estimate. Another amphidromy has its centre in the Caribbean and includes parts of the middle Atlantic east of the USA; a third covers the southern Atlantic.

23.3.2 Tidal range, tidal currents and resonance

The **tidal range**, the height difference between **high tide** and **low tide**, is lowest in the centre of the amphidromy and increases towards its circumference. Enclosed seas that do not have an outlet to the oceans, such as the Mediterranean, the Black Sea and the Baltic, have a very small tidal range, usually only a few decimetres.

Geomorphologically, a low tidal range means that the effect of the breakers is limited both vertically and horizontally to a narrow band of the shore. Where the tidal range is large, the boundary between land and sea shifts over a broad zone during each tidal period, particularly if the coast is flat. The effectiveness of the waves is spread over a much larger area and their impact locally is small. Spring and neap tides also cause a variation in tidal range at any one place. If part of the sea floor near the coast is flooded only when the spring tide is particularly

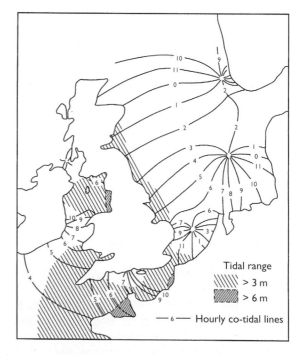

FIGURE 23.5 Amphidromies of the tides in the North Sea (after King, 1972). The co-tidal lines show the position of high water at a particular time in the tidal rhythm.

high, it may dry out enough in the intervening weeks or months for strong winds to move the sand and build dunes that lie above the level of the spring tides and are able to withstand the waves. In this way former sand bars become islands.

Under otherwise equal conditions, the tidal range on continental coasts is, in general, larger than on the open sea or on small oceanic islands. Water moved by the tide near the coast can back up and resonance ocillations may also reinforce the fluctuations in water level.

The rise and fall of the tides is also linked to the horizontal transport of water by the **tidal currents**. Their velocity is much less than that of the tidal waves and they do not advance through the entire circle of the amphidromy but range back and forth over much shorter distances in the rhythm of the tides. The **flood current** occurs when the current advances in the same direction as the tidal wave, and the **ebb current** when the current flows in the opposite direction. The terms flood and ebb refer only to the current and not to the water level of the tides.

The greater the tidal range, the larger the amount of water that must advance or retreat and, given similar conditions, the stronger, therefore, is the tidal current. A small tidal range can result in a strong tidal current if access from the sea to a large bay or lagoon is through a narrow channel and the entire mass of water of the tidal range must advance or retreat through this space during the six and a quarter hours of half a tidal period.

On the open sea and along open, straight ocean coasts the flood current reaches its highest velocity at high tide, the ebb current at low tide. When the current ceases halfway between high and low water, the flow direction reverses and the current turns.

These relationships express the general behaviour of a current in a **progressive tidal** wave in which the water at the wave crest flows forward and the water in the wave trough flows backwards.

In the mouths of tidal rivers (estuaries) and in elongated bays the temporal relationship between water level and current shifts and the advancing wave becomes a **standing wave**: the water level rises as long as the current flows

landwards and falls when it flows seawards. In this case the current turns at high and low water. The maximum velocity of the flood current is reached between low and high water and the maximum ebb current velocity between high and low water. The discharge of a large river can shift this relationship between water level and occurrence of maximum current velocity by a small amount, but does not change it fundamentally.

Large bays develop their own amphidromies. If the length of a bay is about half the critical length of the amphidromy, the centre of the amphidromy lies at the entrance to the bay. Within the bay itself a resonance then develops, similar to the **resonance** of a sound wave in a musical instrument. The tidal wave is strengthened and the tidal range is greater than elsewhere. In accordance with equation (23.14), half the critical length is

$$L/2 = 35\,000t^{0.5} \qquad (23.15)$$

The closer the critical length and the actual length of the bay correspond, the stronger is the resonance and the greater the tidal range. The well-known tides of the Bay of Fundy on the east coast of Canada where the range is up to 17 m, are based on this effect. The bay is about 300 km long and has a mean depth of about 70 m. With this depth the calculated critical half length would be 290 km, so that the prerequisites for resonance are particularly well fulfilled here (Fig. 23.6).

FIGURE 23.6 Sea floor and cliff coast on the Bay of Fundy, Canada, at low water.

23.3.3 Estuaries and estuarine meanders

Estuaries are funnel-shaped river mouths in tidal areas that have been widened by erosion of tidal currents. Sea water constitutes the larger proportion quite far inland. The river water flowing seawards increases the ebb current. Because of its lower specific gravity it is lighter and remains near the surface. The saltier flood current approaching from the sea flows primarily along the floor of the river channel. The mixing of sea and river water gives estuaries their characteristically **brackish water**.

The flow of river water to the sea also alters the tidal curve in the estuary. The ebb current lasts longer because more water flows in the ebb current to the sea than arrives in the flood. The flood current dams the river water arriving in the estuary which, because of the tidal rise and the landward flow of the flood, is prevented from flowing to the sea. The increase in water level during the flood is consequently of shorter duration and steeper.

In many wide estuaries there are several deep channels separated by mud or sand banks and the main flow of ebb and flood often occurs along different channels. Where there are bends in the estuary, centrifugal force pushes the flow of the ebb or flood current to the outside of the bend. The force is strongest in those stretches of the estuary lying downcurrent from each point of greatest curvature. This is similar to the shift in the line of highest velocity in a river meander (section 13.3.1) with the difference that in the estuary, the current's direction alternates with the tide and the undercut bank of the flood current is in another location from that of the ebb.

This spatial shift of lateral erosion with the rhythm of the tides also determines differences in the morphography of river meanders and **estuarine meanders** (Ahnert, 1960a, 1963a). Typical estuarine meanders do not have the parallel banks of river meanders but are narrowest where the paths of the flood and ebb currents cross, that is, where the estuary curves most strongly, and widest between these bends because here the ebb and the flood erode opposite banks (Fig. 23.7).

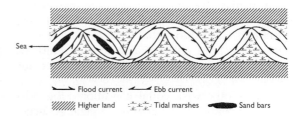

FIGURE 23.7 Schematic diagram of estuarine meanders (after Ahnert, 1960a).

Estuarine meanders are present on many coasts of the world, especially on the east coast of the USA and the coasts of southeast Asia as far as Malaysia and Vietnam and including the deltas of the Ganges and the Irrawaddy (Ahnert, 1963a).

The prerequisites for their development are:

1. that the ebb and flood currents attain their maximum velocity at about the same water level, the ebb current as the tide falls and the flood as it rises;
2. that the maximum velocity of both currents is approximately the same;
3. that both currents are strong enough to erode the banks of the estuary.

The conditions for the first prerequisite are present in the inner part of the estuary where the movement of the water is similar to a standing wave.

The conditions for the second are present more often in the estuaries of small rather than large rivers because the greater the amount of river water flowing into the estuary, the more unequal are the duration and the maximum velocity of both currents. The ebb and the flood usually have similar velocities further downstream in large rivers than in smaller ones, although where the conditions for the first prerequisite are also present, estuarine meanders can develop. The meander at Grays on the Thames estuary in England is an example.

The third prerequisite is most likely to be fulfilled if the estuaries lie in unconsolidated sedimentary material that has been transported and deposited by the river itself. For this reason the banks of most estuarine meanders are formed of marsh deposits.

The estuarine meanders in the area of the Chesapeake Bay in Maryland, USA (Ahnert, 1960a) probably began to develop in a more or less straight, open estuary whose outer limits are now apparent from the straight margins of older Pleistocene terraces bordering the recent tidal marshes. The tidal currents swung back and forth as a result of their own turbulence within the open estuaries, similar to river currents, depositing their loads of sand and silt alternately on the left and right banks in areas not generally reached by the line of high velocity of either the ebb or the flood. These deposits lie opposite the crossing points of the paths of highest velocity of the ebb and flood currents.

In the broad stretches between the crossover points, not much sedimentation takes place because of the lateral shifting of both lines of velocity, which also periodically erode the marsh sediments on the banks. It is possible that these stretches of bank are close to an equilibrium between lateral erosion and sedimentation that is expressed in the existing adjusted form of the curve of the bank. Sand and mud bars are sometimes deposited in the wider stretches of the estuaries between the crossover points because of the lateral separation of highest lines of velocity of the ebb and flood (Figs 23.7 and 23.8).

Suspended load is also deposited at high tide when the current reverses. The particles are deposited in the marsh vegetation. A higher flow velocity is necessary for the erosion of these particles than for their deposition (see Hjulström diagram, Fig. 11.3). The falling ebb current can attain sufficient velocity only when the water hardly covers the marsh or even lies below this level. With the continual raising of the marsh surface by sedimentation, flooding of the marsh becomes less and less frequent and sedimentation decreases.

All these phases of development can be observed in the estuarine meanders of the Chesapeake Bay area. In their lower seaward section the estuaries are funnel shaped and their banks formed of Pleistocene gravel terraces which lie above tide level. There are few marsh areas. The middle sections of the estuaries are dominated by estuarine meanders and recently formed marsh banks. Further inland there is a section in which the tides still have some influence, but in spite of the presence of bends in their courses, the estuaries have largely parallel banks of marsh.

Only in the middle section of the estuary are the ebb and flood currents equally strong and at their maximum at the same water level. In the downstream open section of an estuary, the flood maximum occurs at a higher water level and the ebb maximum at a lower water level. In the upper section, the currents are no longer of similar strength, with the ebb dominating.

This spatial sequence is probably part of a continual sedimentation process that has been taking place since the end of the postglacial rise in sea level 6000 years ago. Initially the estuaries were open funnels free of marsh sediments. Estuarine meanders developed first in the section furthest upstream where the proportion of sea water was then greater than that of river water and the necessary preconditions for the currents were present. With the development of marshes, the estuary cross-section and the sea water proportion of its current budget were progressively reduced. The ebb current, reinforced by the river flow, began to dominate and the stretch of estuarine meanders changed gradually into a stretch of river in which the discharge was pulsated with the tides but flowed largely seawards. Meanwhile, a new stretch of estuarine meanders developed downstream in the previously open estuary. Estuarine meanders can,

FIGURE 23.8 Estuarine meanders on the Patuxent Rivers, near Nottingham, Maryland, USA.

therefore, be seen as a phase in the transformation from an estuary to a river channel. Because the estuaries were former river valleys drowned by the postglacial rise in sea level, this progressive return to their fluvial form is part of the not-yet-completed world-wide adjustment after the last ice age.

A **bore** is a tidal phenomenon in the estuaries of some larger rivers with a large tidal range. The flood tide progresses up the estuary as a steep wall of water with a breaking wave in front, against which the ebb current flows with no reduction in its strength. There is no pause between the occurrence of ebb and flood. Instead, the advancing flood simply stops the ebb by overrunning it. The breaking effect of the ebb current at the foot of the advancing wall of water causes the latter to steepen, become unstable and to collapse progressively like a breaking wave. A large part of the total rise of the tide is concentrated in the narrow steep wall of water of the bore.

The bore generally advances with a moderate velocity. Gierloff-Emden (1980, p. 1014) quotes a velocity of 5 m/s for the Amazon bore. Usually they move more slowly, at little more than a fast walking pace. There are well-developed bores in the estuaries around the Bay of Fundy in Canada, at Moncton, New Brunswick, for example. There is also a large bore in the River Severn in England.

23.3.4 The effect of tides in tidal flats

Tidal flats are areas near the coast, formed mostly of sand and silt, that are flooded and exposed at every tide. In addition to the tidal flats in estuaries, there are tidal flats on the North Sea coast of Europe lying between the coast and the offshore barrier islands and also extensive areas of open tidal flats which are, at most, protected from erosion by a sand bar on their seaward margin. Tidal flats are known as watt in Germany and as wadden in the Netherlands.

The effect of tides on tidal flats is similar to the effect of tides in estuaries. Standing tidal waves dominate, especially in the inner landward edge of the tidal flats. In East Friesland in Germany, for example, the flood flows, with its suspended load, through the deep tidal inlets between the islands into a progressively branching system of channels, until eventually the highest parts of the tidal flat are covered.

The spatial pattern of tidal channels resembles that of a dendritic stream net, although functionally they differ considerably, as is apparent from their form. Not only does the water flow in both directions, on flood and ebb but the very large flow through, compared to that in a stream network, per surface unit of their 'watershed' that must take place in a few hours means that the channels have a high density and branching frequency. For this reason also the width of individual tidal channels increases rapidly seawards.

When the tide turns, the suspended load is deposited on the tidal flats. In the early phase of the falling tide, the ebb current increases in strength. Erosion is most effective, however, after the highest surfaces of the tidal flat are exposed because the ebb's highest flow velocities occur in the tidal channels and not on the flats. Consequently, part of the suspended load brought in on the flood remains as a new layer of sediment on the flat and the height of the flat is raised in the same way that the height of the marsh surface of estuarine meanders (section 23.3.3) and all river and sea marshes in tidal areas that are not dyked is increased.

Sediments on tidal flats are eroded mainly by storm floods. During periods of high wind the sea is driven against the coast which leads to especially high tides, particularly if the storm coincides with spring tides. In the Middle Ages, storm floods destroyed large areas of land on the coasts of Western Europe. Dyking has since reduced the surface area that could be flooded. On a natural undyked lowland coast in the middle latitudes, **salt marshes** or **tidal marshes** succeed the tidal flats on their landward side. They are covered only during extreme spring tides or storm floods. Salt tolerant vegetation (halophytes) colonizes the salt marshes as they develop, especially *Salicornia europaea* which helps to retain the silt and to heighten the level of the marsh. Eventually the *Salicornia* is replaced by salt tolerant grasses.

On tropical coasts, **mangroves**, bushes and low trees, some with stock roots, accumulate the silt very effectively. The mangrove forests are composed of a variety of species whose growth form and physiology have adapted to the special environment of tidal coasts.

23.4 THE SURF AND ITS GEOMORPHOLOGICAL EFFECT

23.4.1 Physical basis of wave movement

The waves produced by the wind on the open sea are continuous transversal oscillations of the boundary area between two moving media, water and air. In contrast to the wind, which advances over great distances, the water rotates in place around a horizontal axis that lies at right angles to the waves' direction of movement. On the front of the wave the water particles move upwards, on the wave crest, forwards, on the back of the wave, downwards and in the trough, backwards, rising again on the front of the next wave. Compared to the depth of the water, wind waves are short. Unlike tidal waves (equation (23.8)) their movement is not influenced by depth in the open sea. The following parameters are required to determine their movement:

1. the wavelength λ (m), measured as the distance from one wave crest to the next;
2. the wave velocity C (m/s);
3. the wave period T, expressed in seconds, between the movement of two successive wave crests past a measuring point.

In water that is deeper than half a wavelength the functional relationship is

$$C = \lambda / T \qquad (23.16)$$

The gustiness and general turbulence of the air produces waves of different lengths simultaneously which superimpose on one another and result in a complex wave spectrum. There is, nevertheless, a characteristic wave size for each wind force that develops when the effective duration of the wind action is sufficient and when the **fetch**, the distance over water that the wind has blown, is larger than the required minimum (Table 23.2).

Waves several hundreds of metres long move out from an area of storm at high velocities as **swell** and produce large breakers on shores in parts of the ocean remote from the storm. Because other, shorter waves do not spread as far from the area of storm, swell, waves have a smoother surface than the original storm waves. The **amplitude**, or height of the swell, decreases considerably as it spreads away from the storm area but the length of the swell waves and their periodicity do not change significantly.

Many wave spectra are heterogenous and include waves from different areas of origin that move in different directions, so that they cross each other. Where two wave crests or two troughs cross, they reinforce one another and create a wave of greater amplitude; where the crest of one wave train meets the trough of another, their amplitudes are diminished. The wave pattern therefore becomes very complex.

TABLE 23.2 Characteristic parameters for fully developed seas

Wind force (Beaufort)	Wind speed (m/s)	Minimum Effective fetch (naut. mile (km))	Effective duration (h)	Maximum Wavelength λ (m)	Period T (s)	$C = \lambda/T$ (m/s)	Maximum Wave height (m)
3	3.6–5.1	6(11)	2.3	18	3.4	5.3	0.36
6	11.3–13.9	140(259)	15	153	9.9	15.4	5.2
9	21.2–24.2	960(1778)	52	490	17.7	27.7	22.2
11	28.2–32.4	2500(4630)	101	900	24.0	37.5	45.0

Source: Dietrich (1963, p. 378).

23.4.2 Refraction

The complexity of a wave spectrum is reduced in shallow water. Shallow in this context is a water depth that is less than half the wavelength of the largest wave. Once this critical threshold value is reached, friction begins on the sea floor and the velocity of the wave movement becomes a function of the water depth (equation (23.8)). One of the consequences of these changes is the phenomenon of **refraction**; others are the progressive integration of the waves and the development of the surf.

Refraction is the change in direction of the progress of waves that arrive at the coast at an angle. It is produced by a change in the velocity of the waves C because of a reduction in the depth t (equation (23.8)). When a wave arrives at the coastline at an angle, the section nearer the coast is more strongly braked by friction and advances more slowly than the other end of the wave further out and still in deeper water. The latter catches up with the slower moving section and swings in towards the shore so that when the waves break they are more or less parallel to the line of the beach.

In front of a projection on a coast, the sea floor slopes away divergently seawards. In this case, refraction causes the waves to turn in and converge on the front and on both sides of the projection.

With the decrease in the wave velocity C, the wavelength also decreases. The wave period T remains constant (equation (23.16)). Because of the reduction in water depth towards the shore, the front of the wave is slowed more than the back part so that the wave crest becomes narrower, steeper and higher as it approaches the shore.

The longest waves of the complex wave spectrum on the open sea are the first, and the furthest out, to be affected because their critical water depth $\lambda/2$ lies furthest from the shore. They are, therefore, the first to swing towards the shore and to steepen and, increasingly, to determine the subsequent advance of all other waves. The smaller waves in the spectrum are affected only passively and become largely integrated in the motion of the larger waves. At the shore, the waves arrive as a sequence of almost parallel wave crests and troughs.

23.4.3 The surf

The surf zone begins where the steepening forward slope of the wave crest becomes unstable and breaks. The wave breaks in different ways depending on the slope of the sea floor. On very gently sloping beaches, a **spilling breaker** (Fig. 23.9) develops in which the water from the crest foams down over the front slope of the wave.

Where the beach slope is steeper, the foot of the front slope of the wave is braked more strongly and the wave crest shoots foward, forming a **plunging breaker** (Fig. 23.10). The energy freed by this type of breaker is very concentrated and particularly effective geomorphologically. The third type of breaker, the **reflection breaker** (Fig. 23.11), is produced when the wave is caught and thrown back, or reflected, from a very steeply sloping beach or cliff shore.

The breakers change the motion of the wave from a **transversal oscillation wave** with a rotational movement of the water particles to a **longitudinal translation wave** in which the water shoots forward as a shallow layer up the beach slope and then flows in an equally shallow or even shallower layer back down the beach.

Sand and gravel particles are set in motion by the power of the breakers in the surf zone. The **swash**, or uprush, the movement to the shore of the translation wave, transports the particles to

FIGURE 23.9 Spilling breakers on the Atlantic coast of Maryland, USA.

FIGURE 23.10 Plunging breakers in Drakes Bay, California, USA.

the beach; the **backwash**, the return movement of the water down the beach, removes them again. The velocity of the backwash is lower and its duration longer than those of the swash. The volume of water in the backwash and its transport capacity are also usually smaller because a part of the water seeps into the unsaturated pores of the upper beach slope. In addition, similar to the sedimentation on tidal flats, some load is deposited when the water movement reverses, because the water velocity necessary for erosion is attained only after the water has

FIGURE 23.11 Reflection breakers on the granite cliff of Land's End, England.

flowed back from the landward margins of deposition. A net transport of unconsolidated material to the shore is characteristic for the swell waves. The steepness and shorter wavelength of storm waves usually has the opposite effect. Because their breaking zone is closer to shore, the slope itself is steepened and the zone of translation waves narrowed. As a result, the net transport is directed seawards and the beach is eroded (King, 1972, p. 240).

The backwash of the translation wave meets the next wave as it advances, and each acts as a brake on the other, producing additional turbulence on the beach slope, often with small secondary breakers. Several zones of breakers, one behind the other and with a diminishing intensity shorewards, can develop on very gently sloping beaches (Fig. 23.12). The fact that the return flow of the translation wave is braked, and partially reversed, by the following wave means that there is a net movement of water in the direction of the beach. This accumulation of water is neutralized by **rip currents** (King, 1972, pp. 121–4) which flow seawards usually several hundreds of metres apart, across the surf zone in channels they have eroded. They cease beyond the surf, joining the currents that flow along the coast.

Rip currents are generally a few tens of metres wide and flow with velocities that are often too strong to swim against. On the beach they are visible as gaps in the line of surf. Their channels

FIGURE 23.12 Series of translation waves on the flat beach at Gwithian in St Ives Bay, Cornwall, England.

are deeper than the rest of the foreshore and the waves break later and closer to the shore in them than on the shore between the channels.

23.4.4 Tsunamis

The tsumami (Japanese *tsu* = harbour, *nami* = wave) is a special form of surf that often has a catastrophic effect on the coasts it reaches. Tsunamis are produced by the shock waves that follow undersea earthquakes or volcanic eruptions, and are most frequent in the Pacific where the crust is tectonically mobile and there are a larger number of volcanoes. Tsunamis have wavelengths of many kilometres, according to King (1972, p. 125) up to 160 km, and because of this, behave like waves in shallow water and move with a velocity that is proportional to the square root of the water depth (equation (23.8)). Since the Pacific is in part more than 6000 m deep, tsunamis can reach speeds of about 800 km/h and cross the entire Pacific in a few hours. An earthquake near the coast of Alaska on 25 March 1964 generated a tsunami which arrived on the Japanese coast 7 h later, the coasts of Ecuador and New Zealand after 14 h and the Antarctic after 21 h.

Because of their great length, the passage of these waves is not perceptible on the open sea. Near the coast their velocity is reduced by friction, their wavelength shortened and their amplitude increased enormously. Wave heights of more than 15 m are not uncommon in tsunamis. Following the Krakatoa eruption in 1883 and the earthquake on the Aleutian Island of Umiak in 1946, tsunami waves with heights of more than 30 m were reported.

Similar to the circular waves produced by throwing a stone into a pond, tsunamis are composed of a series of successive waves. The highest wave crest is often the second or third in the series. The waves follow one another several minutes apart. On the coast this has the effect that the sea retreats far from the land and then returns as an enormous surf wave which crashes on to the shore. The sequence is repeated perhaps a dozen times before the intensity declines.

Tsunamis cause a large amount of damage to coastal settlements and economic activity and, if warnings are insufficient, may cost lives. The geomorphological impact of these catastrophic events is limited mainly to coasts of unconsolidated material which are intensively eroded by the huge waves. Changes on coastal cliffs are limited to the loosening of blocks and the removal of debris.

23.4.5 Bars, beach drift and beach forms

When the largest waves, particularly the swell, reach shallower water on an ocean coast, they are slowed by friction and exert at the same time a thrust on the bottom. If the sea floor is made up of movable particles such as sand grains or gravel, low **offshore bars** or **gravel bars** accumulate at right angles to the direction of the advancing waves. Bars are also formed nearer the shore where shorter waves reach their critical depth and at the breaking line of the surf, these are **breaking point bars** (King, 1972, p. 316).

A positive feedback exists between the growth of the bar and the work of the waves in so far as the increasing height of the bar reduces the water depth and increases friction so that the shorter waves at the bar also become influenced by the friction. On tidal coasts the change in sea level is an additional factor. At low water, a bar that is far from the beach can produce surf, together with translation waves and the associated minor forms, which are then removed by the following high tide.

The position and size of the offshore bars change in several ways depending on the types of waves arriving on the coast. On ocean coasts in the middle latitudes there is a seasonal change in the net sediment transport between the beach and the offshore bar. In summer, the long low swell waves primarily transport sediment from the sea floor to the beach; part of the load originates in the offshore bars, particularly those lying nearest to the beach. By contrast the short steep storm waves of winter break closer to the beach slope and cause a net erosion of the beach (King, 1972, p. 354). Some of the eroded material is carried along the shore. Some is carried to the sea floor in front of the coast, particularly by rip currents, where it is deposited and contributes to the development of the offshore bars. The seasonal sediment flux between beach and offshore

bar is an important component in the mass balance of the littoral system on coasts formed of unconsolidated material.

In the area of the beach slope there is a longitudinal transport of material by the surf. Despite refraction, the translation waves of the surf usually do not arrive exactly at right angles to the beach slope but obliquely, depending on the direction of the waves in front of the coast. The return flow of the translation waves does, however, follow the beach gradient quite closely, although there is often some continuation of the lateral movement. A sand or gravel particle that is not deposited is transported laterally on a zigzag path in the surf zone by the translation waves which results in a net movement along the beach, a process termed **beach drift** (Bird, 1968, pp. 88–92). Beach drift is not to be confused with a longshore current which flows beyond the surf zone.

On beaches little affected by tides, the position and width of the beach drift zone is dependent mainly on the size of waves arriving at the beach. If the translation waves remain in the same general zone on the beach for at least several hours and are sufficiently constant and strong enough to move beach material, their surf can produce **beach dells** and **beach cusps**. Because of their lateral component, the field of movement within the translation waves forms a curve on the beach slope in which the grains of beach material are also moved. If net transport occurs, a shallow depression is produced in the area of the advancing wave into which the following waves flow and deepen the hollow. In cross-section the upper end of these forms resembles a dell; their axes follow the gradient of the beach slope.

The process usually takes place simultaneously along a stretch of beach so that a series of beach dells develops, separated by low beach cusps with convex spurs pointed towards the sea. The cusps are formed in those parts of the beach slope that lie uneroded between the beach dells; some material eroded by the translation waves in the neighbouring dells may also be deposited on the cusp (Fig. 23.13).

The width of the beach dells is determined by the size and periodicity of the translation waves that dominate at a given time. Widths of 10–15 m

FIGURE 23.13 Beach cusps and beach dells on Assateague Island on the Atlantic coast of Maryland, USA.

are common. Where the surf is weak, or on the shores of lakes, they may be smaller. Heavy surf can produce beach dells even on pebble beaches. Often several beach dell systems lie one behind the other, the work of different severe storms. The dells that are higher up the beach are always older than those further seaward.

A **swashbar** or **beach ridge** accumulates on the landward side of the translation waves. In the littoral sequence of coasts formed of unconsolidated material, beach ridges occur where the beach slope joins the flatter beach platform or high shore (Fig. 23.2). They are made up of sediment transported by the translation waves and not removed again when the waves flow back down the beach slope. Seawards of the beach ridge, the beach slope is too steep for material to accumulate as a ridge or bar, and landwards from it, the high shore is not reached by the translation waves. Beach ridges usually appear in summer when there is a net sediment transport from the sea floor to the beach. They develop within a short period and each deposition form is related to an individual water level and a sea with waves of a particular size. Other water levels and waves produce beach ridges elsewhere on the beach slope. For this reason a series of beach ridges may lie one behind the other on a beach. As with beach dells,

no beach ridge survives flooding by surf waves and the ridge closest to the sea is almost always the youngest and that furthest landwards, the oldest.

The coarser the material forming the beach ridge, the steeper its slope. Beach ridges of medium fine sand have an angle of about 5°, gravel about 17° and blocks up to 24° (Figs 23.14 and 23.15; Bird, 1968, p. 85).

The formation of a beach ridge takes a minimum of a few hours. During this period the translation waves have to break on the beach at approximately the same place for their load to be deposited as a ridge. Coasts with a significant tidal range do not generally have beach ridges. On sandy beaches in the tropics and subtropics with very small tides, the sea water in the translation waves evaporates partially on the edge of the high shore and leaves behind its salts, including calcium carbonate and magnesium carbonate, in the sand pores. The sand near the surface hardens to a crust, or **beachrock** (Russell, 1962).

23.4.6 Abrasion platforms and cliffs

The development of the present day coasts in bedrock began at the end of the postglacial rise in sea level about 6000 years ago when the sea, at

FIGURE 23.14 Several gravel beach ridges on Chesil Beach near Wey-mouth, Dorset, England. The highest beach ridge has a height up to 13 m above sea level.

its new level, began to alter the form of the slopes on the land surfaces. On steep slopes, reflection breakers wash out and remove the regolith, exposing the bedrock. Weathering also

FIGURE 23.15 Beach pebbles with their long axes parallel to the shoreline, in the area of breaking translation waves on Chesil Beach, Dorset, England. Compare Fig. 11.2

widens the crevices in the rocks and weakens their resistance. The impact of the surf compresses the air held in the crevices and the pressure developing within the rock structure loosens it. The suction produced by the return of the surf pulls loose blocks out of position and moves them downslope below the water level where they collect as submarine talus. The area from which the blocks have been removed remains as a levelled area in the slope where the translation waves can now be effective. In addition, reflection breakers are replaced by plunging breakers.

Together the surf and the translation waves deliver the energy for the transformation of the small initial flattening on the shore to an **abrasion platform**. The levelled area is smoothed by the abrasive action of the rock debris moved back and forth by the translation waves. During heavy seas the surf continues to impact on the rock on the landward side of the levelled area, producing more rock debris and causing the base of the slope to retreat further inland.

In this way the platform widens. The slope at the back is attacked by waves, steepened, and eventually becomes unstable. Mass movements, slides and rock falls follow and the slope is transformed into a littoral **cliff**. At the base of the cliff a **wave cut notch** may develop as a result of wave attack. Not all wave cut notches are caused by the mechanical removal of rocks by the surf. Solution weathering at the water line can produce similar forms.

The rate at which the cliff retreats depends on a number of factors. Important are the resistance of the rock, the intensity of the surf and the width of the abrasion platform in front of the cliff. Various estimates of rates of retreat have been made. In unconsolidated volcanic material, a rate of retreat of 1 m per day can occur and in glacial deposits a rate of 2–15 m per year. Some cliffs of Mesozoic chalk on the English and French coasts retreat a few decimetres annually, whereas cliffs in Palaeozoic rock in southwest England have receded very little. Most estimated rates are means for a long period of years. The retreat in any one section on an individual coast can vary greatly from any mean (Irving, 1962; Zenkovich, 1967, p. 8; King, 1972, pp. 467–74).

The rock material delivered by the cliff as it retreats falls to the cliff foot on the inner margin of the abrasion platform. Here, it is worked by the surf and used to abrade and further smooth the abrasion platform. The higher the sheer cliff becomes, the more material it delivers and the harder it is for the surf to remove the material and renew its attack on the bedrock at the foot of the cliff. Temporary surpluses of debris accumulate at the foot of the cliff and form a **cliff foot talus** (Fig. 23.16).

In this way, a typical littoral sequence develops on a cliff coast (section 23.1.2). Above the cliff are the prelittoral landforms into which the cliff retreats, cutting across the previous denudation forms on land and creating a geological and structural cross-section. The upper edge of the cliff represents a profile of the previous landscape. Prelittoral valleys whose streams flow into the sea on the coast lose their lowest valley stretch and, if downcutting does not keep pace with the cliff retreat, become **sea cliff hanging valleys** which appear as incisions on the upper edge of the cliff. This type of hanging valley can be seen on many cliff coasts.

Debris that has fallen from the cliff and been worked by the surf often forms a **gravel beach**, with beach ridges, beach dells and beach cusps, on the inner zone of the abrasion platform. Part of this debris is moved seawards but is replaced

FIGURE 23.16 Chalk cliff with abrasion platform near Birling Gap, Sussex, England. On the right in the background is cliff foot talus from a recent rock fall. The perpendicular groove in the cliff is an ancient well which has been exposed by the retreat of the cliff.

by new material from the cliff. On the seaward section of the abrasion platform there is hardly any debris. The transport system of the surf deposits the material that arrives from the cliff as a submarine talus beyond the seaward margin of the abrasion platform.

If the sea level remains constant this development eventually approaches a limit: the further the cliff base retreats, the less often it is reached by the breaking surf and the more energy of the translation waves is expended on the abrasion platform. Debris from the cliff remains for longer and longer periods at the cliff foot before being redeposited by the surf. The width of the abrasion platform cannot therefore exceed a maximum which is determined by the local surf intensities and the tidal range. A cliff that is no longer undercut by the surf is changed gradually into a normal slope by denudation processes. The previously **active cliff** becomes an **inactive cliff**. This transition can also come about when sediment from further along the coast is deposited in the area in front of the cliff, separating it from the active surf zone.

Abrasion platforms on tidal coasts may be wider than on coasts with little or no tide because high water brings the surf further in over the platform, although there are limits to the absolute width possible. Earlier theories concerning the development of large areas of denudation surfaces by marine abrasion by, for example, Ramsay (1863), can only be applied when the abrasion of the surface took place during a long, very slow rise in sea level.

The formation of individual abrasion platforms and cliffs is strongly influenced by the structure of the rock. In unfolded sedimentary rocks that have resistant horizons at sea level, the abrasion platform is very likely adjusted to the surface of one of the resistant layers. On the English Channel coast, for example, a horizon of flints in the chalk at Birling Gap between Eastbourne and Seaford, which is only a few centimetres thick, forms a large part of the surface of the abrasion platform and protects the soft chalk beneath from rapid erosion. The flint also supplies the pebbles that form the beach at the foot of the chalk cliffs (Fig. 23.16).

At Portland Bill, also on the Channel coast, the abrasion platform is stepped with several levels that are adjusted to the bedding planes of the Portland limestone layers which dip at a very low angle to the sea in this area. On the landward side of these abrasion platforms there are accumulations of large rounded limestone boulders (Fig. 23.17).

FIGURE 23.17 Several abrasion platforms adjusted to the bedding planes in Portland limestone (Jurassic) and accumulations of large limestone boulders on the Bill of Portland, England.

Structure and resistance also have a considerable influence on the smaller forms on cliffs. Cliffs in soft material have a more or less smooth surface. In hard rocks, even if they are petrographically uniform, the joints provide surfaces and lines of weakness where denudation by the surf and rock fall can etch out details of the rock structure. The granite at Land's End in England, for example, has two sets of vertical joints running almost at right angles to one another as well as a large number of horizontal pressure release joints, which together give the rock a rectangular structure. As a result, the cliff is made up of right-angled projections and columns, and is stepped everywhere because of the many pressure release joint surfaces (Fig. 23.11).

Where joints are dense, the surf can remove the blocks more easily and **wave cut inlets** develop along the zones of higher joint density. A **marine stack** is formed, standing free on the abrasion platform, if a wave cut inlet cuts off a projecting area of cliff from the cliff itself (Fig. 6.5). **Marine arches** and **sea caves** (Fig. 23.18) develop if the rocks above the wave cut inlet do not collapse but are held up by lateral cohesion with the neighbouring rock and remain as arches or roofs. A good example of a marine arch is Durdle Door near Lulworth on the south coast of England which is formed in a hogback of Portland limestone that rises out of the sea (Fig. 23.19).

23.5 FORM ASSOCIATIONS ON COASTS OF UNCONSOLIDATED MATERIAL AND NEUTRAL SHORELINES

23.5.1 Barrier beaches and spits

A **barrier beach** is an elongated accumulation of unconsolidated material that rises above sea level and on whose seaward side the littoral sequence of unconsolidated material coasts (section 23.1.2) is more or less completely developed. The barrier beach separates a shallow area of water, a **lagoon**, from the sea. The cross-section from beach to the lagoon shows:

1. the beach with the forms described in section 23.4.5;

FIGURE 23.18 Wave cut inlets and caves at Land's End, Cornwall, England.

FIGURE 23.19 Durdle Door near Lulworth Cove, Dorset, England. A marine arch in steeply dipping Portland limestone.

2. the dune belt;
3. marsh, sheltered from the sea by the dunes.

Barrier beaches develop in a number of ways. They occur mostly where the tidal range is small and often begin as a sand reef in shallow water in front of the coast. Once sedimentation has raised the reef sufficiently for it to be flooded only occasionally, the sand above the water may be blown into dunes. If the dunes withstand subsequent flooding, the sand reef gradually becomes a permanent island whose initial shape resembles that of the original reef. While still in the sand reef stage, waves break on the seaward side of the reef and begin to form a beach. On the shore of the sand reef there is a lateral transport of sand. Sand eroded by rip currents either accumulates as new sand reefs in the outer surf zone or is transported by the current in the same direction as the beach drift material, along the coast parallel to the beach. Part of this sand is returned to the beach by the surf. Because of this longshore sand transport, the leeward end of the sand reef is lengthened. The extended area of the reef also develops a beach, dunes and marsh area. In this way **barrier islands** or **free barrier beaches** such as the Lido of Venice in Italy are formed. All beaches are called lidos in Italy even when they are on the mainland coast, the Lido di Ostia near Rome, for example; the word lido cannot, therefore, be used as a term to describe barrier islands, as has sometimes been done.

By contrast, **spits** extend from a projection on the coast and begin their development to leeward of the dominating wave and storm direction. They grow longitudinally as a result of beach drift and the transport of sediment by longshore currents. Eventually the spit may join the next projection along the coast and convert the bay it cuts off into a lagoon. Generally a passage through the spit remains, however, especially on tidal coasts. In the shelter of the barrier beach, whether free or connected to the mainland, a marsh area of organic mud and inorganic silt and clay particles from the suspended load develops on the landward side of the spit.

The end of the spit is affected by wave refraction as it grows longitudinally. The surf waves swing round at the end of the spit and during periods of high seas cause the beach drift to turn into the bay lying behind the spit to form a **hook**. The spit often continues to grow in the original direction during periods of calmer seas. The position of former hook ends of the spit can be seen as curving sand ridges on its lee side.

There are a large number of spits, hooks and barrier beaches on the coasts of Europe. Hooks and barrier beaches have developed in several locations on the south coast of the Baltic Sea, including the barrier beaches of the Kurische Nehrung and the Frische Nehrung on either side of the Samland peninsula near Kaliningrad and the spit of the Hela peninsula on the west side of the Bay of Gdansk. Several spits have developed along the English Channel coast including Hurst Castle spit and Hengistbury Head, both relatively simple, small spits. The largest in the area is Chesil Beach near Weymouth, a 19.2 km long barrier beach of pebbles connecting the Isle of Portland on one end with the mainland on the other (Fig. 23.4). The combination of landforms, an island connected to the mainland by one or several barrier beaches, is known as a **tombolo**. On Chesil Beach the pebble size decreases from east to west, against the predominating wave direction. This apparent paradox has led to a great deal of research into its development (King, 1972, pp. 307–10). Another complex coastal landform development is represented by Dungeness, a triangular **cuspate foreland**, also on the Channel coast, which juts out into the sea between Hythe and Winchelsea. It has been created by spits growing towards each other from opposite directions as a result of alternating wave directions.

On the coasts of the Netherlands, Germany and Denmark there are a series of **barrier island chains**. The East Frisian Islands extend from Texel to Wangerooge and the North Frisian Islands from Amrum and Sylt to Romö and Fanö. The sea inlets (gats) between the islands are much wider than the passages between more typical barrier islands. In Valentin's classification (1952) they form a separate type, the barrier island chain with tidal flats (Table 23.1).

The East Frisian Islands are migrating from west to east. The beach drift and the net coastal current move in this direction so that the islands are eroded at their western ends and extended at their eastern ends by accumulation. In general more has been deposited at the eastern end of the islands than has been eroded from the western end, so that over the past 300 years the length of most islands has increased by several kilometres. The tidal range on the North Sea coast is the primary cause for the particular development of the Frisian Islands. Instead of a lagoon, there are extensive areas of tidal flats behind both the East Frisian and the North Frisian Islands, in the latter case also several marsh islands. The smallest of these, which are unprotected by dykes from the winter storms, are known as **halligen**.

In the USA, there are hundreds of kilometres of barrier beaches and islands along the Atlantic coast from New York to North Carolina, in Florida, and on the coast of Texas in the Gulf of Mexico. The barrier beach at Cape Hatteras, North Carolina is up to 50 km from the mainland. Elsewhere the lagoons are only a few kilometres wide and have often been largely filled by marsh. Sand layers within the marsh sediments indicate that the dune belt on the barrier beach is occasionally flooded by waves from storm floods which spread the sand from the dunes in a layer over the marsh surface.

Storm floods also break through the barrier beaches but the gap is closed again by beach drift, unless humans interfere with this development by building groynes and moles. At Ocean City, Maryland in the USA, a hurricane flood broke through the barrier beach in the early 1930s. The breach was kept open artificially by a mole so that the channel could be used as an access to the ocean, especially by sport fishing boats. The mole interrupted the beach drift, which moves from north to south, so effectively that on the north side of the gap a wide sand beach has accumulated. At the same time, immediately to the south of the channel, the beach has receded several hundred metres because the sand supply from the north is missing in its mass balance (Fig. 23.20).

23.5.2 Local mass balance of barrier beaches

The example at Ocean City shows the importance of beach drift in the mass balance of a barrier beach coast. The sand balance of the barrier beach, excluding the marsh behind the coast and any aeolian sand transport within the dune belt, can be understood as a simple process response system with the following components:

FIGURE 23.20 Topographic map of the barrier beach at Ocean City, Maryland, USA about 30 years after a hurricane had broken through the barrier. On the north side of the mole, a broad beach has developed. The continuation of the barrier beach (Assateague Island) south of the break has retreated. Further south the barrier beach has maintained its old position.

C' the amount of sand present in a cross-section of the barrier beach at the beginning of a time unit;

C the amount of sand present in this cross-section at the end of this time unit;

A_s the supply of sand from beach drift per time unit;

A_b the sand supplied by the surf from the sea floor per time unit;

A_w sand brought by the wind per time unit;

R_s removal of sand by beach drift per time unit;

R_b removal of sand by rip currents to the ocean floor per time unit;

R_w the volume of sand removed by the wind or by flooding into the tidal marshes or lagoons per time unit.

The balance equation is, therefore

$$C = C' + A_s + A_b + A_w - R_s - R_b - R_w \quad (23.17)$$

The mass balance is in equilibrium when

$$A_s + A_b + A_w = R_s + R_b + R_w \qquad (23.18)$$

To simplify the balance concept it is assumed that when equilibrium occurs the position of the shoreline of the beach does not change but that with a positive balance ($C > C'$) the shoreline advances seawards and with a negative balance ($C < C'$) it recedes.

Each change in a component of equation (23.18) causes the shoreline to advance or retreat. For example, north of the mole that kept the channel open at Ocean City, R_s was reduced which led to the accumulation and advance seawards of the shore, while south of the channel, A_s was absent and the shoreline retreated. Subsequent developments on this stretch of coast indicate a return to the former state of equilibrium. Without further human interference, accumulation north of the mole will continue only until the seaward end of the mole is reached. Thereafter, the advance of the shoreline will cease and the transport of sand southward along the coast can begin again. R_s north of the mole will increase, the channel will be closed and the southern continuation of the barrier beach will receive an increase in its sand supply that will slow down and eventually stop its retreat.

On coasts with a general, not only local, positive mass balance, an advance occurs less as a seaward shift of the existing shorelines than as an emergence of sand bars near the shore and the creation of new barrier beaches. The coast behind the present day beach consists then of more or less parallel sand or gravel bars separated by lower lying strips of, in general, wetland. At Nayarit in Mexico, there is a 15 km wide belt of barrier beaches separated by depressions (Gierloff-Emden, 1980, p. 927), which can be termed a **barrier island series coast**. These barriers are considerably higher and broader than beach ridges and should not be confused with them. A similar barrier series occurs on the Atlantic coast of Florida near St Augustine where the present outer barrier beach is separated from the older barrier by a narrow lagoon that is being filled in.

23.5.3 Neutral shorelines

The postglacial rise in sea level covered part of a varied landscape that had been formed by terrestrial fluvial, glacial or aeolian process systems. Most of the coasts that developed had, therefore, bays and inlets.

The waves arriving at the coast changed direction in shallow water, due to refraction, and converged on the headlands. The abrasion effect of the surf on the rocks formed cliffs. Beach drift moved the rock debris to the inner part of the bays on either side of the headlands where it formed a crescent-shaped beach of increasing width.

These changes are continuing today on many bay coasts. While the headlands retreat because of denudation, the shorelines in the bays are advancing and the bays are being filled in with sediment (Fig. 23.21). In the final stage of this development the shoreline of the beach connects the former headlands in a straight line and a neutral shoreline is established on which stretches of cliff alternate with stretches of sand accumulation. On coasts that become straight as a result of barrier beach development, instead of the advance of the beach in the bay, the bay becomes a lagoon that is separated from the sea and is eventually filled in.

23.6 STRUCTURALLY CONTROLLED COASTAL TYPES

The geological structure exercises a considerable influence on the shape and often also on the profile of the coast. The 80 km long north coast of the Nyboe Land peninsula in northern Greenland, for example, is a fault scarp coast. A young fault scarp falls very steeply and at an almost uniform angle into the northern Polar Sea. The steep drop has not been formed by the sea but consists primarily of the surface of the fault (Fig. 20.4). The coasts on both sides of the Red Sea which are part of the African–Near Eastern fault system also belong to this type. The classification of a particular stretch of coast is not always unequivocal, especially as it can also depend on the scale used in the approach.

FIGURE 23.21 A beach accumulated in a bay as a result of refraction, Sennen Cove near Land's End, Cornwall, England.

Coastal forms in folded structures are largely determined by the strike direction and whether the coast runs with the strike of the folds, across the strike or at an oblique angle to it so that either a **longitudinal coast**, a **transversal coast** or a **diagonal (oblique) coast** results. Valentin divided these coasts into young folded and old folded structures. The age of the folding, which in every case is older than the postglacial eustatic rise in sea level and which determines the position of the coast, seems to be morphologically less important, however, than the strike direction relative to the direction of coast.

The Dalmatian coast is a good example of a longitudinal coast, also termed canale coast. The direction of the folds and the coastline are almost parallel. The islands offshore are mostly anticlinal ridges or horsts that rise above sea level. The intervening arms of the sea are synclinal valleys or grabens. The entire west coast of North America from southern Alaska to Mexico can be defined as a longitudinal coastline, although there are some short stretches of diagonal coastline, especially in California. The west coast of Peru and Chile is also a longitudinal coastline. Valentin (1952) classified these coasts between 50°N and 40°S as young folded coasts and poleward of these latitudes as fjord and skerry coasts because of glacial action (Table 23.1).

The south coast of the Peloponnesus in Greece is a transversal coast. Here the folded and faulted structures of the Dinaric–Greek mountain system extend southwards. The ridges form peninsulas and the longitudinal valley zones the bays. On the transversal coast of Galicia in northwest Spain, the bays, which reach far inland, are termed **rias**. In Europe, there are also ria coasts on the west coast of Brittany and in southwest Ireland.

On transversal coasts formed of old folded more or less homogeneous rocks where the surf is strong, the cliffs cut across the folds almost in a straight line. At Hartland Point on the north coast of Devon in England, the carboniferous sandstones and siltstones form straight walls in the cliff but on the abrasion platform to seaward, single resistant sandstone beds have been left standing as narrow rock walls, up to several metres in height, which run in the direction of the strike into the sea.

The coastline between San Francisco and Los Angeles is a diagonal coast: the strike of the outer chains of the Coastal Ranges, such as the Santa Cruz Mountains, the Santa Lucia Range and the Santa Ynez Mountains reach the coast at an acute angle and the valleys that lie behind the ranges form bays, where they reach the coast. Monterey Bay is an example.

23.7 COASTLINES DETERMINED BY CLIMATE

Directly or indirectly climatic factors play a role in the formation of all coasts. They influence rock weathering, the frequencies and strengths of different wind directions and thereby the wave direction, and affect the development of vegetation. In regions of extreme cold or extreme heat, the influence of environmental conditions determined by climate are especially strong, sometimes stronger than the influence of other factors. For example, ice in cold regions and the growth of coral reefs in warm regions often dominate over other non-climatic factors.

23.7.1 Coast influenced by glacial factors

Some coasts are composed entirely of ice, including the several hundred kilometre long rim of the Antarctic ice shelves in the Ross and Weddell Seas and the more than 100 km long ice front of the Humboldt Glacier in northwest Greenland. Valentin (1952) did not classify these coasts because they are not formed of rock, but in their function as a boundary between land and, at least during part of the summer, water, they are a type of coast.

The effect of pressure by grounded icebergs and pack ice on beaches of unconsolidated material in the Arctic and Antarctic is a factor on coasts in these regions. Pack ice pressure results in the formation of ice-pushed ridges on beaches. They are similar to beach ridges on ice-free coasts although more irregular in form.

The shape and form of coasts influenced by glaciation is primarily determined by the action of the glaciers themselves. Coasts in regions of glacial erosion consist mainly of landforms in bedrock. Valentin (1952) distinguishes between linear and areal erosion (Table 23.1). Linear glacial erosion includes the drowned troughs, **fjords**, produced by linear ice streams (Fig. 23.22). The fjords in Greenland were formed in the Pleistocene and are, in part, still being eroded by outlet glaciers flowing from the inland ice. In the Antarctic there are relatively few fjords. The Norwegian fjord coast from Stavanger northwards and the area between Narvik and the North Cape clearly indicates the adjustment of the glacial erosion to the linear structure of the rocks. Many of the fjords are overdeepened in their long profiles because glacial erosion at the end of the glacier was less intensive than further inland where the glacier ice was thicker and moved more rapidly.

Strandflats are characteristic of the Norwegian fjord coast. They are platforms on the outer margin of the coastal area up to several kilometres in width. Their development has not been fully explained (King, 1972, p. 555). They are not abrasion platforms. In the outer areas of the Norwegian and other fjord coasts, **skerries**, which are partially drowned roches moutonnées, rise above the water level. The Scottish equivalent of fjords are the sea lochs, most of which are on the west coast.

In North America, fjord coasts occur in the east in Labrador, Baffin Island and Ellesmere Island and in the west on the coast of British Columbia and Alaska. Structurally, both the west coast of North America and the coastline in Chile from Puerto Montt to Cape Horn are also longitudinal coasts.

Coasts with forms resulting from areal glacial erosion are areas of low relief that were covered by the eroding inland ice. Low longitudinal depressions that were eroded out by the ice are much shallower than fjords. After the postglacial rise in sea level they became rocky elongated bays known as **fjärds**. Because of the low relief, skerries are also common on these coasts. They are widespread in Finland and central Sweden and on the coast of Maine in the USA.

Both predominantly linear forms and areally developed forms are present in regions of glacial accumulation. A **fœrde** is a bay in glacial deposits that is elongated along an axis and formed by the erosion of a glacier tongue on the inland ice margin, often in conjunction with subglacial meltwater. Generally an end moraine from the last glaciation lies at the head of the fœrde. Examples are the Flensburg Fœrde, the Schlei, the Eckernfœrder Bay and the Kiel Fœrde in the western part of the Baltic coast in Germany. **Bodden** are bays in areas of glacial ground

FIGURE 23.22 Romsdal Fjord near Andalsnes, Norway.

moraine deposits occurring a little further east. The undulating relief of the partially flooded moraine is the cause of their complex irregular outlines. The largest are the Saale Bodden, the Jasmund Bodden on Rügen and the Greifswald Bodden. Both the Saale and Jasmund Boddens are separated from the Baltic by barrier beaches and can be termed lagoons with glacially controlled shorelines.

23.7.2 Coral coasts

Coral reefs in tropical oceans range in size from small local calcareous bars to islands and island groups. They are of interest to biologists, ecologists and earth scientists. To geomorphologists they are of interest primarily as organically developed marine coastal forms. Guilcher (1958, pp. 118–36), Wiens (1959), Bird (1968, pp. 190–211), Fairbridge (1968d) and Gierloff-Emden (1980, pp. 955–80) have described the geomorphology of coral coasts. Wiens (1962) has provided the most comprehensive and detailed treatment of the geology, morphology and ecology of coral islands.

Corals are sessile polyps attached to the sea floor. They take up dissolved calcium carbonate from the sea water which is deposited in the form of their skeletons. Reef corals live in colonies and their skeletons build up masses of coral limestone. Calcareous algae live in symbiosis with the corals and participate in the building of the coral limestone layers. The algae draw their nutrients from the corals and also deliver oxygen to them through photosynthesis and, in addition, secrete calcium. **Coral reefs** are large hard masses of living and dead coral which incorporate the coral skeletons and secretions from calcareous algae as well as mussel shells, foraminifera and other calcareous remains of marine forms of life.

In order to live, reef coral requires the following environmental conditions (Guilcher, 1958, p. 118):

1. water temperature over 18°C, optimally between 25°C and 30°C;
2. salinity between 2.7 and 4.0 per cent;
3. water depth of less than 27 m, because of the light requirements of the calcareous algae for photosynthesis;
4. turbulence in the water to increase oxygen and nutrient contents;
5. as little sediment load as possible because fine sand, silt and clay deposits suffocate the coral.

These conditions explain the distribution of coral reefs. They are present mainly on the islands and coasts of the central and western Pacific Ocean, the tropical Indian Ocean, the Red Sea and the east coast of Central and South America as far as eastern Brazil. Their northern boundary in the Atlantic is around Bermuda at about 32°N, in the Pacific at Midway Island at 28°N. Lord Howe Island at 31°30'S between Australia and New Zealand lies at their southern boundary.

Cold ocean currents flowing northward on the west coasts of continents limit the development of corals, especially in South America where the Humboldt Current brings cold water from the south to the equator and in Africa where the water temperature remains high enough only between about 10°N and 10°S because of the Benguela Current flowing from the south and the Canary Current from the north. The cold currents on the west coast of Australia and on the coast of California are apparently not strong enough to have a similar limiting effect in these areas.

Coral reefs do not develop at the mouths of large rivers if the salinity is reduced to less than 2.7 per cent. Even small rivers with a suspended load can prevent coral growth. The turbulence factor becomes apparent in that reefs developed in the surf on the windward sides of islands are wider and more massive than those on the lee side developed in weaker surf (Verstappen, 1968).

Although live corals occur only to a depth of about 27 m, coral reefs are present at greater depths. Several explanations have been put forward to account for this. Daly (1934) in his glacial control theory suggests eustatic changes as the cause. The glacial eustatic drop in sea level during the last ice age was more than 100 m so that coral growth could have taken place at a depth of about 130 m below the present sea level. When the postglacial rise in sea level began, these corals died off and new corals grew as the sea level rose. For this to take place, a growth rate is necessary that corresponds to the rate at which the sea level rose. According to Fairbridge (1961), in the 12 000 years from 18 000 BP to 6000 BP sea level rose 100 m, a mean rate of 0.8 m per 100 years. Growth rates measured on present-day coral reefs approach these rates in only a few instances (Wiens, 1962, pp. 104–5). Most are considerably lower. There is, however, no doubt that the mean rate during the period of sea level rise that ended 6000 years ago is correct and that the growth rate of the coral reefs that now lie just below the ocean surface must have kept pace with the rise in sea level. This leads to the conclusion that the mean growth rates of the corals during the period of rising sea level were greater than those prevailing during the past 6000 years when the sea level was more or less constant. Perhaps the progressive rise in sea level was itself a reason for the larger growth rates in the past because when the sea level is rising the corals are subjected to the destructive effect of the surf, which reduces the growth rate, for a shorter time than if the sea level were constant.

The calcareous reefs of many coral islands reach to a depth at which glacial eustatic changes are not sufficient to explain their location. Bore holes in coral reefs in the Pacific have been drilled to a depth of 339 m on Funafutu, Tuvalu, of 319 m on Oahu, Hawaii and 640 m on Bikini in the Marshall Islands. None of these bore holes reached the bottom of the limestone corals. The deepest corals were probably developed more than 25 million years ago in the Oligocene period (Guilcher 1958, pp. 131–2).

The presence of coral limestone at these depths can only be explained by a long-term sinking of the ocean floor. There is independent evidence that this took place in the form of **guyots**, named after A. H. Guyot (1804–1884) a geologist and geographer. Guyots are extinct volcanoes that rise from the ocean floor. Their summits are flattish and lie more than 200 m below the surface of the ocean (Hess, 1946; Gierloff-Emden, 1980, pp. 393–402). Pebbles of basalt and coral material on their level summits indicate that the summit surfaces are former abrasion platforms (Guilcher, 1958, pp. 242–4).

Guyots in the central Pacific rise 4000 m from the deep sea floor to within 1200 m of sea level. The latter amount apparently corresponds to the

depth they have sunk since their summits were abraded. Reasons for the sinking include a regional sinking trend of the crust caused by plate tectonics and local isostatic lowering of the oceanic crust due to the weight of the volcano.

Darwin (1842) developed a classification of coral reefs and an explanation of their genesis that assumed a lowering of the sea floor which, in its essentials, is valid today. Based on his observations during the voyage of the *Beagle* (1831–1836) around the world, he distinguished the main types of coral reef: the **fringing reef**, immediately in front of a coast, the **barrier reef** which is also linear but more distant from the coast so that between the reef and shore there is a shallow water zone protected from the surf, and the **atoll**, a ring-shaped coral reef or coral island which encloses a lagoon (Fig. 23.23).

The initial stage of Darwin's theory is a volcanic island rising above the level of the ocean, around the coast of which a fringing reef is developed (Fig. 23.23a). As the island sinks, the reef continues to grow upward in its former position so that its surface remains close to sea level. The island becomes smaller as it sinks and its shores lie further from the reef, which then becomes a barrier reef (Figs 23.23b and 23.24). Finally, the island disappears below the sea and the barrier reef has the form of an atoll (Fig. 23.23c).

Darwin's theory is not valid for **shelf reefs** that have developed in shallow water on the continental shelf away from the mouths of large rivers. An example is the Bahama Reef in front of

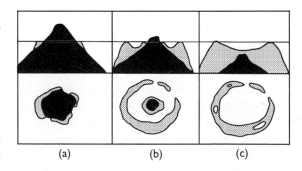

FIGURE 23.23 The development of coral reefs, based on Darwin (1842). (a) fringing reef; (b) barrier reef; (c) atoll (Kelletat, 1989, p. 153).

FIGURE 23.24 Barrier reef around Bora-Bora Island in the Pacific (Photograph by W. Kreisel).

the Florida coast. Because of the shallow conditions in which they develop, these reefs spread in various directions and usually have a more irregular outline than other types of reef. The largest shelf reef in the world is the Great Barrier Reef off the east coast of Queensland in Australia. It is 2000 km long but is not a barrier reef as defined by Darwin.

At various locations on all types of coral reef where coral growth is less intensive, there are passages through the reef which provide channels for ships and boats. The passages in fringing reefs are usually at locations where there is a reduction in salt content or an increase in the suspended load in front of a river mouth. In barrier reefs and atolls the passages are mainly kept open by tidal currents.

When the surf breaks up the coral and calcareous debris accumulates as a beach ridge on the reef, coconuts and other plant seeds that float on to the beach create the beginnings of a vegetation cover on the reef. Resistance to erosion is increased and the height of the beach raised by storm debris until the reef becomes a coral island. On an atoll, there may be a series of islands forming a ring around the central lagoon (Fig. 23.23c). The largest islands usually form spur-like projections on the atoll because the reef

platform is wider in this area than in the stretches of reef in between.

The outline of the atolls varies from almost circular or star shaped with broad points to an elongated oval. Some are horseshoe shaped with broad reefs to windward and little or no reef development on their lee side. For atolls lying in the trade wind belt of the Pacific, the dominating wind direction remains the same throughout the year.

The atoll size depends on the size of the surface of the sunken volcano or group of volcanoes on which the coral reefs have developed. The largest in the world is Kwajalein in the Marshall Islands with a long axis of 120 km and a short axis of 28 km. Another large atoll is Namonuitu in Micronesia which measures 83 km × 50 km (Wiens, 1962, p. 28). In such large lagoons the inside of the reef is also affected by the surf, and broad reefs develop their own small lagoons. Small atolls of this type that form part of the reef of a large atoll are termed **faro**. They are particularly frequent in the northern Maldives, where the word faro originates (Guilcher, 1958, p. 126).

The outer margins of atolls slope steeply into the sea. Angles of 45° are common. The inclination on the inside toward the lagoon is less steep.

Lagoon depths are usually more than 10 m but rarely more than 100 m. This limit strengthens the supposition, based on Daly's theory, that the development of the present-day reefs is controlled by the postglacial rise in sea level. The floor of individual lagoons is irregular because corals also grow in the lagoons, but less densely than on the outer reef washed by the surf.

Coral coasts, islands and reefs are the predominant marine landform in the greater part of the tropics. In detail, their morphology is very complex. Further details of their forms, their development and their functional relationships to the related processes can be found in the references cited at the beginning of the section.

23.8 SHELF FORMS AND SUBMARINE CANYONS

The **continental shelf** is the submerged margin of the **continental platform** and reaches to a depth of about 200 m, beyond which the steeper **continental slope** descends to a depth of 3000–4000 m. The continental slope is the real structural margin of the continental crust with the oceanic crust (*see* hypsographic curve, Fig. 4.1).

The width of the shelf varies greatly. On the longitudinal coasts of western North and South America and Africa, there are mountain chains along the coasts and the shelf is very narrow. In Europe the continental shelf reaches almost 1500 km westward from the German and Danish North Sea coasts. Apart from a deep channel near the coast of southern Norway, the entire North Sea is a shelf sea. The British Isles lie on this shelf. Other large shelf areas are the Yellow Sea and the East China Sea, the southern part of the South China Sea together with the adjoining Java Sea, the Arafura Sea between Australia and New Guinea, the northern Asiatic shelf in the Arctic Ocean, the Great Newfoundland Bank and the Falkland Shelf.

During the glacial eustatic lowering of sea level in the Pleistocene ice ages (section 5.1), much of the shelf area became land and was, therefore, affected by terrestrial geomorphological processes. Rivers continued their courses across this area to the ice age coasts. Their valleys, channels and deposits can be identified on the present sea floor. During the early phases of the Saale ice age, the Rhine and the Meuse flowed northwards in the present area of the North Sea but were forced to flow westward towards the English Channel by the advancing ice (Zonneveld, cited in Flint, 1971, 238; Gierloff-Emden, 1980, pp. 332–47). The ice left glacial forms and deposits on the floor of the glaciated part of the North Sea. During the subsequent rise in sea level, coastal forms such as beach ridges, abrasion platforms and cliffs were developed, all of which are now submerged. Tidal currents affect the development of forms on the present-day continental shelf by shifting shelf sediments. Similarly, the suspended load in river water, which has a lighter specific gravity, is carried far beyond the river mouth before being deposited on the shelf.

Submarine canyons were early recognized as a significant form on the continental slope (Shepard, 1948). They are deeply cut erosion channels, tens of kilometres wide and often more than 100 km long which reach down to the foot of the continental slope. Many submarine canyons are aligned as the continuation of a former river channel on the shelf that originated during a glacial period of lower sea level and is connected to the mouth of a present-day river. The submarine canyons in front of the Hudson River in New York, the Zaire (Congo), the Indus and the Ganges belong to this group. Other canyons have no continuation on the mainland.

Submarine canyons are very probably formed by turbidity currents, gravitational currents loaded with suspended load which, because of their high density, flow down and erode the continental slope, similar to mudflows. Large rivers can deliver such suspended loads directly at their mouths. But elsewhere, too, the continental shelf and its margin is also a storage area for sediment. Submarine mudflows could be triggered by block sliding on the continental slope, or by earthquakes whose shock waves loosen saturated sediments, increasing their pore water pressure and making them fluid (section 8.2.3).

ASPECTS OF APPLIED GEOMORPHOLOGY

24.1 INTRODUCTION

A science is applied when its body of knowledge and methods are used for another purpose than the methodological progress and understanding of the subject matter of the science itself.

In the USA, applied geomorphology was developed in the first half of the 20th century mainly in response to enormous erosion damage caused by water and wind to the cultivated land on the Great Plains and in other areas. The term **soil erosion** had come into use by 1928 (Bennett and Chapline, 1928). Protection of the soil and the introduction of ecologically sound soil-use methods became of public concern and were supported by the Government with new laws and the setting-up of the Soil Conservation Service in the US Department of Agriculture. The 1938 Yearbook, *Soils and Man* (US Dept of Agriculture, 1938), which dealt comprehensively with soil erosion problems and solutions, became a classic in soil erosion literature. In 1933, the US Government had created the **Tennessee Valley Authority (TVA)** to advance soil reclamation and encourage economic development in the catchment area of the Tennessee River. The public engagement meant that applied geomorphology became part of the national programme for improving the environment during the 1930s. The demands of military geography during the Second World War also widened the applications of geomorphology. Thornbury's (1954) introductory text in geomorphology included a chapter

of 38 pages covering such topics as water supply, mineral deposit sites, the geomorphological aspects of the construction of roads, reservoir dams and airfields, and of the use of oil fields. A three-volume series edited by Coates (1972, 1973, 1974) with the title *Environmental Geomorphology and Landscape Conservation* is a comprehensive collection of classical and modern papers concerned with applied geormophology.

In Europe, Bakker (1959) and Tricart (1962) drew attention to the wide range of uses for applied geomorphology. Cooke and Doornkamp (1974) and Cooke *et al.* (1982) have described more recent work. Verstappen (1983a) has emphasized the value of air photographs. Other recent publications include Hails (1977), Gerrard (1984), Costa and Fleisher (1984), Embleton (1988), De Ploey, *et al.* (1991) and Mäckel (1991).

2.4.2 STRUCTURE OF APPLIED GEOMORPHOLOGY

A considerable amount of research is devoted to landforms and processes that are created, or at least influenced, by human activity. These investigations are usually considered as part of applied geomorphology, but they belong to the core of geomorphology as well, since they deal with the effects of humans as a factor in otherwise natural geomorphological systems.

Human geomorphological activity results from other causal relationships than those in a natural process system such as denudation or stream work. They can, however, be expressed statistically in the form of functional probability statements. The application and consequences of stream regulation or the removal and accumulation of material in a lignite open cast pit are examples of human intervention and change. They interfere with the natural systems and interact with them by feedbacks. For example, stream regulation influences the erosion and sediment transport which alter the form of the regulated stream channel.

Of equal importance is research into the actual and potential influences of geomorphological systems on social, cultural and economic circumstances. The use of favourable geomorphological conditions for a settlement or the building of a road are as important as the conscious avoidance of geomorphological hazards such as avalanches, rock falls or floods.

A third significant area of investigation in applied geomorphology is the influence of geomorphological systems on the development, properties and spatial differentiation of the soil, vegetation and local fauna, essentially a geoecological application.

There are several ways to subdivide applied geomorphology. One is by subject matter, land use, forestry etc. (Gellert, 1968); another is to use methodological criteria similar to the subdivision of systematic geomorphology into morphography, functional geomorphology and historic-genetic geomorphology (Ahnert, 1981c):

1. the description of the **static system** of landforms and materials, as a basis for local or regional planning or for the investigation of the relationships between morphographic aspects and other components of land use;
2. the investigation of the landscape changes brought about by humans in the framework of **process reponse systems**, as a basis for the explanation of the interaction between geomorphological systems and human activities;
3. the investigation of **evolution** of the interaction between humans and the geomorphological environment on the basis of the results of 1 and 2.

Contemporary applied geomorphology is often concerned with relationships between geomorphological states and processes and non-geomorphological problems in a cross-section of time in the past or the present. It could relate to the reconstruction of geomorphological environmental conditions in the Neolithic or an examination of past development such as the silting up of an estuary and the decline in its usefulness for navigation (Gottschalk, 1945) or it could predict the future behaviour of a geomorphological system in relation to planning decisions, the landslide risks as a hazard for urban expansion (Hansen, 1984), for example.

24.3 MAPPING

Knowledge of the physical geography of the planning area, especially its landforms, is essential in land-use planning. **Fractional code mapping** was used to map the area covered by the Tennessee Valley Authority to obtain an inventory as a basis for planning (Finch, 1933; Hudson, 1936).

For this method the land area is divided into small units that are as homogeneous as possible. In practice, the units are seldom larger than a hectare so that the mapping is at a large scale. The relevant characteristics are determined for each unit and shown on the map by a code of numbers, arranged on two lines like the numerator and the denominator of a fraction.

Each digit of the code represents a single observed characteristic or parameter of the landscape unit, expressed in a number from 0 to 9. If, for example, the extent of soil erosion has to be mapped in relation to the characteristics of the land form, soil and vegetation, the sequence of digits in a code with five places can show above the line (the numerator), the density, mean length and depth of erosion gullies, the degree of interrill wash denudation and the area or volume of local accumulation. The digit sequence below the line (the denominator), can show the slope angle, soil type, land use and vegetation density. Both quantitative and qualitative variables are used. Of importance is that each digit has the same position in the sequence on a particular map. Where a characteristic is not determined, an x is entered in the sequence. The content and sequence of the digits depends on the purpose of the map so that the method is flexible and widely applicable.

Fractional code mapping allows a high density of information to be shown on a map and should receive increased attention in computer-based recording techniques in association with digital height models (DHM) and geographic information systems (GIS).

In the former Warsaw Pact countries, geomorphological mapping based on air photographs and field observation was used in regional planning. The International Geographical Union produced a handbook of mapping methods and legends (Demek, 1972) which has provided the basis for all later geomorphological mapping. A volume edited by Embleton and Verstappen (1988) discusses later international work in this field.

A programme of detailed geomorphological maps has been carried out in Germany at scales of 1:25 000 and 1:100 000. Most of the 35 maps planned at 1:25 000, representing characteristic landscapes in all parts of Germany, have been completed (Barsch and Stäblein, 1989). One of the goals was the development and use of a standardized legend (Leser and Stäblein, 1975) so that maps of different regions could be compared. This has largely been achieved although the problems of mapping greatly contrasting landscapes in sufficient detail remain.

Problems arise because all of the information has been printed on one sheet in several colours, including landform genesis, structure, hydrology, present-day processes, substrate, bedrock, curvature of the surface, small landforms, slope angle classes and the topographic base. For particular uses such as agriculture, other features than for regional planning or road building need to be emphasized. Similar problems occur with the 1:100 000 map (GMK 100) of which only a few sheets have been prepared. It is possible to transfer the map information of the GMK series on to a computer database (Stäblein, 1981; Dikau, 1988; Strobl, 1988) Some of the maps and their accompanying handbooks are also important contributions to regional geomorphology, for example the map of the Grönenbach area in the Alpine Foreland (Habbe, 1986).

Some geomorphological maps are made for special purposes. A danger map of Grindelwald in Switzerland by Kienholz (1977) shows the probability of catastrophic and other less devastating geomorphological process events in various parts of the Grindelwald area. Different types of dangers are distinguished such as avalanches and rock falls, together with their probable frequency. Maps of this type are particularly useful in relation to planning for building sites and the development of ski routes, paths and other tourist facilities on previously unused slopes. Other maps include a danger map of an active volcanic region in central Java (Verstappen, 1983b) and a model to estimate landslide risk that combines morphography, geology and meteorological data (Dikau 1990).

A number of maps have been produced to estimate the danger of soil erosion on agricultural lands. Rodolfi (1988) has developed a system using several maps to evaluate the agricultural potential of the strongly eroded Tertiary hills in the Chianti region in Tuscany. Landforms and soils were mapped at 1:10 000, supplemented by maps at 1:25 000 of slope angle classes, present-day land use, traces of geomorphological processes and the usefulness of the terrain for irrigated cultivation, particularly significant in the Mediterranean region with its dry summers. Herweg (1988) has mapped landforms and erosion damage in another part of Tuscany using the German GMK concept, but partially at much larger scales, up to 1:2000.

24.4 FUNCTIONAL GEOMORPHOLOGICAL APPLICATIONS

24.4.1 Soil erosion

The term **soil erosion** is applied in geomorphology only to the removal of regolith material when the process has been accelerated by humans. Sheet wash, rill wash and gullying are the most common soil erosion processes. Aeolian removal of soil is less significant and also less investigated. Introductions to soil erosion have been written by G. Richter (1976), Bryan (1979) and Morgan (1979).

Research in soil erosion is particularly important in agricultural areas in which there are dry periods and periods of heavy rainfall, such as the Middle West and southeast of the USA (Fig. 24.1), China, the regions of Mediterranean climate and the subhumid tropics. The temperate middle latitudes are less vulnerable although on some soils, such as loess (Bork, 1983; Pecsi, 1987), on steep slopes and under some types of cultivation (vineyards) severe local damage may occur (G. Richter, 1965, 1977).

Long-term measurements of soil erosion in vineyards on the Moselle, Saar and Ruwer rivers resulted in an estimated soil loss in the measured areas of up to 444 kg/ha in some years (G. Richter, 1978b; 1979, p. 39). Losses of up to several tonnes per hectare on steep slopes with no vegetation have been estimated in central Belgium (Govers 1991).

Soil erosion has historically been a widespread problem in central Europe (Hard, 1976; Machann and Semmel, 1970; Bork, 1988). R.-G. Schmidt (1979) has examined methods to record and quantify soil erosion and J. Schmidt (1991) has developed a mathematical model of soil erosion.

Thousands of years of intensive cultivation in the Mediterranean has lead to extensive soil erosion losses (Bork, 1988; Brückner and Hoffmann, 1992). Forest fires also expose the soil and accelerate erosion (Sala and Rubio, 1994). Hempel (1988) has shown that the bareness of the hill slopes in Greece is not only due to the land use practices but to the changes of climate since the end of the last ice age. Van Asch (1983) found that the soil erosion by wash on freshly ploughed land in southern Italy is transport limited, while on unploughed fallow it is detachment limited and the load transported consisted mainly of particles that were loosened from the soil by splash. The spatial variation of vegetation cover and land use influences soil erosion in the tropical savanna in a similar way (Planchon *et al.* 1987).

Experiments in laboratories and the field can duplicate erosion processes under controlled conditions, measure their components and identify qualitative and quantitative causal relationships (Slaymaker, 1980). De Ploey founded the *Laboratory for Experimental Geomorphology* in Leuven, Belgium. The first experiments dealt with the effect of splash (De Ploey and Savat, 1968), followed by experiments on wash denudation, first in the laboratory and later in a ploughed field (De Ploey, 1972, 1989; De Ploey and Poesen, 1987; Govers, 1987; Govers *et al.* 1990). De Ploey

FIGURE 24.1 Soil erosion resulting from over grazing in the Black Hills, South Dakota, USA.

also developed theoretical estimates of erosion sensitivity for various soils in relation to climatic conditions (De Ploey, 1992; De Ploey *et al.* 1991). Similar erosion experiments have been undertaken in other laboratories (Bryan, 1990, Bryan and Yair, 1982).

Early field experiments by Musgrave (1935) and Zingg (1940) in the USA provided the first empirical functional equations relating to the removal of soil by wash as a function of the slope length and angle. Zingg's equation is still used as a removal equation for wash denudation (equation 8.16). Wischmeier and Smith (1962, 1978) added several variables and developed the Universal Soil Loss Equation (USLE) which has become a frequently used equation for estimating soil erosion. The USLE is

$$A = RKLSCP \qquad (24.1)$$

where

A is soil loss in t/0.4 ha (0.4 ha = 1 acre);
R is erosivity of the precipitation, expressed as kinetic energy of the maximum 30 min intensity of a single precipitation event;
K is erodibility of the soil, a coefficient obtained by empirical comparisons;
L is slope length as a multiple of the standard test field of 22.6 m;
S is slope angle, usually a percentage, or a value relative to the standard test field angle of 9 per cent;
C is a crop management factor, a coefficient of the growth type compared to cultivated fallow;
P is a coefficient of cultivation methods which compares the cultivation method used with the method that would most encourage erosion (ploughing straight downslope).

The particular value of the equation is that it allows estimates of probable changes in the potential for soil erosion with a change in land use. The values for the USLE variables have been obtained from analyses of test fields cultivated in the USA and are not necessarily transferable to other countries without local modification (Bryan, 1979, p. 211). Auerswald (1991) has compared the US values with values obtained in Bavaria in Germany.

The USLE has been adapted for use with digital terrain models to produce high-resolution maps of potential soil erosion (Flacke *et al.*,

(1990). Imeson and Vis (1982) have determined and evaluated variables that affect the erodibility factor *K* along a traverse of the Central Cordillera of Columbia which shows clearly that *K* varies depending on altitude and slope direction.

The values determined for the USLE should be regarded as relative rather than absolute and used for comparing different localities. The dimensions of the value of *A* in equation (24.1), t/0.4 ha, require an extrapolation of the actual values because the standard test fields on which it is based were limited in size to a width of 1.8 m, a length of 22.6 m and had a slope angle of 9 per cent. Overland flow usually occurs over much larger slopes and longer distances so that the values for depth, velocity and turbulence exceed the values obtained on the test fields. Comparability is difficult because many test fields have been laid out with quite different dimensions, ranging from 1 m plots to major slope segments.

There are also some practical problems associated with using the erosivity factor *R* because in order to determine its long-term significance for soil erosion, the maximum rainfall intensity, determined over a duration of 30 min, should be measured at the test site and not at some distant meteorological station. The record of a special recorder near the test field is not usually long enough for the cumulative effect of rainfall over a long period to be estimated. Magnitude-frequency of the daily rainfall as a variable to estimate erosivity has been suggested as an alternative (section 5.1.1, De Ploey *et al.*, 1991).

Artificial rain-making equipment has been used in the field and the laboratory so that the amount and intensity of rainfall can be controlled exactly although the frequency distribution of the drop size often does not resemble natural rainfall (Prahsun and Schaub, 1991; Luk *et al.*, 1993).

To overcome some of the limitations of USLE, Morgan *et al.* (1982) developed a model of the long-term effects of soil erosion which includes additional variables dealing especially with the feedbacks in the system related to wash denudation. The model was developed at two agricultural research stations in Malaysia and has been successfully tested on other cultivated areas in that country. The method is simple and uses data easily available from public sources so that costs are low, important factors in developing countries in the tropics and subtropics where soil erosion is a major problem.

24.4.2 Other applications of geomorphology

Urban geomorphology examines the changes caused by the requirements of urban residential, economic and traffic functions (Cooke, 1976; Cooke *et al.*, 1982). Towns are adjusted to the relief and the relief is also adjusted to the needs of construction and planning. The changes that occur as a result of urban development are also influenced by their interaction with the disturbed geomorphological process response systems, such as weathering on building stone resulting from air pollution (Wolff, 1976; Viles, 1993).

Local landforms have played a central role throughout history in the choice of sites for settlements and their further development has often been influenced by the regional geomorphology. The location of Italian cities on defensive hill sites, the development of the Greek city states in fertile basins surrounded by hills, locations at easy river crossings, on terraces above flood levels, on bays, at the landward end of navigable estuaries or at the exit of large mountain valley systems on to a foreland are some examples. Villages located in spring hollows, along valley floors or on sandy upland bordering marsh areas all once took advantage of the local relief and hydrological conditions. Settlement geography is incomplete without taking into account the morphography and hydrology of the area. The spatial development of the city of Ghent in Belgium was largely limited, until the second half of the 19th century, to the flood-free area of Pleistocene sands, lying above the river marshes. After a drainage project, canal building and the extension of the harbour basin, the city expanded on to the former marshy areas (Heyse and De Moor, 1988). The American capital Washington DC was originally built on the extensive late Pleistocene terraces of the Potomac River. Only in the 20th century has it spread to the older higher terraces and the Piedmont to the northwest of the city (Ahnert, 1958).

Erosion increases in urban areas when, for example, building activity bares the soil and causes extensive wash denudation. The streams become overloaded with sediment as a result and their cross-sections are changed (Wolman, 1967; Wolman and Schick, 1967). Road building has a similar effect, particularly the construction of multi-lane highways. As urban areas expand, more soil surfaces are sealed by buildings, streets, parking areas, industrial and commercial activity so that the surface runoff flows into the sewage system. Groundwater supplies are reduced as a result and flood frequency and the peak flood height are increased (Leopold, 1968; Graf, 1975; Sala and Inbar, 1992). Gupta (1984) has discussed the special geomorphological and hydrological problems of cities in the tropics.

Human activity has increased the runoff maxima in many river basins, and this has been compensated for by the building of reservoirs which store the excess runoff and release it gradually. Reservoir basins are also sediment traps for all the bed load and at least part of the suspended load. Immediately below the dam the stream is practically free of solid load so that it picks up new load and erodes its channel bed downstream from the reservoir (Gregory and Park, 1974; Gregory, 1979). The bed load is smaller than before because the high-water peaks are lower than before the dam was completed. One of the largest reservoirs to act as a sediment trap is Lake Nasser behind the Aswan High Dam in Egypt. Before the dam was built in 1970, the Nile delta received enormous quantities of sediment during the annual flood from the catchment area of the upper Nile in the Sudan, Ethiopia and Uganda. The sediment supply maintained the equilibrium of the delta's coastline between deposition by the Nile and littoral removal by surf and currents of the Mediterranean Sea. Now the Nile transports sediment only from the stretch below Aswan, in which there are no tributaries, to the delta. The amount of sediment is insufficient to compensate for the attack by waves and currents and the coastline is retreating (Hails, 1977b; Bird, 1979).

The high concentration of energy in littoral processes means that all artificial disturbances of the natural dynamic eqilibrium on the sea coast have very rapid and often unforeseen effects.

The building of a breakwater parallel to the shore about 700 m in front of the coast at Santa Monica, California in 1934 had the effect that the waves approaching the shore from the Pacific were turned inwards by refraction at both free ends of the breakwater. As a result, the direction of movement of the translation waves converged on the beach from both sides towards the middle behind the breakwater, which lead to a broad triangular accumulation of sand (Fig. 24.2, Bird 1979, p. 94). In 1983, a storm destroyed the breakwater and ended the development (information from A. Orme, University of California at Los Angeles).

Breakwaters projecting into the sea and even small groynes at regular intervals on the beach interfere with natural sediment movement and disturb the inherent tendency to a dynamic transport equilibrium of the littoral system. An example is the development at Ocean City, Maryland in the USA discussed in section 23.5.1.

The building of houses and roads on steep slopes changes their stability. Roads cut into slopes reduce the critical height and increases the danger of landslides. Slides in road cuts are particularly frequent in climates with periods of high rainfall where the soil pores are filled with water and the specific gravity increased and cohesion decreased. Haigh (1984) has studied the problem in the foothills of the Himalayas. Based on quantitative analysis of the observation data he suggested ways to recognize landslides on air

photographs and estimate landslide risks. The latter are particularly high on slopes that were close to the limit of stability before any interference by man. Even building a house can exceed the stability limit and lead to slides. Examples are discussed in Brunsden and Prior (1984), Selby (1979) and Anderson and Richards (1987).

Geomorphology also plays a role in the exploration of mineral deposits near to the surface and the investigation of mine tips. Toy (1984) has summarized the problem and methods. Mineral deposits close to the surface are mostly of two types and in both cases their origin is related to the geomorphological development of the landform. The first type develops through the enrichment of economically useful mineral substances as **supergene ore** deposits. These are weathering residues in the regolith (Schwarz and Germann, 1994) and include ferricrete (section 7.5.5), bauxite (Schellmann, 1994) and, most importantly, metal ores such as copper, nickel and chrome ores, which are worth mining even at concentrations of low percentages, and precious metals. Gold is mined at concentrations down to <1 ppm. Since enrichment requires a long time, supergene ore deposits are found usually in autochthonous regolith covers produced by the weathering of crystalline rocks on old peneplains (section 7.5.1) on which there has been little or no recent mechanical denudation, making long-term soil development possible. Valeton (1994) has discussed this process for ore and phosphate deposits, Hale and Porto (1994) and Freyssinet (1994) discuss gold enrichment in laterite soils. Clark *et al.* (1967) describe the relationship between the enrichment of copper deposits and the geomorphological development in the Chilean Atacama Desert. The geomorphological mapping of mineral rich regolith in Western Australia by Chan (1988) is an example of this type of mineral exploration.

The second type of mineral enrichment near the surface is the local enrichment of mineral substances in stream deposits as a result of differences in specific gravity compared to the other sediments in the stream. They are known as placers. Most gold and diamond deposits are of

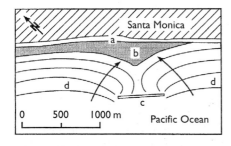

FIGURE 24.2 Beach accumulation due to refraction behind a breakwater at Santa Monica, California, USA (after Bird, 1979): (a) old beach until 1934; (b) beach increase since 1934; (c) breakwater; (d) wave crests moving towards the shore.

this type. Geomorphological methods help to find them (Sutherland, 1984). Thomas and Thorp (1993) have compared diamond and gold placers in West Africa, northwest Australia and Indonesia and identified the most important geomorphological conditions of such deposits. They lie for the most part in tectonically and topographically low areas, grabens for example, with high eroded flanks, in sedimentary basins and in areas of crustal downwarping near coasts. Placers form in stream gravels because the large pores between the gravel particles act as sediment traps for the smaller but heavy gold and diamond grains.

The storage of waste material from mining in the form of mine tips creates considerable safety problems because of the potential instability of the tip slopes. In 1966, a mine tip on a slope in Wales slid without warning into a school and killed many children. Haigh (1980) has dealt with the question of stability and geomorphological development of mine tips in Great Britain.

In the area of the Ranger uranium mines in the Northern Territory in Australia, East *et al.* (1993) have investigated the stability of different types of natural landforms in relation to their use as a base for mine waste. Riley (1994) has developed detailed models of the soil formation, the expected water budget and the possible erosion of the mine tip. Because part of the waste contains radioactive material and the entire operation lies in the Kakadu National Park, every effort has to be made to avoid contamination of the surrounding area. This research is of great practical importance for the preservation of the ecological systems in the region and for tourism.

The evaluation of the recreational potential and the aesthetic impact of the landscape for purposes of tourism are other aspects of applied geomorphology. Leopold (1968) and Morisawa (1971) have written appraisals. More recently, Hamann (1988) has made a geomorphological-aesthetic evaluation with maps of three cirque landscapes in the Austrian Alps. The application of this approach in other landscapes would be useful for tourism planning.

The role of geomorphology in history and prehistory has not often been discussed. Many historical events might be better understood if the contemporary conditions in the landscape and physical environment had been taken account of. Butzer (1964) used geomorphology to interpret the archaeological evidence relating to settlements and traces of human activity by reconstructing the prehistoric conditions that existed for the location of settlements and also the local and regional environmental conditions. Dating of artifacts can also help geomorphologists to date the development of landforms.

25

BRIEF HISTORICAL SUMMARY OF GEOMORPHOLOGY

The advance of a science through time is expressed in its history. The 200-year history of geomorphology is made up of a series of observations, explanations and methodological approaches, many of which have influenced the progress of geomorphology and which are still relevant today. An attempt is made here to summarize the development of geomorphology briefly in order to provide a historical framework.

Geomorphology began to evolve at the same time as modern geology at the end of the 18th century, particularly in the work of Hutton (1795) and Playfair (1802), whose concept of uniformitarianism (section 1.4.6) became widely accepted and recognized with the publication of Lyell's three-volume *Principles of Geology* (1830–1833). A major impetus to this new geoscience was given by A. von Humboldt's account of his South American Journey (34 volumes, 1805–1834) in which empirical observations and, where possible, measurements were emphasized. The word morphology in relation to the earth's surface was first used by Naumann (1858).

One of the first to investigate the forms of the land surface was Ramsay (1863) who suggested that peneplains might be explained by marine abrasion (section 19.1); he also discussed stream erosion and the influence of different rock types on the development of landforms. During the exploration and opening up of the western United States, Powell (1876) developed the theory, later expanded by Davis (1899), that peneplains are the result of long periods of denudation. Gilbert (1877) was the first to recognize that the functional relationship between the process components of a process response system, in this case, weathering, slope denudation and the work of streams, tended toward the establishment of a dynamic equilibrium.

The first half of the 20th century was dominated by the cycle theory (Davis, 1899, 1909), particularly in English speaking countries and France. Its seeming logic, detailed terminology and apparent applicability to a variety of systems and landforms led to the theory being widely accepted for many years. For Davis, the cycle theory was, above all, a general verbal model of landform development and, as such, has not entirely lost its validity even today. It was not his fault that many of his supporters and also some of his opponents transformed his concepts into inflexible schemes that in the end did not lead anywhere. A disadvantage of the cycle theory's popularity was that the importance of Gilbert's approach was not generally recognized until after 1950 (e.g. Hack, 1960).

In Germany, von Richthofen (1886) developed a systematic geomorphology in his *Führer für Forschungsreisende*. The following year Hettner (1887) included Gilbert's ideas in his geomorphological investigation of Saxon Switzerland in Germany. However, the first standard text of morphographic, functional and genetic systematic geomorphology was *Morphologie der Erdoberfläche* (*Morphology of the Earth's Surface*) by A.

Penck (1894). Interestingly, Penck included a large number of quantitative functional equations for geomorphological processes, including a variation of Coulomb's law (Vol. I, p. 223), the Chezy equation (Vol.1, p. 274; *see* equation (8.15)) and an equation for the critical shear stress for the fluvial transport of various gravel sizes (Vol. 1, p. 279; *see* equation (11.3)). Unfortunately, these quantitative beginnings remained largely unnoticed and unused for half a century.

Questions of climatic geomorphology were discussed by Walther (1900, 1924) and in the comprehensive investigation by A Penck and Brückner of *The Alps in the Ice Age* (*Die Alpen im Eiszeialter* 1901–1909; sections 22.7–22.8). Morphological research in subhumid tropical regions advanced other aspects of climatic geomorphology, including the morphology of inselbergs (Bornhardt, 1900; section 19.3–19.5).

A. Penck's son, W. Penck published a stimulating book on process-oriented geomorphology in 1924 that owed much to Gilbert. It was translated into English in 1953. The fact that it contains a much criticized theory of slope development (section 18.3) is of less importance than his emphasis on the principle that the key to the explanation of the development of forms lay in an understanding of the process mechanisms involved. However, the historic-genetic development of the landscape and its formation in different climatic zones began to dominate the research interest in Germany soon after and W. Penck's ideas were largely ignored (Passarge, 1926; Behrmann, 1927; Schmitthenner, 1927; Thorbecke, 1927a,b).

Büdel (1937) introduced a climato-genetic approach to geomorphology with his discovery and explanation of the presence of Pleistocene periglacial solifluction deposits on the high surfaces of the central uplands in Germany. He later turned to the question of denudation surface development in the tropics (section 19.3) and eventually also to the climatic morphological and climato-genetic zonation of the entire earth (1957, 1963, 1977).

In France, de la Noë and de Margerie (1888) and especially de Martonne (1909) based their geomorphology largely on systems of explanation similar to the Davisian cycle of landform development. Baulig (1949, 1952) investigated slope development and denudation surfaces and Birot (1958, 1960) was concerned with structural morphology and climatically determined variations of the Davis cycle. Tricart and Cailleux (1962–1969) published a comprehensive re-examination of climatic geomorphological problems which was based partly on their work on the influence of Pleistocene periglacial conditions on the morphological development of Europe (Cailleux, 1942; Tricart 1949). In addition to his fundamental geomorphological text (1965), Tricart laid the foundations for the development of applied geomorphology, roughly at the same time as the Dutch geomorphologist Bakker (1959; Tricart, 1962). In both France and Germany at this time geomorphology remained largely qualitative.

Modern quantitative geomorphology had begun in Europe with work on the theory of slope development and gravel transport in streams (O. Lehmann, 1933; Bakker and Le Heux, 1947; Hjulström 1935), but the impetus for the widespread use of quantitative methods came from North America. A significant role was played by hydrologists who were also geomorphologists and who concentrated on fluvial geomorphology. From hydrology they brought a quantitative functional approach that was not concerned with the historic-genetic development of landforms but tried to explain the present functional relationships between form, material and process characteristics.

The work of Horton (1945) and Strahler (1952, 1957, 1964) on the characteristics of stream networks and river catchment areas and Leopold and Maddock (1953) on the hydraulic geometry related to the work of streams were of fundamental importance. Hack (1957, 1960) and Schumm (e.g. 1956, 1977) developed these ideas further and used them empirically in the field. Leopold *et al.* (1964) published the first, and still relevant, text that incorporated this research. Chorley (1962) introduced the concept of general systems theory into geomorphology (Chapter 2). A stimulus to further research was Scheidegger's (1961) attempt to give theoretical geomorphology a formal structure by transferring physical laws and concepts directly to it.

From the middle of the 1960s onward, system-oriented functional geomorphology has progressively become part of all branches of geomorphology, together with a trend towards the use of quantitative data and their statistical evaluation. Many of the data are obtained in field and laboratory experiments. They are for the most part carried out to investigate present-day functional relationships between landform, process and material, but are relevant also for understanding the process response mechanisms of past landform development.

Geomorphological theories are expressed increasingly in quantitative terms, to the extent to which this is possible. The use of computers makes it possible to design theoretical process response models to analyse the functional rela-tionships between particular components of geomorphological systems, to test hypotheses concerning the development of certain landform types, or to simulate the evolution of particular landforms. Model experiments which test the effect of different premises on geomorphological systems constitute a third important type of experiment in addition to experiments in the field or in the laboratory.

Detailed accounts of the history of geomorphology have been made by Tinkler (1985) and in a three-volume history by Chorley *et al.* (1964, 1973, 1991), of which volume one covers the 19th century, volume two is essentially a biography of W.M. Davis (1850–1934) and volume three deals with historical and regional morphology from 1890 to 1950.

BIBLIOGRAPHY

Abele, G. 1981: *Trockene Massenbewegungen, Schlammströme and rasche Abflüsse*. Mainzer Geographische Studien 23.

Abele, G. 1994: Felsgleitungen im Hochgebirge und ihr Gefahrenpotential. *Geographische Rundschau*, **46**: 414–20.

Abrahams, A.D. 1972: Drainage densities and sediment yield in E. *Australia. Australian Geographical Studies*, **10**: 19–41.

Abrahams, A.D. 1984: Channel networks, a geomorphological perspective. *Water Resources Research*, **20**: 161–188.

Abrahams, A.D. (ed.) 1986: *Hillslope processes*. Boston Allen & Unwin.

Ahnert, F. 1954: Zur Frage der rückschreitenden Denudation und des dynamischen Gleichgewichts bei morphologischen Vorgängen. *Erdkunde*, **8**: 61–63.

Ahnert, F. 1955: Die Oberflächenformen des Dahner Felsenlandes. *Mitteilungen der Pollichia*, III, **3**.

Ahnert, F. 1958: Washington, DC – Entwicklung und Gegenwartsbild der amerikanischen Hauptstadt. *Erdkunde*, **12**: 1–26.

Ahnert, F. 1960a: Estuarine Meanders in the Chesapeake Bay area. *Geographical Review*, **50**: 390–401.

Ahnert, F. 1960b: The influence of Pleistocene climates upon the morphology of cuesta scarps on the Colorado Plateau. *Annals of the Association of American Geographers*, **50**: 139–56.

Ahnert, F. 1963a: Distribution of estuarine meanders. US Office of Naval Research, Final Report, Project No. 388–U69.

Ahnert, F. 1963b: The terminal disintegration of Steensby Gletscher, North Greenland. *Journal of Glaciology*, **4**(35): 537–45.

Ahnert, F. 1964: Quantitative models of slope development as a function of waste cover thickness. *In Abstracts of papers, 20th Geographical Congress*, London, 118.

Ahnert, F. 1967: The role of the equilibrium concept in the interpretation of landforms of fluvial erosion and deposition. In Macar, P. (ed.), *L'évolution des versants*. Université de Liége, 23–41.

Ahnert, F. 1970a: Functional relationships between denudation, relief and uplift in large mid-latitude drainage basins. *American Journal of Science*, **268**: 243–63.

Ahnert, F. 1970b: An approach towards a descriptive classification of slopes. *Zeitschrift für Geomorphologie, Supplement-Band*, **9**: 71–84.

Ahnert, F. 1973a: Inhalt und Stellung der funktionalen Methode in der Geomorphologie. *Geographische Zeitschrift*, Plewe-Festschrift. 105–13.

Ahnert, F. 1973b: *COSLOP2 – A comprehensive model program for simulation of slope profile development*. Geocom Programs 8.

Ahnert, F. 1976a: Darstellung des Struktureinflusses auf die Oberflächenformen im theoretischen Modell. *Zeitschrift für Geomorphologie, Supplement-Band* **24**: 11–22.

Ahnert, F. 1976b: Brief description of a comprehensive three-dimensional process-response model of landform development. *Zeitschrift für Geomorphologie, Supplement-Band*, **25**.

Ahnert, F. (ed.) 1976c: Quantitative slope models. *Zeitschrift für Geomorphologie, Supplement-Band*, **25**.

Ahnert, F. 1977: Some comments on the quantitative formulation of geomorphological processes in a theoretical model. *Earth Surface Processes*, **2**: 191–201.

Ahnert, F. 1978: Gegenwärtige Forschungstendenzen der Physischen Geographie. *Die Erde*, **109**: 49–80.

Ahnert, F. 1980: A note on measurements and experiments in geomorphology. *Zeitschrift für Geomorphologie, Supplement-Band*, **35**: 1–10.

Ahnert, F. 1981a: Stone rings from random walks. *Transactions of the Japanese Geomorphological Union*, **2**: 301–12.

Ahnert, F. 1981b: Über die Beziehung zwischen quantitativen, semiquantitativen and qualitativen Methoden in der Geomorphologie. *Zeitschrift für Geomorphologie, Supplement-Band*, **39**: 1–28.

Ahnert, F. 1981c: Entwicklung, Stellung and Aufgaben der angewandten Geomorphologie. *Aachener Geographische Arbeiten*, **14** (Festschrift für Felix Monheim): 39–65.

Ahnert, F. 1982: Untersuchungen über das Morphoklima und die Morphologie des Inselberggebietes von Machakos, Kenia. *Catena Supplement*, **2**: 1–72.

Ahnert, F. 1983: Einige Beobachtungen über Steinlagen im südlichen Hochland von Kenia. *Zeitschrift für Geomorphologie, Supplement-Band*, **48**: 65–77.

Ahnert, F. 1984: Local relief and the height limits of mountain ranges. *American Journal of Science*, **284**: 1035–55.

Ahnert, F. 1987a: An approach to the identification of morphoclimates. In Gardiner, V. (ed.), *International Geomorphology 1986*, Part II, John Wiley & Sons, Chichester: 159–88.

Ahnert, F. 1987b: Process-response models of denudation at different spatial scales. *Catena Supplement*, **10**: Cremlingen, 31–50.

Ahnert, F. (ed.) 1987c: Geomorphological models – theoretical and empirical aspects. *Catena Supplement*, **10**: Cremlingen.

Ahnert, F. 1987d: Approaches to dynamic equilibrium in theoretical simulation of slope development. *Earth Surface Processes and Landforms,* **12**: 2–15.

Ahnert, F. 1988a: Das Morphoklima und seine Bedeutung für die Hangentwicklung in Trockengebieten. *Abhandlungen der Akademie der Wissenschaften in Göttingen, Mathematisch, Physikalische Klasse,* 3. Folge, **41**: 229–44.

Ahnert, F. 1988b: Modelling landform change. In Anderson, M.G. (ed.), *Modelling geomorphic systems.* Chichester, John Wiley & Sons, 375–400.

Ahnert, F. (ed.) 1989: Landforms and landform evolution in West Germany. *Catena Supplement,* **15**.

Ahnert, F. 1994a: Randomness in geomorphological process response models. In Kirkby, M.J. (ed.), *Process models and theoretical geomorphology.* Chichester: John Wiley & Sons, 3–21.

Ahnert, F. 1994b: Equilibrium, scale and inheritance in geomorphology. *Geomorphology* **11**: 125–40.

Ahnert, F. 1994c: Modelling the development of non-periglacial sorted nets. *Catena,* **23**: 43–63.

Ahnert, F. 1996: The point of modelling geomorphological systems. In McCann, B. and Ford, D.C. (ed.), *Geomorphology sans frontières.* Chichester: John Wiley & Sons, 81–113.

Anderson, M.G. (ed.) 1988: *Modelling geomorphological systems.* Chichester: John Wiley & Sons.

Anderson, M.G. and Richards, K.S. (eds.) 1987: *Slope stability – geotechnical engineering and geomorphology.* Chichester: John Wiley & Sons.

Andersson, J.G. 1906: Solifluction, a component of subaerial denudation. *Journal of Geology* **14**: 91–112.

Andre, M.F. 1986: Dating slope deposits and estimating rates of rock wall retreat in northwest Spitsbergen by lichenometry. *Geografiska Annaler,* **68A**: 65–75.

Andres, W. Radtke, U. and Mangini, A. 1988: Quartäre Strandterrassen an der Küste des Gebel Zeit (Golf v. Suez/Ägypten). *Erdkunde,* **42**: 7–16.

Arbeitsgemeinschaft Bodenkunde 1971: *Kartieranleitung. Anleitung und Richtlinien zur Herstellung der Bodenkarte 1:25 000.* Hrsg. von der Bundesanstalt für Bodenforschung und den Geologischen Landesämtern der Bundesrepublik Deutschland. 2.Aufl. Hannover; 4. Aufl 1994.

Auerswald, K. 1991: Significance of erosion determining factors in the USA and in Bavaria. *Zeitschrift für Geomorphologie, Supplement-Band,* **83**: 155–60.

Bagnold, R.A. 1941: *The physics of blown sand and desert dunes.*London: Methuen & Co., (2nd edn 1953.)

Bakker, J.P. 1959: Dutch applied geomorphological research. *Revue de Géomorphologie Dynamique,* **10**: 67–84.

Bakker, J.P. and Le Heux, J.W.N. 1947: Theory of central rectilinear recession of slopes. *Kon. Nederl. Akad. van Wetensch.,* **50** (8): 9; 1953: **54** (7): 8.

Barsch, D. 1969: Studien und Messungen an Blockgletschern in Macun, Unterengadin. *Zeitschrift für Geomorphologie, Supplement-Band,* **8**: 11–30.

Barsch, D. 1996: *Rockglaciers.* Berlin: Springer-Verlag.

Barsch, D. and Flügel, W.-A. 1989: Hillslope hydrology – Data from the Hollmuth Test Field near Heidelberg. *Catena Supplement,* **15**: 211–27.

Barsch, D., Gude, M., Mäusbacher, R., Schukraft, G., Schulte, A. and Strauch, D. 1993: Slush stream phenomena – process and geomorphic impact. *Zeitschrift für Geomorphologie, Supplement-Band,* **92**: 39–53.

Barsch, D. and Hell, G. 1975: Photogrammetrische Bewegungsmessungen am Blockgletscher Murtel I, Oberengadin, Schweizer Alpen. *Zeitschrift für Gletscherkunde and Glazialgeologie,* **11**: 111–42.

Barsch, D. and Stäblein, G. 1989: Geomorphological mapping in the Federal Republic of Germany – the GMK 25 and the GMK 100. *Catena Supplement,* **15**: 343–7.

Barth, T.F.W., Correns, C.W. and Eskola, P. 1939: *Die Entstehung der Gesteine* (ed. C.W.Correns), reprinted 1970. Berlin-Heidelberg-New York: Springer-Verlag.

Basu, S.R. and Sarkar, S. 1990: Development of alluvial fans in the foothills of the Darjeeling Himalayas and their geomorphological and pedological characteristics. In Rachocki, A.H. and Church, M. (eds), *Alluvial fans – a field approach.* Chichester: John Wiley & Sons, 321–33.

Bauer, F. 1962: Karstformen in den österreichischen Kalkhochalpen. *Act. 2. International Congress Speleology,* 1958, I, 299–329.

Baulig, H. 1949: Le profil d'équilibre des versants. *Annales de Géographie,* **49**: 81–97.

Baulig, H. 1952: Surfaces d'aplanissement. *Annales de Géographie,* **52**: 161–83, 245–62.

Baulig, H. 1957: Peneplains and pediplains. *Bulletin of the Geological Society of America,* **68**: 913–39.

Baumgartner, A. and Liebscher, H.-J. 1990: *Allgemeine Hydrologie – quantitative Hydrologie. Lehrbuch der Hydrologie,* Band 1. Berlin: Gebrüder Borntraeger, 673.

Baver, L.D., Gardner, W.H. and Gardner, W.R. 1972: *Soil physics.* 4th ed. New York. John Wiley.

Beaty, C.B. 1963: Origin of alluvial fans, White Mountains, California and Nevada. *Annals of the Association of American Geographers,* **53**: 516–35.

Beaty, C.B. 1990: Anatomy of a White Mountain debris flow – the making of an alluvial fan. In Rachocki, A.H. and Church, M. (eds), *Alluvial fans – a field approach.* Chichester: John Wiley & Sons, 69–89.

Behrmann, W. 1927: Die Oberflächenformen in den feuchtwarmen Tropen. Düsseldorfer Geographische Vorträge and Erörterungen, 3. Teil: Morphologie der Klimazonen. Breslau, 4–9.

Bennet, H.H. and Chapline, W.R. 1928: *Soil erosion – a national menace.* US Department of Agriculture, Circular 33.

Bertalanffy, L.v. 1951: An outline of general systems theory. *Journal of British Philosophical Science,* **1**: 124–65.

Bertalanffy, L.v. 1962: General systems theory – a critical review. *General Systems* (Yearbook of the Society for General Systems Research), VII, 1–19.

Besler, H. 1987: Entstehung und Dynamik von Dünen in warmen Wüsten. *Geographische Rundschau,* **39**: 422–8.

Besler, H. 1992: *Geomorphologie der ariden Gebiete.* Erträge der Forschung, Band 280. Darmstadt: Wissenschaftliche Buchgesellschaft.

Besler, H. Blümel, W.D., Heine, K., Hüser, K., Leser, H., and Rust, U. 1994: Geomorphogenese and Paläoklima Namibias. Eine Problemskizze. *Die Erde* **125**: 139–65.

Bibus, E. 1980: Zur Relief-, Boden- and Sedimententwicklung am unteren Mittelrhein. Frankfurter Geowiss. Arbeiten, Serie D – Physische Geographie, Bd.1.

Billi, P., Hey, R.D., Thorne, C.R. and Tacconi, P. (eds.) 1992: *Dynamics of gravel-bed rivers.* Chichester: John Wiley & Sons.

Bird, E.C.F. 1968: *Coasts.* Canberra: Australian National University Press.

Bird, E.C.F. 1979: Coastal processes. In Gregory, K.J. and Walling, D.E. (eds.), *Man and environmental processes*. Folkestone: Wm. Dawson & Sons, 82–101.

Birot, P. 1958: *Morphologie structurale*, 2 vols. Paris: Orbis.

Birot, P. 1960: Le cycle d'érosion sous les different climats. Universidad do Brasil, Rio de Janeiro. *The cycle of erosion in different climates*. London: Batsford.

Blair, T.C. 1987: Sedimentary processes, vertical stratification sequences and geomorphology of the Roaring River alluvial fan, Rocky Mountain National Park. *Journal of Sedimentary Petrology*, **57**: 1–18.

Blume, H. 1971: *Probleme der Schichtstufenlandschaft*. Darmstadt: Wissenschaftliche Buchgesellschaft.

Blume, H. (ed.) 1976: Strukturbetonte Reliefs. *Zeitschrift für Geomorphologie, Supplement-Band*, 24.

Blume, H. 1991: *Das Relief der Erde, ein Bildatlas*. Stuttgart: F. Enke Verlag.

Blume, H. and Remmele, G. 1989: Schollengleitungen an Stufenhängen des Strombergs (Würtembergisches Keuperbergland). *Jahresberect und Mitteilungen der oberrheinischen geologischen Vereigng N.F.*, **71**: 225–46. Stuttgart.

Blümel, W.D. 1982: Calcretes in Namibia and SE-Spain – relations to substratum, soil formation and geomorphic factors. *Catena Supplement*, **1**: 67–82.

Bögli, A. 1964: Mischungskorrosion – ein Beitrag zum Verkarstungsproblem. *Erdkunde*, **18**: 83–92.

Bögli, A. 1970: Das Hölloch and sein Karst. *Stalactite*, Supplement 4a.

Bork, H.R. 1983: Die holozäne Relief- und Bodenentwicklung in Lößgebieten – Beispiele aus dem südlichen Niedersachsen. *Catena Supplement* 3: 1–93.

Bork, H.R. 1988: Bodenerosion und Umwelt. Verlauf und Folgen der mittelalterlichen und neuzeitlichen Bodenerosion. Bodenerosionsprozesse, Modelle und Simulationen. *Landschaftsgenese und Landschaftsökologie*, **13**: Braunschweig.

Bornhardt, W. 1900: *Zur Oberflächengestaltung und Geologie Deutsch-Ostafrikas*. Berlin.

Braun, U. 1969: Der Felsberg im Odenwald. *Heidelberger Geographische Arbeiten*, 26.

Bremer, H. 1977: Reliefgenerationen in den feuchten Tropen. *Würzburger Geographische Arbeiten*, **45**: 25–38.

Bremer, H. 1989a: *Allgemeine Geomorphologie*. Berlin: Gebr. Borntraeger.

Bremer, H. 1989b: On the geomorphology of the South German Scarplands. *Catena Supplement*, **15**: 45–67.

Broili, L. 1967: New knowledge on the geomorphology of the Vaiont slip surfaces. *Felsmechanik und Ingenieurgeologie*, **5**: 38–88.

Bronger, A. 1980: Zur neuen Soil Taxonomy der USA aus bodengeographischer Sicht. *Petermanns Geographische Mitteilungen* (4/1980): 253–62.

Bronger, A. and Bruhn, N. 1989: Relict and recent features in tropical alfisols from south India. *Catena Supplement*, **16** 107–28.

Brown, R.W., Summerfield, M.A. and Gleadow, A.J.W. 1994: Apatite fission track analysis: its potential for the estimation of denudation rates and implications for models of long-term landscape development. In Kirkby, M.J.(ed.), *Process models and theoretical geomorphology*. Chichester: John Wiley & Sons, 23–53.

Brückner, H. and Bruhn, N. 1992: Aspects of weathering and peneplanation in southern India. *Zeitschrift für Geomorphologie, Supplement-Band*, **91**: 43–66.

Brückner, H. and Hoffmann, G. 1992: Human-induced erosion processes in mediterranean countries. *Geoöko plus*, **III**: 97–110.

Brunotte, E. 1978: Zur quartären Formung von Schichtkämmen und Fußflächen im Bereich des Markoldendorfer Beckens und seiner Umrahmung (Leine-Weser-Bergland). *Göttinger Geographische Abhandlungen* **72**.

Brunotte, E. and Garleff, K. 1989: Structural landforms and planation surfaces in southern Lower Saxony. *Catena Supplement* **15**: 151–64.

Brunsden, D. 1979: Weathering (ch.4). Mass movement (ch.5) in Embleton, C. and Thornes, J. (eds.): *Process in geomorphology*, London: Edward Arnold, 73–186.

Brunsden, D. and Prior, D.B. (eds.) 1984: *Slope instability*. Chichester: John Wiley & Sons.

Bryan, K. 1923: Erosion and sedimentation in the Papago country, Arizona. *US Geological Survey Bulletin*, **730**: 19– 90.

Bryan, R. 1979: Soil erosion and conservation. In Gregory, K.J.and Walling, D.E. (eds.), Man and environmental processes. Folkestone: Dawson & Sons, 207–221.

Bryan, R. (ed.) 1990: Soil erosion: experiments and models. *Catena Supplement*, **17**.

Bryan, R. and Yair, A. 1982: *Badland geomorphology and piping*. Norwich: Geo Books.

Buch, M.W. 1988: Zur Frage einer kausalen Verknüpfung fluvialer Prozesse and Klimaschwankungen im Spätpleistozän und Holozän – Versuch einer geomorphodynamischen Deutung von Befunden von Donau und Main. *Zeitschrift für Geomorphologie, Supplement-Band*, **70**: 131–62.

Buch, M.W. and Heine, K. 1988: Klima- oder Prozeß-Geomorphologie. *Geographische Rundschau*, **40**: 16–26.

Buch, M.W. and Zoeller, L. 1992: Pedostratigraphy and thermoluminescence – Chronology of the western margin- (lunette-) dunes of Etosha Pan/Northern Namib. *Würzburger Geographische Arbeiten*, **84**: 361–84.

Buchanan, F. 1807: *A journey from Madras through the countries of Mysore, Canara and Malabar*. London: East India Company, vol. 2.

Büdel, J. 1937: Eiszeitliche und rezente Verwitterung und Abtragung im ehemals nicht vereisten Gebiet Mitteleuropas. *Petermanns Geographische Mitteilungen, Ergänzungsheft*, **229**.

Büdel, J. 1957: Die "Doppelten Einebnungsflächen" in den feuchten Tropen. *Zeitschrift für Geomorphologie N.F.*, **1**: 201–28.

Büdel, J. 1963: Klima-genetische Geomorphologie. *Geographische Rundschau* 15: 269–286.

Büdel, J. 1969: Das System der klima-genetischen Geomorphologie. *Erdkunde*, **23**: 165–83.

Büdel, J. 1977: *Klima-Geomorphologie*. Arbeiten aus der Kommission f. Geomorphologie der Bayer. Akad. d. Wiss. 1. Berlin.Stuttgart.

Büdel, J. and Hagedorn, H. (ed.) 1975: Reliefgenerationen in verschiedenen Klimaten. *Zeitschrift für Geomorphologie, Supplement*, **23**: 162.

Bull, W.B. 1964: Geomorphology of segmented fans in western Fresno County, California. *US Geological Survey Professional Paper 532–F*, 89–128.

Bull, W.B. 1977: The alluvial fan environment. *Progress in Physical Geography*, **1**: 222–70.

Burger, D. 1992: Quantifizierung quartärer subtropischer Verwitterung auf Kalk. Das Beispiel Travertinkomplex von Antalya (Südwesttürkei). *Relief Boden Paläoklima*, Bd. 7, Berlin: Gebr. Borntränger.

Busche, D. and Sponholz, B. 1988: Karsterscheinungen in nichtkarbonatischen Gesteinen der Republik Niger. *Würzburger Geographische Arbeiten*, **69**: 9–43.

Butzer, K. 1964: *Environment and Archeology*. London: Methuen, 524.

Cailleux, A. 1942: Les actions périglaciaires en Europe. *Mémoires Géologiques de France*, **21**: Paris.

Caine, N. 1969: A model for alpine talus slope development by slush avalanching. *Journal of Geology* **77**: 92–100.

Carson, M.A. 1971: *The mechanics of erosion*. London: Pion.

Carson, M.A. and Kirkby, M.J. 1972: *Hillslope form and process*. Cambridge: Cambridge University Press.

Chan, R.A. 1988: Regolith terrain mapping for mineral exploration in Western Australia. *Zeitschrift für Geomorphologie, N.F.*, **68**: 205–21.

Chezy, A. de 1775: Mémoire sur la vitesse de l'eau conduite dans une rigole. *Nachgedruckt in Annales des Ponts et Chaussées*, **60**: 1921. (cited in Richards 1982)

Chorley, R.J. 1962: *Geomorphology and general systems theory*. US Geological Survey Professional Paper 500-B.

Chorley, R.J. and Kennedy, B.A. 1971: *Physical geography – a systems approach*. London: Prentice Hall International.

Chorley, R.J., Dunn, A.J. and Beckinsale, R.P. 1964, 1973, 1991: *The history of the study of landforms or the development of geomorphology*. vol. 1 1964: *Geomorphology before Davis*. London: Methuen, New York: John Wiley, Vol.2 Chorley, R.J., Beckinsale, R.P. and Dunn, A.J.: *The life and work of William Morris Davis*. London: Methuen, New York: Harper & Row. Vol. 3, R.P.Beckinsale and R.Chorley 1991: *Historical and regional geomorphology 1890–1950*. London: Routledge.

Chorley, R.J., Schumm, S.A. and Sugden, D.E. 1984: *Geomorphology*. London: Methuen & Co.

Chow, V.T. (ed.) 1964: *Handbook of applied hydrology*. New York: McGraw-Hill.

Clapperton, C. 1993: *Quarternary geology and geomorphology of South America*. Amsterdam: Elsevier.

Clark, A.H., Cooke, R.U., Mortimer, C. and Sillitoe, R.H. 1967: Relationships between supergene mineral alteration and geomorphology, southern Atacama Desert. *Transactions of the Institute of Mineralogy and Metallurgy, B (Applied Earth Science)*, **76**: B89–B96.

Clark, M.J. (ed.) 1988: *Advances in periglacial geomorphology*. Chichester: John Wiley & Sons.

Coates, D.R. 1968: Finger lakes. In Fairbridge, R.W. (ed.), *The encyclopedia of geomorphology*. New York: Reinhold Book Corporation, 351–7.

Coates, D.R. (ed.) 1972–1974: *Environmental geomorphology and landscape conservation*. Vol. I 1972: Prior to 1900. Vol.II, 1973: *Urban areas*, Vol.III, 1974: *Non-urban*. Benchmark papers in Geology, Stroudsburg: Dowden, Hutchinson & Ross.

Cooke, R.U. 1976: Urban geomorphology. *Geographical Journal*, **142**: 59–65.

Cooke, R.U., Brunsden, D., Doornkamp, J.C. and Jones, D.K.C. 1982: *Urban geomorphology in drylands*. Oxford: Oxford University Press, 324.

Cooke, R.U. and Doornkamp, J.C. 1974: *Geomorphology in environmental management*. Oxford: Clarendon Press.

Cooke, R.U. and Warren, A. 1973: *Geomorphology in deserts*. B.T. London: Batsford. (New edition: Cooke, R.U., Warren, A. and Goudie, A. 1992: *Desert geomorphology*. London: UCL Press.)

Corbel, J. 1959a: Vitesse de l'érosion. *Zeitschrift für Geomorphologie, N.F.*, **3**: 1–28.

Corbel, J. 1959b: Erosion en terrain calcaire. *Annales de Géographie* **68**: 97–120.

Costa, J.E. and Fleisher, P.J. 1984: *Developments and applications of geomorphology*. Berlin: Springer Verlag.

Coulomb, C.A. 1776: Essai sur une application des régles des maximis et minimis à quelques problémes de statique, relatifs à l'architecture. *Mémoires présentés à l'Academie royale des Sciences*, vol. VII. Paris: Nachdruck durch Éditions Science et Industrie, 1971.

Coutard, J.P., Van Vliet-Lanoe, B. and Auzet, A.V. 1988: Frost heaving and frost creep on an experimental slope. Results for soil structure and sorted stripes. *Zeitschrift für Geomorphologie, Supplement Band*, **71**: 13–23.

Cox, N.J. 1992: Precipitation statistics for geomorphologists: variations on a theme by Frank Ahnert. *Catena Supplement* **23**: 189–212.

Credner, W. 1931: Das Kräfteverhältnis morphogenetischer Faktoren und ihr Ausdruck im Formenbild Südost-Asiens. *Bulletin of the Geological Society of China*, **11**: 13–34.

Curtis, C.D. 1976: Chemistry of rock weathering: fundamental reactions and controls. In Derbyshire, E. (ed.), *Geomorphology and climate*. London: John Wiley & Sons, 25–57.

Cvijic, J. 1893: Das Karstphänomen. *Pencks Geographische Abhandlungen*, **5**: 217–329.

Cvijic, J. 1960: La géographie des terrains calcaires. *Académie Monographics des Sciences et des Arts* **141**, Belgrad.

Daly, R.A. 1934: *The changing world of the Ice Age*. New Haven: Yale University Press.

Darwin, C. 1842: *The structure and distribution of coral reefs*. London.

Davis, W.M. 1899: The geographical cycle. *Geographical Journal*, **14**: 481–504.

Davis, W.M. 1909: *Geographical Essays. Nachdruck durch* New York: Dover Publications.

Davies, J.L. 1964: A morphogenetic approach to world shorelines. *Zeitschrift für Geomorphologie N.F.*, **8**: 127–42.

De Geer, G. 1912: A geochronology of the last 12000 years. *In Comptes Rendus, 11. Internationaler Kongreß, Stockholm 1910*, **I**: 241–58.

De Martonne, E. 1909: *Traité de géographie physique*. vol. 2, *Le relief du sol*. Paris: 8th edn 1948.

De La Noë, G. and De Margerie, E. 1888: *Les formes du terrain*. Paris: Service géographique de l'armée

Demek, J.1968: Cryoplanation terraces in Yakutia. *Byuletin Periglacjalny*, **17**: 91–116.

Demek, J. 1969: Cryogene processes and the development of cryoplanation terraces. *Byuletin Periglacjalny*, **18**: 115–25.

Demek, J., 1972: *Manual of detailed geomorphological mapping*. Prague: Akademia.

De Ploey, J. 1972: Enkele bevindingen betreffende erosieprocessen en hellingsevolutie op zandig substraat. *Tijdschrift von de Belgische Vereeniging voor Aardrijkskundige Studies*, **41**: 43–67.

De Ploey, J. 1989: Erosional systems and perspectives for erosion control in European loess areas. *Soil Technology*, **1**: 93–102.

De Ploey, J. 1992: Gullying and the age of badlands: an application of the erosional susceptibility model Es. *Catena Supplement*, **23**: 31–45.

De Ploey, J., Haase, G. and Leser, H. (eds) 1991: Geomorphology and geoecology – geomorphological approaches in applied geography. *Zeitschrift für Geomorphologie, Supplement-Band*, **83**: 259 pp.

De Ploey, J., Kirkby, M.J. and Ahnert, F. 1991: Hillslope erosion by rainstorms – a magnitude-frequency analysis. *Earth Surface Processes and Landforms*, **16**: 399–409.

De Ploey, J.J. and Poesen, J. 1987: Some reflections on modelling Hillslope processes. *Catena Supplement*, **10**: 67–72.

De Ploey, J. and Savat, J. 1968: Contribution a l'étude de l'érosion par le splash. *Zeitschrift für Geomorphologie, N.F.*, **12**: 174–93.

Derbyshire, E. (eds.) 1976: *Geomorphology and climate*. London: John Wiley & Sons, 512.

Dietrich, G. 1963: *General Oceanography*. New York: John Wiley & Sons, 588.

Dietrich, G. Kalle, K. Krauss, W. and Siedler, G. 1975: *Allgemeine Meereskunde*. Berlin: Gebr.Borntraeger.

Dietrich, W.E. and Dunne, T. 1978: Sediment budget for a small catchment in mountainous terrain. *Zeitschrift für Geomorphologie, Supplement-Band* **29**: 191–206.

Dietrich, W.E. and Smith, J.D. 1984: Bedload transport in a river meander. *Water Resources Research*, **20**: 1355–80.

Dietz, R.S. and Holden, J.D. 1970: The breakup of Pangaea. *Scientific American*, October Reprinted 1972 in *Continents adrift*. Readings from Scientific American. San Francisco: W.H. Freeman & Co.

Dikau, R. 1988: Case studies in the development of derived geomorphic maps. *Geologisches Jahrbuch A*, **104**: 329–38.

Dikau, R. 1990: Derivatives from detailed geoscientific maps using computer methods. *Zeitschrift für Geomorphologie, Supplement-Band*, **80**: 45–55.

Dongus, H. 1980: *Die geomorphologischen Grundstrukturen der Erde*. Stuttgart: B.G. Teubner, 200.

Douglas, I. 1980: Climatic geomorphology. Present day processes and landform evolution. Problems of interpretation. *Zeitschrift für Geomorphologie, Supplement-Band*, **36**: 27–47.

Dreybrodt, W. 1988: *Processes in karst systems*. Heidelberg: Springer-Verlag.

Du Boys, M.P. 1879: Études du régime et l'action exercée par les eaux sur un lit à fond de graviers indéfinitement affouillable. *Annales des Ponts et Chaussées*, Serie 5, **18**: 141–95.

Dunne, T., Dietrich, W.E. and Brunengo, M.J. 1978: Recent and past erosion rates in semi-arid Kenya. *Zeitschrift für Geomorphologie, Supplement-Band*, **29** 130–40.

Dury, G.H. 1969: Hydraulic geometry. In Chorley, R.J. (ed.), *Water, earth and man*. London: Methuen & Co., 319–29.

East, T.J., Nanson, G.C. and Roberts, R.G. 1993: Geomorphological stability of sites for the long-term containment of uranium mining wastes in the seasonally wet tropics, Northern Australia. *Zeitschrift für Geomorphologie, Supplement-Band*, **87**: 171–82.

Einstein, H.A., Anderson, A.G. and Johnson, J.W. 1940: A distinction between bedload and suspended load in natural streams. *Transactions of the American Geophysical Union*, **21**: 628–33.

Embleton, C. 1979: Nival processes. In Embleton, C. and Thornes, J. (eds), *Process in geomorphology*. London: Edward Arnold, 307–24.

Embleton, C. (ed.) 1988: Applied geomorphological mapping: methodology by example. *Zeitschrift für Geomorphologie, Supplement-Band*, **68**.

Embleton, C. and King, C.A.M. 1968: *Glacial and periglacial geomorphology*. London: Edward Arnold.

Embleton, C. and King, C.A.M. 1975: *Glacial geomorphology*. London: Edward Arnold.

Embleton, C. and Thornes, J. 1979: *Process in geomorphology*. London: Edward Arnold.

Embleton, C. and Verstappen, H.Th. 1988: The nature and objectives of applied geomorphological mapping. *Zeitschrift für Geomorphologie, Supplement-Band*, **68**: 1–8.

Ergenzinger, P. 1987: Chaos and order – the channel geometry of gravel bed braided rivers. *Catena Supplement*, **10**: 85–98.

Ergenzinger, P. and Conrady, J. 1982: A new tracer technique for measuring bedload in natural channels. *Catena*, **9**: 77–80.

Ergenzinger, P. and Schmidt, K.H. 1990: Stochastic elements of bed load transport in a step-pool mountain river. In *Hydrology in mountainous regions. II Artificial reservoirs; water and slopes*. Proceedings of two Lausanne Symposia, August 1990. IAHS Publication no.194, 39–46.

Fairbridge, R.W. 1961: Eustatic changes in sea level. In Ahrens, L.H. *et al.* (eds.), *Physics and chemistry of the earth*, Vol. 4. London: Pergamon Press.

Fairbridge, R.W. 1968a: Coral reefs – morphology and theories. In Fairbridge, R.W. (ed.), *The encyclopedia of geomorphology*. New York: Reinhold Book Corporation, 186–97.

Fairbridge, R.W. (ed.) 1968b: *The encyclopedia of geomorphology*. New York: Reinhold Book Corporation.

Feldman, S., Harris, S.A. and Fairbridge, R.W. 1968: Drainage patterns. In Fairbridge, R.W. (ed.), *The encyclopedia of geomorphology*. New York: Reinhold Book Corp., 284–91.

Finch, V.C. 1933: Geographical Surveying. Geographical Society of Chicago Bulletin 9, 3–11 and 15–44.

Fischer, K. 1965: Murkegel, Schwemmkegel und Kegelsimse in den Alpentälern, unter besonderer Berücksichtigung des Vinschgaus. *Mitt. d. Geogr. Gesellschaft München*, 127–60.

Fischer, K. 1989: The landforms of the German Alps and the alpine foreland. *Catena Supplement*, **15**: 69–83.

Fitzpatrick, E.A. 1980: *Soils. Their formation, classification and distribution*. London: Longman.

Flacke, W., Auerswald, K. and Neufang, L. 1990: Combining a modified universal soil loss equation with a digital terrain model for computing high resolution maps of soil loss resulting from rain wash. *Catena*, **17**: 383–97.

Fleisher, P.J. 1984: Maps in applied geomorphology. In Costa, J.E. and Fleisher, P.J. (eds.), *Developments and applications of geomorphology*. Heidelberg: Springer-Verlag, 171–202.

Flint, R.F. 1971: *Glacial and quaternary geology*. New York: John Wiley & Sons.

Ford, D. and Williams, P.W. 1989 *Karst geomorphology and hydrology*. London: Unwin Hyman.

Francis, P. 1976: *Volcanoes*. Harmondsworth: Penguin Books.

French, H. 1976: *The periglacial environment*. Longman, London.

Freyssinet, PH. 1994: Gold mass balance in lateritic profiles from savanna and rain forest zones. *Catena*, **21**: 59–172.

Gellert, J.F. 1968: Vom Wesen der angewandten Geomorphologie. *Petermanns Geographische Mitteilungen*, **112**: 256–264.

Gerlach, T. 1967: Évolutions actuelles des versants dans les Carpathes, d'après l'exemple d'observations fixes. In Macar, P. (eds), *L'évolution des versants*. Les congrès et colloques de l'Universite de Liège, vol. 40, 129–38.

Gerrard, J. (ed.) 1984: Applied geomorphology. *Zeitschrift für Geomorphologie, Supplementband* **51**.

Gerstenhauer, A. 1977: Kritische Anmerkungen zu den Vorstellungen von der Genese der Korrosionspoljen. *Abhandlungen zur Karst-und Höhlenkunde, Reihe A*, **15** (Festschrift für Alfreed Bögli).12–25.

Gierloff-Emden, H.G. 1980: *Geographie des Meeres – Ozeane and Küsten*. 2 Bände. Berlin De Gruyter.

Gilbert, G.K. 1877: *Report on the geology of the Henry Mountains*. Geographic and geologic survey of the Rocky Mountain Region, US. Washington DC.

Goetz, R., Müller, S. and Wacker, F. 1971: Zusammenhänge zwischen den Tagesrhythmen der Bodentemperatur und der Bodenprofilprägung. *Mitteilungen des Vereins für Forstliche Standortkunde und Forstpflanzenzüchtung*, Oktober.

Goldich, S.S. 1938: A study in rock weathering. *Journal of Geology*, **46**: 17–58.

Gottschalk, L.C. 1945: Effects of soil erosion on navigation in upper Chesapeake Bay. *Geographical Review*, **35**: 219–38.

Goudie, A. 1973: *Duricrusts in tropical and subtropical landscapes*. Oxford University Press.

Goudie, A. (ed.) 1981: *Geomorphological techniques*. London: George Allen & Unwin.

Govers, G. 1987: Spatial and temporal variability in rill development processes at the Huldenberg experimental site. *Catena Supplement* **8**: 17–33.

Govers, G. 1991: Rill erosion on arable land in central Belgium: rates, control and predictability. *Catena* **18**: 133–55.

Govers, G. Everaert, W., Poesen, J., Rauws, G., De Ploey, J. and Lautridou, J.P. 1990: A long flume study of the dynamic factors affecting the resistance of a loamy soil to concentrated flow erosion. *Earth Surface Processes and Landforms*, **15**: 313–28.

Gradmann, R. 1931: *Süddeutschland*. 2 Bände. Stuttgart.

Graf, W.L. 1975: The impact of suburbanization on fluvial geomorphology. *Water Resources Research*, **11**: 690–2.

Graf, W.L. 1988: *Fluvial processes in dryland rivers*. Berlin-Heidelberg, Springer-Verlag.

Grahmann, R. 1932: Bemerkungen über die Begriffe Diluvium, Eiszeit and Vereisung. *Zeitschrift für Gletscherkunde* **20**: 470–4.

Gregory, K.J. 1979: River channels. In Gregory, K.J. and Walling, D.E. (eds.), *Man and environmental processes*. Folkestone: Dawson & Sons, 123–43.

Gregory, K.J. and Park, C.C. 1974: Adjustment of river channel capacity downstream from a reservoir. *Water Resources Research*, **10**: 870–83.

Gregory, K.J. and Walling, D.E. 1973: *Drainage basin form and process – a geomorphological approach*. London: Edward Arnold.

Grund, A. 1914: Der geographische Zyklus im Karst. *Zeitschrift der Gesellschaft für Erdkunde Berlin*, 621–40.

Gudehus, G. 1981: *Bodenmechanik*. Stuttgart: Ferdinand Enke.

Guilcher, A. 1958: *Coastal and submarine geomorphology*. (übersetzt von B.W Sparks und R.H.W. Kneese). London: Methuen.

Gumbel, E.J. 1958: *Statistics of extreme values*. New York: Columbia University Press.

Gundestrup, N.S., Dahl-Jensen, D., Johnsen, S.J. and Rossi, A. 1993: Bore-hole survey at Dome GRIP 1991. *Cold Regions Science and Technology*, **21**: 399–402.

Gupta, A. 1984: Urban hydrology and sedimentation in the humid tropics. In Costa, J.E. and Fleisher, J.P. (eds.), *Developments and applications of geomorphology*. Berlin: Springer-Verlag, 240–67.

Habbe, K.A. 1986: Zur geomorphologischen Kartierung von Blatt Grönenbach (I) – Probleme, Beobachtungen, Schlußfolgerungen. *Erlanger Geographische Arbeiten*, **47**: 365–479.

Habbe, K.A. and Rögner, K. The Pleistocene Iller Glaciers and their outwash fields. *Catena Supplement*, **15**: 311–28.

Häberle, D. 1911a: Über Kleinformen der Verwitterung im Hauptbuntsandstein des Pfälzerwaldes. *Verhendlungender Naturwissenschaflicker and Heidelberg*.

Häberle, D. 1911b: Über die Meßbarkeit der Fortschritte der Verwitterung. *Jahresberichte der Oberrheinschen Geologischen Vereinigung*: 53–54.

Hack, J.T. 1957: *Studies of longitudinal stream profiles in Virginia and Maryland*. US Geological Survey Professional Paper 294B, 53–63.

Hack, J.T. 1960: Interpretation of erosional topography in humid temperate climate. *American Journal of Science*, **258–A**: 80–97.

Hagedorn, H. 1968: Über äolische Abtragung und Formung in der Südost-Sahara. Ein Beitrag zur Gliederung der Oberflächenformen in der Wüste. *Erdkunde*, **22**: 257–69.

Hagedorn, H. 1988: Äolische Abtragungsformen im Massiv von Termit (NE-Niger). *Würzburger Geographische Arbeiten*, **69**: 277–87.

Hagedorn, J. 1964: Geomorphologie des Uelzner Beckens. *Göttinger Geographische Abhandlungen* **31**: 200.

Hagedorn, J. 1989: Glacial and periglacial morphology of the Lüneburg Heath. *Catena Supplement*, **15**: 85–94.

Haigh, M. 1977:The use of erosion pins in the study of slope evolution. *British Geomorphological Research Group Technical Bulletin*, no. **18**: 31–49.

Haigh, M. 1980: Slope retreat and gullying on revegetated mine dumps, Waun Hoscyn, Gwent. *Earth Surface Processes*, **5**: 77–9.

Haigh, M. 1984: Landslide prediction and highway maintenance in the Lesser Himalayas, India. *Zeitschrift für Geomorphologie, Supplement-Band*, **51**: 17–37.

Hails, J.R. (ed.) 1977: *Applied geomorphology*. Amsterdam. Elsevier.

Hails, J.R. 1977b: Applied geomorphology in coastal-zone planning and management. In Hails, J.R. (ed.), *Applied geomorphology*. Amsterdam: Elsevier, 317–62.

Hale, M. and Porto, C.G. 1994: Geomorphological evolution and supergene gold ore at Posse, Goias State, Brazil. *Catena* **21**: 145–57.

Hamann, C. 1988: Geomorphologische Karten zur Beurteilung der optischen Qualität von Landschaften. *Zeitschrift für Geomorphologie, Supplement-Band*, **68**: 125–41.

Hansen, A. 1984: Engineering geomorphology: the application of an evolutionary model of Hong Kong's terrain. *Zeitschrift für Geomorphologie, Supplement-Band*, **51**: 39–50.

Hard, G. 1976: Exzessive Bodenerosion vor und nach 1800. In Richter, G. (ed.), *Bodenerosion in Mitteleuropa. Wege* *der Forschung*, 430. Darmstadt: Wissenschaftliche Buchgesellschaft, 195–239.

Harvey, A.M. 1991: The influence of sediment supply on the channel morphology of upland streams: Howgill Fells, Northwest England. *Earth Surface Processes and Landforms*, **16**: 675–84.

Hastenrath, S. 1967: The barchans of the Arequipa region, southern Peru. *Zeitschrift für Geomorphologie N.F.* **11**: 300–31.

Hedges, J. 1969: Opferkessel. *Zeitschrift für Geomorphologie N.F.*, **13**: 22–55.

Heine, K. 1992: On the age of humid Late Quaternary phases in southern African arid areas (Namibia, Botswana). *Palaeoecology of Africa*, **23**: 149–64.

Hempel, L. 1981: Die Tendenzen anthropogen bedingter Reliefformung in den Ackerländereien Europas. *Zeitschrift für Geomorphologie N.F.*, **15**: 312–29.

Hempel, L. 1988: Jungquartäre Erosion und Akkumulation im Landschaftshaushalt Griechenlands. *Geographische Rundschau*, **40**(4): 12–19.

Hermes, K. 1964: Der Verlauf der Schneegrenze. *Geographisches Taschenbuch 1964/65*, 58–71.

Herrmann, R. 1977: *Einführung in die Hydrologie*. Stuttgart. B.G.Teubner.

Herweg, K. 1988: The applicability of large-scale geomorphological mapping to erosion control and soil conservation in a research area in Tuscany. *Zeitschrift für Geomorphologie, Supplement-Band*, **68**: 175–88.

Hess, H.H. 1946: Drowned ancient islands of the Pacific Basin. *American Journal of Science*, **244**: 772–91.

Hettner, A. 1887: Gebirgsbau und Oberflächengestaltung der Sächsischen Schweiz. *Forschungen zur deutschen Landes-and Volkskunde*, **2**: 249–355.

Heyse, I. and De Moor, F.R. 1988: The influence of the natural environment on the site and urban development of Ghent in historical times: a study in applied geomorphology. *Zeitschrift für Geomorphologie, Supplement-Band*, **68**: 143–53.

Hjulström, F. 1935: Studies of the morphological activity of rivers as illustrated by the river Fyris. *Bulletin of the Geological Institute, University of Uppsala*, **25**: 221–527.

Höllermann, P. 1967: Zur Verbreitung rezenter periglazialer Kleinformen in den Pyrenäen and Ostalpen. *Göttinger Geographische Abhandlungen*, **40**: 219.

Hormann, K. 1963: Torrenten in Friaul und die Längsprofilentwicklung auf Schottern. *Münchener Geographische Hefte*, **26**: 80.

Hormann, K. 1965: Das Längsprofil der Flüsse. *Zeitschrift für Geomorphologie N.F*, **9**: 437–56.

Horton, R. 1945: Erosional development of streams and their drainage basins hydrophysical approach to quantitative morphology. *Bulletin of the Geological Society of America*, **56**: 275–370.

Hövermann, J. 1987: Neues zur pleistozänen Harzvergletscherung. *Eiszeitalter and Gegenwart 37*, 99–107.

Hövermann, J. and Hövermann, E. 1991: Pleistocene and Holocene geomorphic features between the Kunlun Mountains and the Taklamakan Desert. *Die Erde, Ergänzungsheft*, **6**: 51–72.

Hudson, G.D. 1936: The unit method of land classification. *Annals of the Association of American Geographers*, **26**: 99–112.

Humboldt, A. von 1805–1834: *Voyage aux régions équinoxiales du Nouveau Continent, fait en 1799, 1800, 1801, 1802, 1803 et 1804, par A. de Humboldt et A.Bonpland*. 34 vols. Paris.

Hunt, R.E. 1984: *Geotechnical Engineering Investigation Manual*. New York: McGraw-Hill Book Co.

Hutchinson, J.N. 1969: A reconsideration of the coastal landslides of Folkestone Warren, Kent. *Geotechnique* **20**: 6–38.

Hutton, J. 1795: *Theory of the earth with proofs and illustrations*. 2 vols. Edinburgh.

Imeson A.C. and Vis, M. 1982: Factors influencing the credibility of soils in natural and semi-natural ecosystems at different altitudes in the Central Cordillera of Columbia. *Zeitschrift für Geomorphologie, Supplement-Band*, **44**, 91-106.

Irving, E.G. 1962: Coastal cliffs: report of a symposium. *Geographical Journal*, **128**: 303–20.

Ives, J.D. 1974: Permafrost. In Ives and Barry (eds.), *Arctic and alpine environments*. London. Methuen, 159–194.

Jennings, J.N. 1985: *Karst geomorphology*. London: Basil Blackwell.

Jessen, O. 1936: *Reisen und Forschungen in Angola*. Berlin: Heimer.

Judson, S. and Ritter, D.T. 1964: Rates of regional denudation in the United States. *Journal of Geophysical Research*, **69**: 3395–401.

Jungerius, P.D. and Van Der Meulen, F. 1989: The development of dune blowouts, as measured with erosion pins and sequential air photos. *Catena*, **16**: 369–76.

Jungerius, P.D. and Dekker, L.W. 1990: Water repellency in the dunes with special reference to the Netherlands. In Bakker, T.H.W., Jungerius, P.D. and Klijn, J.A. (eds.), *Dunes of the European coasts. Catena Supplement*, **18**: 173–83.

Karte, J. 1979: Räumliche Abgrenzung und regionale Differenzierung des Periglaziärs. *Bochumer Geographische Arbeiten*, **35**: 211.

Kaufmann, H. 1929: *Rhythmische Phänomene der Erdoberfläche*. Braunschweig.

Keller, R. 1961: *Gewässer und Wasserhaushalt des Festlandes*. Berlin.

Kellersohn, H. 1952: Untersuchungen zur Morphologie der Talanfänge im mitteleuropäischen Raum. *Kölner Geographische Arbeiten* **1**.

Kelletat, D. 1989: *Physische Geographie der Meere and Küsten*. Stuttgart: Teubner,

Kesel, R.H. 1985: Alluvial fan systems in a wet-tropical environment, Costa Rica. *National Geographic Research*, **1**: 450–69.

Kienholz, H. 1977: Kombinierte geomorphologische Gefahrenkarte von Grindelwald. *Catena*, **3**: 265–94.

Kim, J.W. 1989: Funktionale Morphologie der Kall. *Aachener Geographische Arbeiten*, **21**: 190.

King, L.C. 1953: Canons of landscape evolution. *Bulletin of the Geological Society of America*, **64**: 721–52.

King, C.A.M. 1972: *Beaches and Coasts*. 2nd edn. London: St. Martin's Press.

Kirkby, M.J. 1971: Hillslope process-response model based on the continuity equation. In Brunsden, D. (ed.), *Slope form and process*. Institute of British Geographers Special Publication 3, 15–30.

Kirkby, M.J. 1986: Mathematical models for solutional development of landforms. In Trudgill, S.T. (ed.), Solute processes. Chichester: John Wiley & Sons, 440–95.

Kirkby, M.J. 1992: An erosion-limited hillslope evolution model. *Catena Supplement* **23**: 157–87.

Kirkby, M.J. (ed.) 1994: *Process models and theoretical geomorphology*. Chichester: John Wiley & Sons.

Kirkby, M.J. and Morgan, R.P.C. (eds) 1980: *Soil erosion*. Chichester: John Wiley & Sons.

Klaer, W. 1962: Die periglaziale Höhenstufe in den Gebirgen Vorderasiens. *Zeitschrift für Geomorphologie N.F.,* **6**: 17–32.

Knighton, D. 1984: *Fluvial forms and processes*. London: Edward Arnold.

Kossack, H.P. 1953: *Die Polargebiete. Geographisches Taschenbuch*. Stuttgart; Franz Steiner Verlag,

Kuhle, M. 1984: Zur Geomorphologie Tibets, Bortensander als Kennform semiarider Vorlandvergletscherung. *Berliner Geographische Abhandlungen*, **36**: 127–38.

Langbein, W.B. and Schumm, S.A. 1958: Yield of sediment in relation to mean annual precipitation. *Transactions of the American Geophysical Union*, **39**: 1076–84.

Lautridou, J.P. 1971: Conclusions générales des experiences de gélifraction expérimentale. Centre de Géomorphologie, CNRS Caen, Bulletin no. 10, 63–84.

Lautridou, J.P. and Ozouf, J.C. 1982: Experimental frost shattering: 15 years of research at the Centre de Géomorphologie du CNRS. *Progress in Physical Geography* **6**: 215–32.

Lawes, E.F. 1974: *An analysis of short duration rainfall intensities*. Technical Memorandum No. 23, East African Meteorological Department, Nairobi.

Lecce, S.A. 1990: The alluvial fan problem. In Rachocki, A.H. and Church, M. (eds): *Alluvial fans – a field approach*. Chichester, John Wiley & Sons, 3–24.

Lehmann, H. 1954: Das Karstphänomen in den verschiedenen Klimazonen, *Erdkunde* **8**: 112–22.

Lehmann, O. 1933: Morphologische Theorie der Verwitterung von Steinschlagwänden. *Vierteljahresschrift d. Naturforsching Gesellschaft Zürich*, **78**: 83–126.

Lehmkuhl, F. and Rost, K.T. 1993: Zur pleistozänen Vergletscherung Ostchinas und Nordosttibets. *Petermanns Geographische Mitteilungen* **137**: 67–78.

Leopold, L.B. 1968: *Hydrology for urban land planning – a guidebook on the hydrological effects of urban land use*. US Geological Survey Circular 554.

Leopold, L.B. and Langbein, W.B. 1962: *The concept of entropy in landscape evolution*. US Geological Survey Professional Paper 500-A.

Leopold, L.B. and Maddock, T. 1953: *The hydraulic geometry of stream channels and some physiographic implications*. US Geological Survey Professional Paper 252.

Leopold, L.B., Wolman, M.G. and Miller, J.P. 1964: *Fluvial processes in geomorphology*. San Francisco: W.H. Freeman,

Leser, H. 1993: *Geomorphologie*. 2. Aufl. Braunschweig: Westermann,

Leser, H. and Stäblein, G. (eds) 1975: Geomorphologische Kartierung. Richtlinien zur Herstellung geomorphologischer Karten 1:25 000 ("grüne Legende"). *Berliner Geographische Abhandlungen, Sonderheft*, **39**.

Liedtke, H. 1984: Die nordischen Vereisungen in Mitteleuropa. *Forschungen z. Deutschen Landeskunde* **204**: 307 (1. Aufl.1975).

Liedtke, H. 1989: The landforms in the north of the Federal Republic of Germany and their development. *Catena Supplement*, **15**: 11–24.

Linton, D. 1955: The problem of tors. *Geographical Journal*, **121**, 470–86.

Liu Tungsheng 1988: *Loess in China*. Berlin: Springer-Verlag.

Lobeck, A.K. 1939: *Geomorphology*. New York: McGraw-Hill,

Löffler, E. 1977: *Geomorphology of Papua New Guinea*. Canberra: *Commonwealth Scientific and Industrial Research Organization* (CSIRO) and Australian National University Press,

Lohnes, R.A. and Handy, R.L. 1968: Slope angles in friable loess. *Journal of Geology* **76**: 247–58 (cited in Carson, 1971).

Louis, H. and Fischer, K. 1979: *Allgemeine Geomorphologie*. Vol. 4. Berlin: Walter De Gruyter,

Luk, S-H., Abrahams, A.D. and Parsons, A.J. 1993: Sediment sources and sediment transport by rill flow and interrill flow on a semi-arid piedmont slope, southern Arizona. *Catena*, **20**: 93–111.

Lyell, C. 1830–1833: *Principles of geology*, 3 vols. London.

Machann, R. and Semmel, A. 1970: Historische Bodenerosion auf Wüstungsfluren deutscher Mittelgebirge. *Geographische Zeitschrift*, **58**: 250–66.

Mackay, H.R. 1978: Contemporary pingos: a discussion. *Biuletyn Periglacjalny*, **27**: 133–54.

Mackay, H.R. 1994: Pingos and pingo ice of the western Arctic coast, Canada. *Terra*, **106**: 1–11.

Mäckel, R. 1991: Aktuelle Geomorphodynamik und angewandte Geomorphologie. *Zeitschrift für Geomorphologie, Supplement-Band*, **89**: 155.

Mainguet, M. 1976: Propositions pour une nouvelle classification des édifices sableux éoliens d'après les images des satellites Landsat I, Gemini, Noaa 3. *Zeitschrift für Geomorphologie N.F.*, **20**: 275–96.

Mandelbrot, B.B. 1982: *The fractal geometry of nature*. New York. Freemann.

Mangelsdorf, J., Scheurmann, K. and Weiss, F.-H. 1990: River morphology – a guide for geoscientists and engineers. Berlin-Heidelberg-New York: Springer-Verlag, (Ursprüngliche deutsche Ausgabe 1980: *Flußmorphologie*. München R. Oldenbourg-Verlag.)

Manning, R. 1889: On the flow of water in open channels and pipes. *Transactions of the Institution of Civil Engineering of Ireland*, **20**: 161–207.

Melton, M.A. 1965: Debris-covered hillslopes of the southern Arizona desert – considerations of their stability and sediment contribution. *Journal of Geology* **73**: 715–29.

Mensching, H. 1949: Talauen und Schotterfluren im Niedersächsischen Bergland. *Göttinger Geographische Abhandlungen*, **4**: 60.

Mensching, H. 1951: Akkumulation und Erosion niedersächsischer Flüsse seit der Rißeiszeit. *Erdkunde*, **5**: 60–70.

Mensching, H. 1978: Inselberge, Pedimente and Rumpfflächen im Sudan (Republik). Ein Beitrag zur morphogenetischen Sequenz in den ariden Subtropen and Tropen Afrikas. *Zeitschrift für Geomorphologie, Supplement-Band*, **30**: 1–19.

Meyer, W. 1986: *Geologie der Eifel*. Stuttgart: E.Schweizerbarth'sche Verlagsbuchhandlung.

Millot, G. 1970: *Geology of clays*. Paris-Berlin-London: Masson/Springer/Chapman,.

Milne, G. 1935: Some suggested units of classification and mapping particularly for East African soils. *Soil Research*, **4**: 183–95.

Miotke, F.-D. 1975: Bedeutung und Grenzen von Verkarstungsprozessen. *Zeitschrift für Geomorphologie, Supplement-Band*, **23**: 107–17.

Moeyersons, J. 1975: An experimental study of pluvial processes on granite grus. *Catena*, **2**: 289–308.

Moeyersons, J. 1983: Measurements of splash-saltation fluxes under oblique rain. *Catena Supplement*, **4**: 9–13.

Moeyersons, J. and De Ploey, J. 1976: Quantitative data on splash erosion, simulated on unvegetated slopes. *Zeitschrift für Geomorphologie, Supplement-Band*, **25**: 120–31.

Montgomery, D.R. and Dietrich, W.E. 1989: Source areas, drainage density, and channel initiation. *Water Resources Research*, **26**: 1907–18.

Morgan, M.A. 1969: Overland flow and man. In Chorley, R.J. (ed.), *Water, earth and man*, London: Methuen & Co., 239–55.

Morgan, R.P.C. 1979: *Soil erosion*. London: Longman,

Morgan, R.P.C., Hatch, T., and Wan Sulaiman Wan Harun. 1982: A simple procedure for assessing soil erosion risk: a case study for Malaysia. *Zeitschrift für Geomorphologie, Supplement-Band*, **51**: 69–89.

Morisawa, M. 1968: *Streams – their dynamics and morphology*. New York: McGraw-Hill,

Morisawa, M. 1971: *Rivers*. London and New York: Longman.

Mortensen, H. 1953: Neues zum Problem der Schichtstufenlandsachaft. Einige Ergebnisse einer Reise durch dem Südwesten der USA, Sommer und Herbst 1952. Nachrichten der Akademie d. Wissenschaften Göttingen, Mathematisch-Physikalische Klasse IIa, Mathematisch-Physikalisch-Chemische Abteilung, Nr.2.

Musgrave, G.W. 1935: Some relationships between slope-length, surface-runoff and the silt-load of surface-runoff. *Transactions of the American Geophysical Union*, **16**: 472–8.

Naumann, K.F. 1858: *Lehrbuch der Geognosie*. 2. Auflage. Leipzig.

Ollier, C.D. 1965: Some features of granite weathering in Australia. *Zeitschrift für Geomorphologie*, **9**: 285–304.

Ollier, C.D. 1969: *Weathering*. New York: American Elsevier,

Ollier, C.D. 1976: Catenas in different climates. In Derbyshire, E. (ed.), *Geomorphology and climate*. London: John Wiley & Sons, 137–69.

Pardé, M. 1964: *Fleuves et riviéres*, vol. 4. Paris: Colin,

Parizek, E.J. and Woodruff, J.F. 1957: Description and origin of stone layers in soils of the southeastern states. *Journal of Geology*, **65**: 24–34.

Parsons, A.J. 1988: *Hillslope form*. London: Routledge,

Passarge, S. 1926: Geomorphologie der Klimazonen oder Geomorphologie der Landschaftsgürtel. *Petermanns Geographische Mitteilungen*, **72**: 173–5.

Pecsi, M. (ed.), 1987: Loess and environment. *Catena Supplement*, **9**.

Penck, A. 1894: *Morphologie der Erdoberfläche*. 2 Bände. Stuttgart: Engelhorn.

Penck, A. and Brückner, E. 1901–1909: *Die Alpen im Eiszeitalter*. 3 Bände, Leipzig: Tauchnitz.

Penck, W. 1924: *Die morphologische Analyse*. Stuttgart: Engelhorn. English translation, 1953: *The morphological analysis of landforms*. London: Macmillan.

Pfeffer, K.H. 1978: *Karstmorphologie*. Erträge d. Forschung Bd. 79. Darmstadt: Wissenschaftl. Buchgesellschaft,

Pfeffer, K.H. 1989: The karst landforms of the northern Franconian Jura between the rivers Pegnitz and Vils. *Catena Supplement*, **15**: 253–60.

Philipp, H. 1931: *Das ONO-System in Deutschland und seine Stellung innerhalb des saxonischen Bewegungsbildes*. Abhandlung Nr. 17 der Heidelberger Akademie der Wissenschaften, Mathematisch-Naturwissenschaftliche Klasse.

Pirsson, L.V. and Knopf, A. 1958: *Rocks and rock minerals*. New York: John Wiley & Sons.

Pissart, A. 1977: Apparition et évolution des sols structuraux périglaciaires de haute montagne. Expériénces de terrain au Chambeyron (Alpes, France). In *Abhandlungen der Akademie der Wissenschaften in Göttingen, Mathematisch-Physikalische Klasse*, Folge III, **31**: 142–56.

Planchon, O., Fritsch, E. and Valentin, C. 1987: Rill development in a wet savannah environment. *Catena Supplement*, **8**: 55–70.

Playfair, J. 1802: *Illustrations of the Huttonian theory of the Earth*. Edinburgh.

Poser, H. 1948: Boden-und Klimaverhältnisse in Europa während der Würmeiszeit. *Erdkunde*, **2**: 53–68.

Powell, J.W. 1876: *Report on the Geology of the Uinta Mountains*. Washington, D.C.

Prasuhn, V. and Schaub, D. 1991: The different erosion dynamics of loess and clay soils and the consequences for soil erosion control. *Zeitschrift für Geomorphologie, Supplement-Band*, **83**: 127–34.

Price, W.A. 1955: *Correlation of shoreline types with offshore bottom conditions*. A. & M. College of Texas, Dept. of Oceanography, Project 63. (cited in King, 1972).

Priesnitz, K. (1988): Cryoplanation. In Clark, M.J. (ed.), *Advances in periglacial geomorphology*. Chichester: John Wiley & Sons, 49–67.

Quitzow, H.W. 1974: *Das Rheintal und seine Entstehung*. Conference de la Societé Geologique Belge: *L'Evolution Quaternaire des Bassins Fluviaux de la Mer du Nord Meridionale*. Liege, 53–104.

Rachocki, A.H. 1990: The Leba River alluvial fan and its palaeogeomorphological significance. In Rachocki, A.H. and Church, M. (eds), *Alluvial fans – a field approach*. Chichester: John Wiley & Sons, 305–17.

Rachocki, A.H. and Church, M. (eds); 1990: *Alluvial fans – a field approach*. Chichester: John Wiley & Sons,

Radtke, U. and Brückner, H. 1991: Investigations on age and genesis of silcretes in Queensland (Australia) – preliminary results. *Earth Surface Processes and Landforms*, **16**: 547–54.

Ramsay, A.C. 1863: *The physical geology and geography of Great Britain*. London.

Rapp, A. 1960: Recent development of mountain slopes in Kärkevagge and surroundings, northern Scandinavia. *Geografiska Annaler*, **42**: 71–200.

Ratzlaff, J. 1974: *Effects of environmental characteristics upon storm runoff and sediment production in selected Middle Atlantic Watersheds*. Ph.D. Dissertation, University of Maryland,

Reneau, S.L. and Dietrich, W.E. 1991: Erosion rates in the southern Oregon Coast Range: Evidence for an equilibrium between hillslope erosion and sediment yield. *Earth Surface Processes and Landforms*, **16**: 307–22.

Richards, K. 1982: *Rivers – form and process in alluvial channels*. London: Methuen & Co.

Richter, G. 1965: Bodenerosion – Schäden und gefährdete Gebiete in der Bundesrepublik Deutschland. *Forschungen zur Deutschen Landeskunde*, **152**: 592.

Richter, G. (ed.) 1976: *Bodenerosion in Mitteleuropa* (unter Mitarbeit von W. Sperling). Wege der Forschung, Band 430. Darmstadt: Wiss. Buchgesellschaft.

Richter, G. 1977: *Bibliographie für Bodenerosion und Bodenerhaltung 1965–1975*. Universität Trier, Forschungsstelle Bodenerosion.

Richter, G. 1978a: Bodenerosion – Bodenschutz. In Olschowy, G. (ed.), *Naturschutz und Umweltschutz in der Bundesrepublik Deutschland*. Hamburg: Verlag Paul Parey, 98–111.

Richter, G. 1978b: Bodenerosion in den Reblagen an Mosel-Saar-Ruwer. Formen, Abtragungsmengen, Wirkungen. *Verhandlungen des Deutschen Geographentages Mainz 1977*, Wiesbaden.

Richter, G. 1979: Bodenerosion in Reblagen des Moselgebiets. Ergebnisse quantitativer Untersuchungen 1974–1977. Forschungsstelle Bodenerosion der Universität Trier, Mertesdorf (Ruwertal), 3. Heft,

Richter, M. 1987: Die Starkregen und Massenumlagerungen des Juli-Unwetters 1987 im Tessin und Veltlin. *Erdkunde*, **41**: 261–74.

Richthofen, F. von 1886: *Führer für Forschungsreisende*. *Hannover*. Nachdruck 1973, Darmstadt: Wissenschaftliche Buchgesellschaft,

Riley, S.J. 1994: Modelling hydrogeomorphic processes to assess the stability of rehabilitated landforms, Ranger Uranium Mine, Northern Territory, Australia – a research strategy. In Kirkby, M.J. (ed.), *Process models and theoretical geomorphology*. Chichester: John Wiley & Sons, 357–88.

Rodda, J.C. 1969: The flood hydrograph. In Chorley, R.J. (ed.), *Water, earth and man*, London: Methuen & Co., 405–18.

Rodolfi, G. 1988: Geomorphological mapping applied to land evaluation and soil conservation in agricultural planning: some examples from Tuscany (Italy). *Zeitschrift für Geomophologie, Supplement-Band*, **44**: 91–106.

Rögner, K. 1987: Temperature measurements of rock surfaces in hot deserts (Negev, Israel). In Gardiner, V. (ed.), *International Geomorphology 1986*, Part II, 1271–86. Chichester: John Wiley & Sons,

Römer, W. 1993: Die Morphologie des Alkalikomplexes von Jacupiranga und seiner Umgebung. *Aachener Geographische Arbeiten*, **26**: 300.

Rohdenburg, H. 1989: Landschaftsökologie – Geomorphologie. Cremlingen: Catena Verlag.

Rost, K.T. 1993: Die jungpleistozäne Vergletscherung des Quinlin Shan (Provinz Shanxi). *Erdkunde*, **47**: 131–42.

Roth, E.S. 1965: Temperature and water contents as factors in desert weathering. *Journal of Geology*, **73**: 454–68.

Ruhe, R.V. 1959: Stone lines in soils. *Soil Science*, **87**: 223–31.

Russell, I.C. 1898: *Rivers of North America*. New York: Putnam.

Russell, R.J. 1962: Origin of beachrock. *Zeitschrift für Geomorphologie, N.F.*, **6**: 1–16.

Sala, M. and Inbar, M. 1992: Some effects of urbanization in Catalan rivers. *Catena*, **19**: 345–61.

Sala, M. and Rubio, J.L. (eds.) 1994: *Soil erosion as a consequence of forest fires*. Logrono: Geoforma Ediciones.

Scheffer/Schachtschabel 1976: *Lehrbuch der Bodenkunde*. 9.Aufl. bearb. von P. Schachschabel, H.-P. Blume, K.H. Hartge and U. Schwertmann. Stuttgart: Enke-Verlag.

Scheidegger, A.E. 1961: *Theoretical geomorphology*. Berlin. Springer-Verlag, (2nd edn 1990, 3rd edn 1991.)

Schellmann, W. 1994: Geochemical differentiation in laterite and bauxite formation. *Catena*, **21**: 131–43.

Schick, A.P. 1974: Alluvial fans and desert roads – a problem in applied geomorphology. *Abhandlungen der*

Akademie der Wissenschaften in Göttingen, Mathematisch-Physikalische Klasse, III. Folge, **29**: 418–25.

Schick, A.P. 1979: Fluvial processes and settlement in arid environments. *Geojournal*, **3**(4): 351–60.

Schick, A.P. , Hassan, M.A. and Lekach, J. 1987: A vertical exchange model for coarse bedload movement: numerical considerations. *Catena Supplement*, **10**: 73–83.

Schmidt, J. 1991: A mathematical model to simulate rainfall erosion. *Catena Supplement*, **19**: 101–9.

Schmidt; K: and Walter; R: 1990: *Erdgeschichte*. Berlin. Walter de Gruyter, 4. Aufl.

Schmidt, K.-H. 1984: *Der Fluß und sein Einzugsgebiet*. Wiesbaden: Franz Steiner Verlag.

Schmidt, K.-H. 1987: Factors influencing structural landform dynamics on the Colorado Plateau – about the necessity of calibrating theoretical models of empirical data. *Catena Supplement*, **10**: 51–66.

Schmidt, K.-H. 1988: Die Reliefentwicklung des Colorado Plateau. *Berliner Geographische Abhandlungen*, **49**: 183.

Schmidt, K.-H. 1992: Stepped pediments in the Henry Mountains, Utah: Gilbert's equilibrium concept and historical geomorphology. *Catena Supplement*, **23**: 135–50.

Schmidt, K.-H. and Ergenzinger, P. 1990: Radiotracer und Magnettracer – die Leistungen neuer Meßsysteme für die fluviale Dynamik. *Die Geowissenschaften*, **8**(4): 96–102.

Schmidt, R.-G. 1979: Probleme der Erfassung und Quantifizierung von Ausmaß und Prozessen der aktuellen Bodenerosion (Abspülung) auf Ackerflächen. *Physiographica* (Basler Beiträge zur Physiographie), Band 1.

Schmitthenner, H. 1926: Die Dellen und ihre morphologische Bedeutung. *Zeitschrift für Geomorphologie*, **1**: 3–28.

Schmitthenner, H. 1927: Die Oberflächengestaltung im außertropischen Monsunklima. *Düsseldorfer Geographische Vorträge und Erörterungen*, 3. Teil: *Morphologie der Klimazonen*. Breslau, 26–36.

Schmitthenner, H. 1954: Die Regeln der morphologischen Gestaltung im Schichtstufenland. *Petermanns Geographische Mitteilungen*, **98**: 3–10.

Schröder, R. 1991: Test of Hack's slope to bed material relationship in the southern Eifel uplands, Germany. *Earth Surface Processes and Landforms*, **16**: 731–6.

Schultz, J. 1995: *Die Ökozonen der Erde*, 2. Auflage. Stuttgart: Ulmer (UTB).

Schumm, S.A. 1956: The evolution of drainage systems and slopes in badlands of Perth Amboy, New Jersey. *Geological Society of America, Bulletin*, **67**: 597–646.

Schumm, S.A. 1960: *The shape of alluvial channels in relation to sediment type*. United States Geological Survey Professional Paper 352B, 17–30.

Schumm, S.A. 1961: *Effect of sediment characteristics on erosion and sedimentation in small stream channels*. US Geological Survey Professional Paper 352C, 31–70.

Schumm, S.A. 1967: Rates of surficial rock creep on hillslopes in western Colorado. *Science*, **155**: 560–1.

Schumm, S.A. 1977: *The fluvial system*. New York: John Wiley & Sons.

Schumm, S.A. and Chorley, R.J. 1964: The fall of Threatening Rock. *American Journal of Science*, **262**: 1041–54.

Schumm, S.A. and Lichty, R.W. 1965: Time, space and causality in geomorphology. *American Journal of Science*, **263**: 110–19.

Schunke, E. 1975: Die Periglazialerscheinungen Islands in Abhängigkeit von Klima und Substrat. Abhandlungen der Akademistie de Wissenschatten in Göttingen, Mathematist-Physikalische Klasse, III. Folge, **30**: 273.

Schunke, E. 1988: Die Fußflächen- und Schichtkammlandschaften der Richardson Mountains (NW-Kanada). *Nachrichten der Akademie der Wissenschaften in Göttingen, II.Mathematisch- Physikalische Klasse.*, **5**: 79–110.

Schwarz, T. and Germann, K. (eds) 1994: Lateritization processes and supergene ore formation. *Catena*, **21** (2/3), special issue.

Schwarzbach, M. 1974: *Das Klima der Vorzeit. Eine Einführung in die Paläoklimatologie.* 3.Auflage. Stuttgart: Enke Verlag.

Seidl, M.A. and Dietrich, W.E. 1992: The problem of channel erosion into bedrock. *Catena Supplement*, **23**: 101–24.

Selby, M.J. 1979: Slopes and weathering. In Gregory, K.J. and Walling, D.E. (eds), *Man and environmental processes.* Folkestone: Wm. Dawson & Sons, 105–22.

Selby, M.J. 1982: *Hillslope materials and processes.* Oxford: Oxford University Press.

Selli, R. and Trevisan, L. 1964: La frana del Vaiont. *Annali del Museo Geologico di Bologna*, **32**: 68.

Semmel, A. 1977: *Grundzüge der Bodengeographie.* Stuttgart: B.G. Teubner.

Semmel, A. 1989: The importance of loess in the interpretation of geomorphological processes and for dating in the Federal Republic of Germany. *Catena Supplement*, **15**: 179–88.

Seppälä, M. 1988: Palsas and related forms. In Clark, M.J. (ed.), *Periglacial geomorphology* 247–78. Chichester: John Wiley & Sons.

Sharpe, C.F.S. 1938: *Landslides and related phenomena.* New York: Columbia University Press.

Shepard, F.P. 1948: *Submarine geology.* New York: Harper. (New edn 1973.)

Shreve, R.L. 1966: Statistical law of stream numbers. *Journal of Geology*, **74**: 17–37.

Shreve, R.L. 1967: Infinite topologically random channel networks. *Journal of Geology*, **75**: 178–86.

Shreve, R.L. 1968: *The Blackhawk landslide.* Geological Society of America, Special Paper 108.

Simons, D.B. 1969: Open channel flow. In Chorley, R. (ed.), *Water, earth and man.* London: Methuen & Co., 297–318.

Simons, D.B. and Richardson, E.V. 1963: Forms of bed roughness in alluvial channels. *Transactions of the American Society of Civil Engineers*, **128**: 184–302.

Slaymaker, O. 1980: Geomorphic field experiments – inventory and prospect. *Zeitschrift für Geomorphologie, Supplement-Band*, **35**: 183–94.

Slaymaker, O. 1991: Mountain geomorphology: a theoretical framework for measurement programmes. *Catena*, **18**: 427–37.

Slaymaker, O. and Balteanu, D. (eds) 1986: Geomorphology and land management. *Zeitschrift für Geomorphologie, Supplement-Band*, **58**.

Slaymaker, O., Dunne, T. and Rapp, A. (eds) 1980: Geomorphic field experiments on hillslopes. *Zeitschrift für Geomorphologie, Supplement-Band*, **35**.

Sölch, J. 1935: Fluß- and Eiswerk in den Alpen zwischen Ötztal and St. Gotthard. *Petermanns Geographische Mitteilungen*, Ergänzungsheft **219**: 143, Ergänzungsheft **220**: 184.

Souchez, R. and Lorrain, R.D. 1991: *Ice composition and glacier dynamics.* Berlin: Springer-Verlag.

Spönemann, J. 1977: Die periglaziale Höhenstufe Ostafrikas. *Abhandlungen der Akademie der Wissenschaften in Göttingen, Mathematisch-Physikalische Klasse*, III. Folge, **31**: 300–32.

Spönemann, J. 1983: Die Mesoformen der periglazialen Höhenstufen Ostafrikas. *Abhandlungen der Akademie der Wissenschaften in Göttingen, Mathematisch-Physikalische Klasse*, III.Folge, **35**: 388–402.

Spönemann, J. 1989: Homoclinal ridges in Lower Saxony. *Catena Supplement*, **15**: 133–49.

Sponholz, B. 1989: Karsterscheinungen in nichtkarbonatischen Gesteinen der östlichen Republik Niger. *Würzburger Geographische Arbeiten*, **75**: 268.

Sponholz, B. 1994: Silicate karst associated with lateritic formations (examples from eastern Niger). *Catena*, **21**: 269–78.

Stäblein, G. 1981: Geomorphologische Standortaufnahme mit dem EDV-Symbolschlüssel, Beispiele von Dateien and Auswertungen. *Zeitschrift für Geomorphologie, Supplement- Band*, **39**: 39–49.

Statham, I. 1981: Slope processes. In Goudie, A.S. (ed.), *Geomorphological Techniques.* London: George Allen & Unwin, 156–77.

Stickel, R. 1927: Zur Morphologie der Hochflächen des linksrheinischen Schiefergebirges und angrenzender Gebiete. *Beiträge zur Landeskunde der Rheinlande*, **5**: 104.

Stingl, H. 1969: Ein periglazialmorphologisches Nord-Süd-Profil durch die Ostalpen. *Göttinger Geographische Abhandlungen*, **49**: 134.

Stingl, H. 1974: Zur Genese und Entwicklung von Strukturbodenformen. Abhandlungen der Akademie der Wissenschaften in *Göttingen, Mathematisch-Physikalische Klasse, III. Folge*, **29**: 250–62.

Stingl, H., Garleff, K. and Brunotte, E. 1983: Pedimenttypen im westlichen Argentinien. *Zeitschrift für Geomorphologie, Supplement-Band*, **48**: 213–24.

Stocking, M.A. 1978: Interpretation of stone-lines. *The South African Geographical Journal*, **60**: 121–34.

Strahler, A.N. 1952: Hypsometric (area-altitude) analysis of erosional topography. *Bulletin of the Geological Society of America*, **63**: 1117–42.

Strahler, A.N. 1957: Quantitative analysis of watershed geomorphology. *Transactions of the American Geophysical Union*, **38**: 913–20.

Strahler, A.N. 1964: Quantitative geomorphology of drainage basins and channel networks. Section 4-II in Chow, V.T. (ed.), *Handbook of applied hydrology.* New York: McGraw-Hill Book Co., 4–39–4–76.

Strahler, A.N. 1975: *Physical Geography.* 4th edn. New York: John Wiley & Sons.

Strobl, J. 1988: Einige Aspekte geomorphologischer Analyse und Dokumentation aus der Sicht der EDV. *Zeitschrift für Geomorphologie, Supplement-Band*, **68**: 9–19.

Strunk, H. 1988: Episodische Murschübe in den Pragser Dolomiten – semiquantitative Erfassung von Frequenz und Transportmenge. *Zeitschrift für Geomorphologie, Supplement- Band*, **70**: 163–86.

Strunk, H. 1991: Frequency distribution of debris flows in the Alps since the "Little Ice Age". *Zeitschrift für Geomorphologie, Supplement-Band*, **83**: 71–81.

Summerfield, M.A. 1991: *Global geomorphology.* Harlow: Longman. New York: John Wiley.

Sutherland, D.G. 1984: Geomorphology and mineral exploration for diamondiferous placer deposits. *Zeitschrift für Geomorphologie, Supplement-Band,* **51**: 95–108.

Sweeting, M.M. 1990: The Guilin Karst. *Zeitschrift für Geomorphologie, Supplement-Band* 77, 47–65

Sweeting, M.M. 1993: Reflections on the development of Karst Geomorphology in Europe and a comparison with its development in China. *Zeitschrift für Geomorphologie, Supplement-Band,* **93**, 127–36.

Tarling, D.H. 1971: *Principles and applications of palaeomagnetism.* London: Chapman & Hall.

Terzaghi, K. 1943: *Theoretical soil mechanics,* 2nd ed. New York: John Wiley.

Terzaghi, K. and Peck, R.B. 1967: *Soil mechanics in engineering practice,* 2nd ed. New York: John Wiley.

Thomas, M.F. 1974: *Tropical geomorphology.* London: Macmillan.

Thomas, M.F. 1978: The study of inselbergs. *Zeitschrift für Geomorphologie, Supplement-Band,* **31**: 1–41.

Thomas, M.F. 1994: *Geomorphology in the tropics.* Chichester: John Wiley & Sons.

Thomas, M.F. and Thorp, M.B. 1993: The geomorphology of some Quaternary placer deposits. *Zeitschrift für Geomorphologie, Supplement-Band,* **87**: 183–94.

Thorbecke, F. (ed.) 1927a: Morphologie der Klimazonen. *Düsseldorfer Geographische Vorträge und Erörterungen,* 3.Teil. Breslau.

Thorbecke, F. (ed.) 1927b: Der Formenschatz im periodisch trockenen Tropenklima mit überwiegender Regenzeit. *Düsseldorfer Geographische Vorträge und Erörterungen,* 3.Teil Breslau, 11–17.

Thorn, C.E. and Welford, M.R. 1994: The equilibrium concept in geomorphology. *Annals of the Association of American Geographers,* **84**: 666–96.

Thornbury, W.L. 1954: *Principles of geomorphology.* New York: John Wiley & Sons, (2nd edn 1969.)

Thornes, J.B. and Brunsden, D. 1977: *Geomorphology and time.* London: Methuen & Co.

Tinkler, K.J. 1985: *A short history of geomorphology.* Boston: Unwin Hyman.

Toy, T. 1984: Geomorphology of surface-mined land. In Costa, J.E. and Fleisher, P.J. (eds), *Developments and applications of geomorphology.* Berlin: Springer- Verlag, 133–70.

Tricart, J. 1949: Les phénomènes périglaciaires dans les Vosges gréseuses. *C.R.S. de la Societé Géologique de France,* **15**: séance du 5 décembre 1949.

Tricart, J. 1962: *L' Épiderme de la Terre. Esquisse d_une géomorphologie appliquée.* Paris.

Tricart, J. 1965: *Principe et méthodes de la géomorphologie.* Masson, Paris.

Tricart, J. and Cailleux, A. 1962–69: *Traité de géomorphologie.* Paris.Bd.1: *Introduction à la géomorphologie climatique* (1965). Bd.2: Le modèle des 5régions périglaciaires (1967). Bd.3: *Le modèle glaciaire et nival* (1962). Bd.4: *le modèle des régions sèches* (1969). Bd.5:*Le modèle des régions chaudes, forêts et savanes* (1965).

Troll, K. 1924: Der diluviale Inn-Chiemsee-Gletscher. *Forschungen zur deutschen Landes und Volkskunde,* **23**(1): 121.

Troll, K. 1926: Die jungglazialen Schotterfluren im Umkreis der deutschen Alpen. *Forschungen zur deutschen Landes-und Volkskunde,* **24**(4): 158–256.

Trudgill, S.T. 1976: Rock weathering and climate: quantitative and experimental aspects. In Derbyshire, E. (ed.), *Geomorphology and climate.* London: John Wiley & Sons 59–100.

Trudgill, S.T. 1988: Hillslope solute modelling. In Anderson, M.G. (ed.), *Modelling geomorphological systems.* Chichester: John Wiley & Sons, 309–39.

Twidale, C.R. 1993: The research frontier and beyond: granitic terrains. *Geomorphology,* **7**: 187–225.

US Department of Agriculture 1938: *Soils and men. Yearbook of agriculture 1938.* Washington, DC: US Government Printing Office.

US Department Of Agriculture 1975: *Soil taxonomy.* Agricultural Handbook no. 436.

Valentin, H. 1952: Die Küsten der Erde. *Petermanns Geographische Mitteilungen,* Ergänzungsheft Nr.246: 118.

Valeton, I. 1994: Element concentration and formation of ore deposits by weathering. *Catena,* **21**: 99–129.

Van Ash, T. 1983 Water erosion on slopes in some land units in a mediterranean area. *Catena Supplement* **4**: 129–40.

Verstappen, H.Th. 1968: Coral reefs – wind and current growth control. In Fairbridge, R.W. (ed.), *The encyclopedia of geomorphology.* New York: Reinhold Book Corporation, 197–202.

Verstappen, H.Th. 1983a: Applied geomorphology – geomorphological surveys for environmental development. Amsterdam: Elsevier.

Verstappen, H.Th. 1983b: Geomorphological surveys and natural hazard zoning, with special reference to volcanic hazards in central Java. *Zeitschrift für Geomorphologie, Supplement-Band,* **68**: 81–102.

Viles, H.A. 1993: The environmental sensitivity of blistering of limestone walls in Oxford, England: a preliminary study. In Thomas, D.S.G. and Allison, R.J. (eds.), *Landscape sensitivity.* Chichester: John Wiley & Sons 309–26.

Vogt, J. 1966: Le complexe de la stone-line. Mise au point. Terrains d'alteration et de recouvrement en zone intertropicale. *Bulletin B.R.G.M,* **4**: 3–51.

Wagner, G. 1960: *Einführung in die Erd- und Landschaftsgeschichte.* Öhringen: Verlag d. Hohenlohe'schen Buchhandlung F.Rau.

Wagner, G.A. and Zöller, L. 1989: Neuere Datierungsmethoden für geowissenschaftliche Forschungen. Unter besonderer Berücksichtigung der Thermolumineszenz. *Geographische Rundschau,* **9**, 507–12.

Walter, R. 1992: *Geologie von Mitteleuropa.* 5.Aufl. Stuttgart; E. Schweizerbart'sche Verlagsbuchhandlung.

Walther, J. 1900: *Das Gesetz der Wüstenbildung in Gegenwart und Vorzeit.* 1. Auflage Berlin. 4. Auflage (1924), Leipzig.

Ward, W.H. 1945: The stability of natural slopes. *Geographical Journal,* **111**: 170–91.

Warwick, G.T. 1964: Dry valleys in the southern Pennines, England. *Erdkunde,* **18**: 116–23.

Washburn, A.L. 1979: *Geocryology.* London: Edward Arnold.

Wayland, E.J. (1934): Peneplains and some other erosional platforms. *Bulletin of the Geological Survey of Uganda, Annual Report, Notes 1,* **74**: 366–77.

Wechmann, A. 1964: *Hydrologie.* München: R. Oldenbourg.

Wegener, A. 1915: *Die Entstehung der Kontinente und Ozeane.* Braunschweig. (4. Aufl. 1929.)

Wenzens, G. 1975: Synsedimentär entstandene Kalkkrusten als morphologische Zeugen quartärer Kaltzeiten in Nordmexiko und ihre Bedeutung für die Datierung der Sedimente. *Würzburger Geographische Arbeiten,* **43**: 164–73.

Wenzens, G., Wenzens, E., and Schellmann, G. 1994: Number and types of piedmont glaciations east of the Central Southern Patagonian Icefield. *Zentral blatt für Geoloyie und Paläontologie*, Teil 1: 779–90.

Wiens, H.J. 1959: Atoll development and morphology. *Annals of the Association of American Geographers*, **49**: 31–54.

Wiens, H.J. 1962: *Atoll environment and ecology*. New Haven: Yale University Press.

Wilhelm, F. 1966: *Hydrologie – Glaziologie*. Braunschweig: Georg Westermann Verlag.

Wilhelm, F. 1975: *Schnee- und Gletscherkunde*. Berlin: Walter Du Gruyter.

Wilhelmy, H. 1958: *Klimamorphologie der Massengesteine*. Braunschweig.

Williams, M.A.J. 1968: Termites and soil development near Brocks Creek, Northern Territory. *Australian Journal of Science*, **31**: 153–4.

Williams, P.W. 1972: Morphometric analysis of polygonal karst in New Guinea. *Bulletin of the Geological Society of America*, **83**: 761–96.

Williams, P.W. 1983: The role of the subcutaneous zone in karst hydrology. *Journal of Hydrology*, **61**: 45–67.

Williams, P.W. 1993: Climatological and geological factors controlling the development of polygonal karst. *Zeitschrift für Geomorphologie, Supplement-Band*, **93**: 159–73.

Willis, B. 1895: *The northern Appalachians*. National Geographic Society Monograph 1.

Wilson, L. 1972: Seasonal sediment yield patterns of United States rivers. *Water Resources Research*, **8**: 1470–1779.

Winchester, V. 1984: A proposal for a new approach to lichenometry. *British Geomorphological Research Group, Technical Bulletin* no. 33: 3–20.

Wischmeier, W.H. and Smith, D.D. 1962: Soil loss estimation as a tool in soil and water management planning. *International Association of Scientific Hydrology*, Publication 59: 148–59.

Wischmeier, W.H. and Smith, D.D. 1978: *Predicting rainfall erosion losses – a guide to conservation planning*. US Department of Agriculture Handbook 537.

Wissmann, H.V. 1959: Die heutige Vergletscherung und Schneegrenze in Hochasien. *Abhandlungen der Mathematisch- Naturwissenschaftlichen Klasse der Akademie der Wissenschaften und der Literatur Mainz*, Nr.14.

Witter, J.V., Jungerius P.D. and Ten Harkel, M.J. 1991: Modelling water erosion and the impact of water repellency. *Catena*, 18: 115–24.

Woldenberg, M.J. (ed.) 1985: *Models in geomorphology*. Boston: Allen & Unwin.

Woldstedt, P. and Duphorn, K. 1974: *Norddeutschland und angrenzende Gebiete im Eiszeitalter*. 3. Aufl. neu bearbeitet von K.Duphorn. Stuttgart; K.F.Koehler Verlag.

Wolff, A. 1976: *Gefahr für den Kölner Dom. Bild-Dokumentation zur Verwitterung*. 2.Auflage. Köln: Metropolitankapitel.

Wolman, M.G. (1967): A cycle of sedimentation and erosion in urban river channels. *Geografiska Annaler*, **49 A**: 385–95.

Wolman, M.G. and Brush, L.M. 1961: *Factors controlling the size of stream channels in coarse, non-cohesive sands*. US Geological Survey Professional Paper 282G, 183–210.

Wolman, M.G. and Miller, J.P. 1960: Magnitude and frequency of forces in geomorphic processes. *Journal of Geology*, **68**: 54–74.

Wolman, M.G. and Schick, A.P. 1967: Effects of construction on fluvial sediment: urban and suburban areas of Maryland. *Water Resources Research*, **3**: 451–64.

Yair, A. 1990: Runoff generation in a sandy area. The Nizzana Sands, Western Negev, Israel. *Earth surface Processes and Landforms*, **15**: 597–609.

Yair, A. 1992: Climate change and environment at the desert fringe, Northern Negev, Israel. *Catena Supplement*, 23: 47–58.

Yair, A. and Berkowicz, S.M. 1989: Climatic and nonclimatic controls of aridity. The case of the northern Negev of Israel. *Catena Supplement*, **14**: 145–58.

Yatsu, E. 1988: *The nature of weathering*. Tokyo: Sozosha.

Young, A.1960: Soil movement by denudational processes on slopes. *Nature*, **188**: 120–2.

Young, A. 1972: *Slopes*. Edinburgh: Oliver & Boyd.

Young, A. 1978: A twelve-year record of soil movement on a slope. *Zeitschrift für Geomorphologie, Supplement-Band*, **29**: 104–10.

Zaruba, Q. and Mencl, V. (1969): *Landslides and their control*. Amsterdam: Elsevier. Prague: Academia.

Zenkovich, V.P. 1967: *Processes of coastal development*. English. edition, ed. J.A. Steers. Edinburgh: Oliver & Boyd.

Zernitz, E.R. 1932: Drainage patterns and their significance. *Journal of Geology*, **40**: 498–521.

Zienert, A. 1989: Geomorphological aspects of the Odenwald. *Catena Supplement*, **15**: 199–210.

Zingg, A.W. 1940: Degree and length of land slope as it affects soil loss in runoff. *Agricultural Engineering*, **21**: 59–64.

Index

A horizon 81, 82
Ablation moraine 272
Ablation zone 263, 264, 265, 276
Abrasion platforms 220, 288, 302, 304, 305
AC horizon 82
Accumulation terraces 183–6, 276, 277
Accumulation valley floor 159, 183–5, 217
Acid igneous rocks 52
Active margin of continent 30
Aeolian accumulation forms 119, 120
Aeolian denudation 88
Aeolian process response system 117
Aeolian processes 117, 118, 119
    accumulation forms 120
    cover sands 120
    deflation 119
    desert pavement 119
    dunes 120
    dynamic equilibrium 119
    feedbacks 120
    loess 120
    reptation 117
    saltation 117
    sandstorms 118
    transport 120
    wind abrasion 119
    wind ripples 120
    wind shear 118
Aeolian sandstones 54
Aeolian system 49, 50
Aeolian transport 120
Albite 78
Alimentation zone 263, 264, 265, 276
Alluvial cones 191
Alluvial fans 191–4, 277
    alluvial cones 191
    branching 192

dissection 193
dynamic equilibrium 193
gradient 192
headward accumulation 193
natural levees 191, 192
new channel braiding 192
pluvial periods 193
size and growth 192
terracing 193
Alpine foreland 187, 199, 207
Alpine mountain belt 34
Alpine orogeny 34, 237
Alps 1, 6, 267, 282
Amphidromies 291, 292
Andesite 52
Angara shield 33
Angle of internal friction 89
Anhydrite 56, 76
Annular drainage pattern 208
Anorthite 78
Antarctic ice cap 265, 281
Antecedent valleys 201, 202
Anticlines 16, 25, 237
Aphanitic texture 52
Appalachian Mountains, U.S.A. 177, 202–4
Applied geomorphology 9, 317–24
Aquicludes 136, 139, 239
Aquifers 136, 139, 239
Arcuate delta 197
Areal downwearing 48
Arkose 53
Asthenosphere 27, 29
Atolls 314, 315
Avalanche denudation 88
    avalanche trenches 98
Avalanches 98
Avulsion 165

B horizon 81, 82
Back scarps 241
Back swamp 160, 165
Backwash 299

Badlands 116
Baltic Shield 2, 32, 33
Barchans 121
Barrier beaches 306–10
    dunes 307
    dynamic equilibrium 308, 309
    lagoons 307, 308
    marsh 307
    mass balance 308–10
Barrier islands 307, 308
Barrier reefs 314, 315
Basalt 52
Basalt columns 232
Base level 43, 174, 200, 202, 225, 235
    changes in base level 175
    headward erosion 175
    local base level 174
    regional base level 174
    vertical erosion 175
Basic igneous rocks 52
Batholith 36, 42
Bauxite 79, 85
Beach cusps 301
Beach dells 301
Beach drift 301, 308, 309
Beach forms 301
    cusps 301
    dells 301
    drift 301
    ridges 302, 303, 304
    feedbacks 301
    gravel bars 301
    high shore 302
    off shore bars 301
    swales 288
Beach pebbles 288, 305
Beach platforms 288
Beach ridges 288, 302, 303, 304
Beaches 286
Beachrock 302
Bedding structures 236
    facies 236

Bedding structures (*cont*)
  sedimentation 237
  tectonic movements 237
Bergschrund 267
Bifurcation ratio 205
Biotite 51
Bioturbation 86
Bird's foot delta 196
Black earth 82
Blind valleys 254
Block disintegration 71
Block fall 88, 93
Block fields 111
Block slides 88
Block streams 88, 110, 209–11
Bornhardts 223
Boulder clay 273
Braided streams 162–5
  avulsion 165
  erosional braiding 162
  in-channel braiding 162
  new channel braiding 162, 165
Branching 192
Breakers 298
Breccia 53
Brown earth 82
Bubnoff unit (B) 6
Butte 242

C horizon 81
Caesium 138–9
Calcite 59, 77
Calcium carbonate 54, 77
Calcrete (caliche) 84, 85
Calderas 40
  collapse calderas 40
  explosion calderas 40
Caledonian orogeny 33
Cambering 97
Canadian Shield 2
Carbonation 77
Capture 188
Cascade waterfalls 180
Causal relationships 5
Caves 257–60
Cementing agents in sedimentary
  rocks 54
Central African graben 35
Chalk 55, 62
Chelation 79
Chemical elements 51
Chemical weathering, *see*
  Weathering
Chernozem 82
Chezy equation 114, 146
Cinder and ash eruptions 37
Cinder and ash volcanoes 37
Circum Pacific mountain belt 34

Cirque glaciers 267
Cirques (cwms, corries) 267, 270
  cirque lake 270
  cirque threshold 270
  snowline 270
Clastic sedimentary rocks 53
Clay 53, 62, 81
Claystone 53
Cliff coasts 288, 302–5
  abrasion platforms 288, 302, 304,
    305
  active cliffs 305
  beach pebbles 288, 305
  cliff foot talus 288, 304
  cliff retreat 304
  cliffs 302, 303
  inactive cliffs 305
  marine arches 305
  marine stacks 306
  sea caves 305
  sea cliff hanging valleys 304
  submarine talus 288, 305
  wave cut notch 288, 303
Cliff foot talus 288, 304
Climatic change and processes 285
Climatic geomorphology 9, 326
Climatic snowline 263
Closed systems 10
Coastal classifications 288–90
Coastal dunes 288
Coastal geomorphology 9
Coasts 287, 288
  cliff coasts 288
  length of coasts 287
  unconsolidated material coasts
    288
Cockpits 255
Collapse sinks 253
Complexing 79
Cone and tower karst 256, 257
Confluence spurs 186
Conglomerates 53
Contact metamorphism 58, 60
Contact springs 241
Contact zones of magma 59
Continental crust 25
Continental drift 29
Continental platform 25
Continental shelf 27, 316
Continental slope 25, 316
Continuous creep 92, 100
Contraction joints 230
Coral coasts 313
  calcium carbonate 313
  coral reefs 313
  glacio-eustatic changes in sea
    level 313, 314

Coral reefs 314, 315
  atolls 314, 315
  barrier reefs 314, 315
  distribution of coral reefs 313
  fringing reefs 314, 315
  growth rates 314
  guyots 314
Coulomb's Law 89
Coupling points 213
Crater Lake, U.S.A. 41
Creep 100, 101
Creep as a result of swelling and
  shrinking 102
Creep denudation 88
Creep on the moon and planet
  Mars 103
Cretaceous 30
Crevasses 262, 268
Critical regolith thickness 67
Cross-bedding 54
Crustal blocks 34
Crustal movements 1, 6, 16, 25,
  43–4, 130, 186, 209, 210, 230
Crusts 84
Cryostatic pressure 65
Cryoturbation 88
Crystalline rocks 69
Cuestas 33, 238–48
  back scarps 241
  crenulation index for cuesta
    scarps 248
  development of scarps 239
  dip slope 238
  fluvial process response system
    239
  front scarps 241
  in Europe and North America
    242, 243, 244
  model of cuesta scarps 248
  morphometric characteristics 247
  residual outliers 242
  retreat of cuesta scarps 242
  scarp forming rocks 239
  slope 238, 241
Cuspate delta 196

Davis cycle 220–2, 325
Dead ice 264, 274
Debris slides 88, 97
Decay equilibrium 13
Deep sea trenches 25, 29
Deflation 88, 119
Dells 105, 218, 219
  cryoplanation terraces 106
  nivation hollows 106
Deltas 194–8
  age and distribution 198
  arcuate delta 197

barriers 197
bars 196
bird's foot delta 196
bottomset beds 194
cuspate delta 196
delta arms 195
delta bedding structure 194
delta lakes 197
delta outline 195
dynamic equilibrium 196
estuaries 197
estuarine delta 197
foreset beds 194
Mississippi delta 196
natural levees 195
Niger delta 197
Nile delta 197
Rhine-Maas delta 197
topset beds 195
winged delta 196
Dendrochronology 8
Denudation 1, 17, 48, 88, 103, 111
    aeolian denudation (deflation) 88
    avalanche denudation 88, 98, 99
    block fall 88, 94
    block slides 88
    creep denudation 88, 100–103
    cryoturbation 88
    debris slides 88, 97
    denudation levels 228
    earthflows 88, 99
    gelifluction (solifluction) 88,
        103–5
    glacial erosion 88, 181, 269–78
    landslides 88, 93, 94, 96
    mass movements 88–115
    mudflows 88, 99
    rock fall 88, 93
    rock glaciers 88, 109
    splash 88, 102
    wash denudation 88, 112
Denudation rates 123–4
    determination 123
    erosion pins 123
    Gerlach troughs 124
    mean denudation rate of a
        catchment basin 124
    measurement of rapid mass
        movements 125
    measuring lowering of land
        surface 123
    sediment traps 124
    measuring slow mass movement
        125
    Young pits 125
Denudation terraces 183, 244

Grand Canyon of the Colorado
    River, U.S.A. 244
    theoretical models 244
Depression of snowline 283
Deranged drainage pattern 208
Desert varnish 85
Desilification 78
Detersion 269, 276
Detraction 269
Diagenesis 53, 212
Diagonal coasts 311
Diagonal joints 232
Diapirs 57
Diorite 52
Dip 237
Discharge regimes 142, 143
    complex regimes 142, 143
    Rhine River 142
    simple regimes 142, 143
Dissolved load 148
    rate of dissolved load transport
        148
Dolines 253, 254
Dolomite limestone 55, 250
Double planation 220, 222
Drag force 115
Drainage basins 18
Drift 273
Dripstone 260
Drumlins 275, 278
Dry valleys 250, 251
Dunes 54, 120, 123, 157, 288
    coastal dunes 288
    angle of internal friction 120
    barchans 121
    kupsten dunes 123
    longitudinal dunes 122
    parabolic dunes 122
    rates of movement 122
    sand grain movement 121
    star-shaped dunes 122
Dynamic equilibrium 12–14, 21, 69,
    127, 160–1, 214–18, 263–5, 308,
    309, 325

Earth pillars 116
Earth's mantle 27
Earthflows 88, 99
East African graben 35
Eastern and central African rift
    valleys 32
Eifel, Germany 37
Eksystemic 10, 209
Electron spin resonance analysis 8
Empirical observation 14
End moraine (terminal moraine)
    274, 278
Endogenic 16, 209

Endogenic crustal movements 43
Endogenic energy supply 209
Endogenic process response system
    25–42
Endorheic streams 142
Energy transfers 11
Englacial moraine 272, 275
Ensystemic 10, 209
Epiphreatic caves 259
Epirogeny 28
Erosion 88, 149–52
Erosion pins 123
Erosional braiding 162, 163, 179, 180
    bed load 162
    feedbacks 163
    vertical erosion 162
    waterfalls and rapids 163
Erratics 273
Eskers 276
Estuaries 197, 294
    ebb current 294
    flood current 294
    tidal bores 296
Estuarine meanders 294–6
    ebb current 294–5
    flood current 294–5
    ice age 296
    lateral erosion 295
    line of highest velocity 295
    sedimentation 295
    tidal currents 294–5
Eustatic change 43, 186, 189
    deposition eustasy 43
    glacio-eustasy 43, 283, 284
    tectonic eustasy 43
Evaporites 56
Exfoliation 74
Exogenic 17
Exogenic energy supply 209
Exogenic process response systems
    48
Exorheic streams 143
Exotic streams 143
Expansion and contraction creep
    100

Fall Line, eastern U.S.A. 180
FAO World Soil Map 83
Fault block mountains 34, 35, 236
Fault block uplift 36
Fault blocks 1, 34
Fault breccia 234
Fault line scarps 234, 235, 236
    inversion of relief 236
    obsequent fault line scarp 236
    resequent fault line scarp 236
Fault scarps 234
    hanging valleys 235

Fault scarps (*cont*)
  Rhine Graben 234
  triangular facets 235
Fault structures 233, 234
  grabens 233
  horsts 234
  tilted blocks 234
  vertical block movements 233
Faults 25, 233
  nappes (sheet overthrust) 233
  normal faults 233
  overthrusts 233
  step faults 234
  thrust faults 233
  transcurrent faults 233
Feedbacks 5, 17, 52, 67, 157, 161
Feldspar 52
Ferrallitic soils 83
Ferralsols 83, 84, 85
Fetch 297
Firn 261
Fission track analysis 8
Fjords 281, 312
Flaking (thermal exfoliation) 73
Flandrian transgression 286
Flat-bottomed valley floor 217
Flint 78
Floodplain 159, 160
  alluvial loam 160
Flow velocity 113, 114
Fluvial deposition 211
Fluvial erosion 211
Fluvial geomorphology 9
Fluvial hydraulics 135–46
  channel cross-section 146
  Chezy equation 146
  continuity equation 147
  discharge equation 146
  feedbacks 146
  flow velocity 145
  Froude number 145
  hydraulic geometry 146
  kinematic viscosity 145
  laminar flow 144
  line of highest velocity 145
  Manning equation 146
  plunging flow 145
  Reynolds number 145
  shooting flow 145
  stream width 147
  streaming flow 145
  turbulent flow 144, 145
Fluvial process response system
  209–11, 215–18
  accumulation valley floor 217
  dells 218, 219

eksystemic energy supply 209,
  210
endogenic energy supply 209, 210
ensystemic components 209
exogenic energy supply 209, 210
flat-bottomed valley floor 217
fluvial deposition 211
fluvial erosion 211
fluvial transport 211
form components 209, 210
glacial erosion 214
gorges 214
groundwater 135
headward erosion of contact
  springs 218, 219
magnitude–frequency analysis
  209, 210, 212
material components 210, 211
models of slope development
  214
process components 209, 211
regional relief 210
sediment delivery ratio 213
seepage sapping 218, 219
slope development 214
slope form and height 210
steady state 210
stream discharge 135
stream load 211
V-shaped valleys 211
valley cross-section (cross
  profile) 214, 217
valley heads 218, 219
vertical erosion 214
water balance 135
water budget 135
Fluvial system 48
Foerden 271, 312
Fold mountains 1, 33–4, 246
  anticlines 16, 25, 140, 238, 246
  Appalachians 244–6
  cluse 244
  synclines 16, 25, 140, 238, 246
Foliated metamorphic rocks 57
Foliation 73
Foliation planes 57
Folkestone Warren, England 97
Form and material systems 11
Form generations 36
Form systems 11
Fossil soils 229
Fractional code mapping 318
Free meanders 166, 168, 171
  channel cross-section 168
  channel form 167
  chute cutoff 169
  helical secondary current 167

lateral erosion 167
line of highest velocity 167
meander spur 169
meander wave length 167, 168
neck cutoff 169
ox-bow lake 169
riffle pool sequences 168
sinuosity 166
slip-off bank 168
stream bed composition 167
undercut bank 168
Freeze–thaw 64, 101–2, 103, 104, 106
  frequency 64, 65
Fringing reefs 314, 315
Front scarps 241
Frost heave 65, 106–7
Frost shattering 65, 69–71, 91
Frost weathering 64–6, 270
Froude number 145
Functional association relationships
  6
Functional geomorphology 4, 326
Functional process relationships 5
Functional relationships 5

Gabbro 52
Gelifluction (solifluction) 88, 105–6
  dells 105
  tongues 105
General systems theory 10
Geochronological dating 8
Geocratic periods 43
Geologic time 7
Geomorphodynamic system 16, 17
Geomorphological mapping 319
Geosynclines 34
Geothermal energy 262, 276
Geysers 139
Gibbsite 79, 85
Glacial coasts 311–13
  bodden 312
  fjords 312
  foerde 271, 312
  linear glacial erosion 312
  skerries 312
  strandflats 312
Glacial deposits 272
  boulder clay 273
  drift 273
  erratics 273
  moraine 272
  till 273
Glacial erosion 181, 269–78
  detersion 269, 276
  detraction 269
  glacial striations 269
  glacier rock flour 269
  plucking 269

Glacial geomorphology 261–86
Glacial lakes 272
Glacial landforms 261, 269
    cirques 266, 269
    end moraines 274
    foerde 271, 312
    glacial lakes 272
    glacier trough valley 270, 271
    hanging valleys 271
    kettle holes 274
    knob and kettle landscape 274
    lateral moraines 274
    push moraines 274
    rock basins 269
    shields 269
    valley steps 271
Glacial meltwater 275
    bed load 276
    dissolved load 276
    glacier flour 276
    suspended load 276
Glacial sequence 278
Glacial spillways (Urstromtäler) 278,
    282, 283
Glacial striations 269
Glacial system 49, 261
Glacial tongue basin 274, 278
Glacials 279, 285
Glacier advance 264
Glacier flour 269, 276
Glacier ice 261, 262, 263
    crevasses 262
    firn 261
    flow movement of glaciers 261,
        262
    geothermal energy 262
    ice temperature 262
    melting point of ice 262, 269
Glacier milk 276
Glacier retreat 264, 278
Glacier trough valley 270
    drumlins 275, 278
    gorges 271
    hanging valleys 271
    trough basin lakes 270
    trough shoulder 270
    valley steps 271
Glacier types 265
    cirque glaciers 267
    fjord glaciers 268
    ice shelf 268
    ice stream nets 267
    inland ice 265, 281
    outlet glaciers 265, 275
    piedmont glaciers 268, 274
    plateau glaciers 265, 266, 274
    valley glaciers 265, 274

Glaciers 261–5
    ablation 263
    ablation zone 263, 264, 265, 276
    alimentation zone 263, 264, 265,
        276
    bergschrund 267
    crevasses 262
    dead ice 264
    dynamic equilibrium 264
    glacier advance and retreat 264
    mass budget 262, 263
    rate of flow of glaciers 264
    snowlines 263
    steady state 264
Glacio-eustatic change 43, 283, 284
    Flandrian transgression 286
    regression of sea level 286
    terraces 44
Glacio-isostatic change 289
Glacio-isostatic crustal movements
    283, 284
Glacio-isostatic subsidence 28, 33
Glacio-isostatic uplift 28
Glaciofluvial deposits 275–9
    alluvial fans 278
    eskers 276
    kame terraces 276
    kames 276, 277
    outwash plains 277, 278
    varves 278
Glaciofluvial processes 275
Gley 77, 83
Global water balance and water
    budget 135
Gneiss 58
    ortho-gneiss 58
    para-gneiss 58
Goethite 77, 81
Gondwana 30, 33
Gorges 214, 272
Graben zones 35
Grabens (Rift valleys) 35, 233
Grain size classes of soils 81
Grand Canyon of the Colorado
    River, U.S.A. 244
Granite 51, 52
Granular disintegration 69
Gravel 53
Gravel bars 156, 163, 288, 301
Great Basin of the U.S.A. 35, 223
Great Salt Lake, U.S.A 28
Greenland ice cap 28, 265, 281
Greywacke 53
Ground moraine 269, 272, 273, 275,
    278
Groundwater 137, 141
    aquicludes 137

aquifers 136
catchment area 138
chemical weathering 138
Darcy's Law 137
hydraulic conductivity 137
hydraulic gradient 136
mechanical weathering 138
groundwater recharge 137
seepage outflow 138
springs 138
water table 137
Grus 81, 111
Gullies 115
Guyots 314
Gypsum 56, 76, 250

Haematite 77
Halite 56
Hanging valley waterfalls 181, 214
Hanging valleys 178, 181, 214, 271
Headward accumulation 193
Headward erosion 175, 204, 241
Headward sedimentation 176
High mountains 25
High shore 288
Historic-genetic geomorphology 4
Historical time 7
History of geomorphology 325–7
Hogbacks 246
    anticlines 246
    Appalachians 240–7
    back slopes 246
    cluse 246
    front slopes 246
    homoclinal ridges 246
    morphometric characteristics 246
    ramp scarps 246
    ridge and valley topography
        246, 247
    synclines 246
Homoclinal ridges 246
Honeycomb (hole) weathering 70
Hooks 307, 308
Hornblende 52
Horst 234
Horton overland flow 112, 113
Horton stream order system 205
Hydration 76
Hydraulics, *see* Fluvial hydraulics
Hydro-static subsidence 28
Hydro-static uplift 28
Hydrofracturing 65
Hydrolysis 77
Hypsographic curve 25

Ice ages 6, 187, 278, 279, 280, 282,
    289, 296, 316
    area glaciated 280, 282

Ice ages (*cont*)
    causes of ice age 280, 281
    climatic zones beyond glaciated
        areas 283, 284
    periglacial climates 283, 284
    pluvials 285
Ice islands 268
Ice lenses 65
Ice shelf 268, 311
Ice wedge nets 108
Icebergs 268, 311
Idiographic 4
Igneous rocks 51
Illite 78
In-channel braiding 163
    branching 164
    dynamic equilibrium 164
    sand and gravel bars 163
Incised meanders 169
    cutoff spur 170
    ingrown meander 169
    inherited meander 169, 170
    lateral erosion 169
    underfit streams 170
    valley cross-section 169
Inclination of sedimentary rocks 237
    anticlines 237
    dip 237
    strike 237
    synclines 237
Inherent tendency to equilibrium
    14, 21
Inland ice 265, 281
Inselbergs 220, 222, 224, 225–7
    azonal inselbergs 227
    bornhardts 223
    chemical weathering 222, 223
    double planation 222
    dynamic equilibrium 222
    regolith cover 222
    shield inselbergs 223
    tors 223
    wash denudation 222
    zonal inselbergs 227
Interception 175
Interflow 83, 115, 116, 135, 136
Interglacials 187, 188, 279, 285
Iron oxide 54
Island arcs 29
Isostasy 25, 27

Joint-determined forms 230–3
Joints 71, 139, 180, 230–3, 255–6,
    258, 305
    contraction joints 230
    diagonal joints 231
    joint sets 231
    joint systems 230, 231

longitudinal joints 231
orientation of joints 231, 232
pressure release joints 74, 91,
    231, 305
tectonic joints 231
transversal joints 231
Jordan rift valley 32
Jurassic 30

Kame terraces 276, 277
Kames 276, 277
Kaolinite 78
Karst 250–60
    calcium carbonate 250
    climatic conditions for formation
        257
    cockpits 255, 256
    collapse sinks 253
    cone and tower karst 256, 257
    dolines 253
    dry valleys 250, 251
    headward erosion 251
    hole lapies 251
    in dolomite 250
    in gypsum 250
    in limestone 250
    in rock salt 250
    joint lapies 252
    karst chimneys 253
    lapies 251
    model of karst development 258
    perviousness of rock 250
    poljes 254
    polygon karst 255, 256
    ponor 254
    rill lapies 251
    solubility of rock 250
    solution sinks 253
    springs 259
    uvalas 253
    water table 251
Karst caves 257–60
    corrosion by mixing 259
    dripstone 260
    epiphreatic caves 259
    joints 258
    karst springs 259
    karst water table 259
    phreatic caves 259
    stalactites 260
    stalagmites 260
    vadose zone 258
Karst geomorphology 9, 250–60
Karst springs 259
Karst water table 259
Kettle hole 274
Knick points 173, 177, 178, 271
    concave knick points 178

convex knick points 178
    hanging valleys 178
    waterfalls 179
Knob and kettle landscape 274
Kupsten dunes 123

Laach Lake, Eifel, Germany 40
Laccolith 41
Lake Bonneville 28
Landform associations 5
Landform size and duration 4
Landforms 1
Landslides 88, 94, 95, 96
Landslips 94, 95
Lapies 252, 253
    hole lapies 252
    joint lapies 252
    rill lapies 252
    silicate karst 253
    solution pits (opferkessel) 253
Lapilli 37
Lateral erosion 152, 165–7, 184, 294,
    295
Lateral moraine 272, 273
Laterite 83, 85
Latosols 83
Laurasia 30
Laurentian Shield 32, 33
Lava 25, 36, 230
Lichenometry 8
Limestone 55, 62, 250
Linear expansion coefficient of rock
    64
Lithosphere 27
Littoral system 49, 286
Loam 81
Loess 120, 284, 285
Longitudinal coasts 311
Longitudinal dunes 122
Longitudinal joints 232
Longitudinal stream profile (long
    profile) 173, 176, 177
    base level 174
    concave knick points 178
    convex knick points 178
    erosion equilibrium 177
    feedbacks 177
    Rhine River 173
    tendency to equilibrium 176
    transport deficit 177
    transport equilibrium 176
Lunda Rise 33

Maars 37
Magma 25, 36, 52
Magnitude–frequency in process
    response systems 212
    coupling points 213

dynamic equilibrium 213
    spatial magnitude–frequency 213
Magnitude–frequency analysis 44,
    210
    of precipitation regimes 44
    of temperature regimes 48
    of wind regimes 48
Magnitude–frequency index 47
    of stream discharge 143
Mangroves 296
Manning equation 112, 146
Marble 58, 59
Marine abrasion 220
Marine arches 306
Marine stacks 306
Marl 55
Mass balance 5, 11, 14
Mass movement 88–115
    angle of internal friction 89, 91
    apparent cohesion 91
    avalanches 98
    block fall 92
    block streams 88, 109
    bulk density 92
    cementing material 91
    cohesion 90
    continuous creep 91, 101
    Coulombs's law 89, 91
    creep as a result of swelling and
        shrinking 102
    creep denudation 100
    critical shear stress 89, 90, 91
    critical shearing resistance 89
    critical tangential stress 90
    cuesta scarps 96
    debris slides 97
    detachment fissure 92
    earthflows 99
    expansion and contraction creep
        101
    freeze–thaw frequency 101
    gravity 89
    joints 91
    landslide debris 95
    landslide scar 95
    landslides 93–5
    landslips 94, 95
    lateral pressure release 95
    mudflow cones 99
    mudflow tracks 99
    mudflows 99
    needle ice 102
    negative pore water pressure 91
    normal force 89, 90
    normal stress 89, 90, 91
    plastic flow 89, 91
    positive pore water pressure 91

regolith 101
retrograde movement 101
rock fall 93
shear box 90
shear force 89
shear plane 93
slope angle 89
slope stability 92
slumps 96, 97
splash creep 100, 102
talus 93
three layer clay minerals 102
undercutting of slopes 94
vaulted overhang 93
Material systems 11
Material transfers 11
Mauna Kea, Hawaii 40
Mauna Loa, Hawaii 40
Maximum possible height of
    mountain ranges 22–3
Mean denudation rate 18
Mean relief 18
Mean slope angle 18
Meanders, *see* River meanders
    estuarine meanders 294–6
Mechanical weathering, *see*
    Weathering
Medial moraine 272, 273
Mediterranean–Mjösa graben zone
    35
Mesozoic 1
Metamorphic rocks 51, 57
Metastable equilibrium 13
Mica 51
Mica schist 58
Mid-ocean ridges 25
Mine tips 323, 333
Mineral exploration 323
Minerals 51
Mississippi delta 196
Model of cuesta scarps 248
Model of denudation terraces of the
    Grand Canyon 244
Model of karst development 257
Model of relief development 20–3,
    215–18
Model of slope profile with mass
    movement and wash denudation
    133, 134
Model of slope profile with slow
    mass movement 130, 131
Model of slope profile with wash
    denudation slopes 132
Models 14
Monadnocks 36
Montmorillonite 66, 78, 79
Moraine 272–5

ablation moraine 272
drainage in moraine areas 282
end moraine (terminal moraine)
    274, 275
englacial moraine 272, 274
ground moraine 272, 273, 275,
    278
ice core moraine 273
lateral moraine 272
medial moraine 272
old moraine zone 282
push moraine 274
rate of accumulation 274
retreat moraine 274
young moraine zone 282
Morphoclimates 44, 62, 64, 86, 87,
    105, 129
Morphogenesis 6
Morphography 4
Morphostructural units of
    continents 32
Mt St Helens, U.S.A. 39
Mudflows 88, 99
    mudflow cones 99
    flow tracks 99
Murram 85
Mylonite 234

Nappes 233
Natural levees 159, 165, 191, 192
    back swamp 160
    overflow channels 165
    return flow channels 165
Needle ice 181
Negative feedbacks 12
Net bifurcation ratios 205
Neutral shorelines 310
New channel braiding 165, 192
    back swamp 165
    branching 165
    crevasse channels 165
    natural levees 165
    overflow channels 165
    return flow channels 165
Niagara Falls, Canada/U.S.A. 179,
    180
Niagara type waterfalls 179, 180
Niger delta 197
Nile delta 197
Nomothetic 4
Normal faults 233
Nunataks 270

Obsidian 52
Ocean floor 25
Oceanic crust 25
Old fold mountains 33, 36
Olivine 52

Open systems 10
Orogenies 28, 36, 230
Orthoclase feldspar 78
Outlet glaciers 265
Outwash plains 277, 278
Overthrusts 233, 237
Oxidation 77, 81

Paired and unpaired terraces 190
Palsas 108, 109
Pangea 29, 32
Parabolic dunes 120
Parabrown earth 82
Parallel stream networks 207
Passive margin of continent 32
Pediments 220–5
    Great Basin of the U.S.A. 223
    inselbergs 224
    lateral planation 224
    pediment pass 224
    pediplain 224
    peneplains 225
    playas 224
    valley side pediments 225
Peneplains 220, 221, 224–9, 325
    abrasion platforms 220
    Davis cycle 220–2
    double planation 220
    marine abrasion 220
    pedimentation 220
Periglacial climates 283, 284
Periglacial denudation 103–12
    altitudinal zones 103, 106
    gelifluction (periglacial
        solifluction) 105
    wash denudation 107
Periglacial processes 284
    cryostatic pressure 109
    frost cracks 108
    frost heave 107
    gelifluction 112
    glacial denudation 110
    ice wedge nets 108, 112
    injection ice 110
    nivation 111
    palsas 108, 109
    permafrost 103
    pingos 109
    polygon stone nets 10
    rock glaciers 109, 110, 111
    stone nets 107
    stone polygons (stone rings) 107
    stone stripes 107, 111
    wash denudation 107
Permafrost 103
    active layer 103
    discontinuous permafrost 104
    seasonal permafrost 104

pH value 76
Phaneritic texture 52
Phonolite 52
Phreatic caves 259
Phreatic eruptions 37
Phyllite 58
Physical time 6, 7
Piedmonts 225–8
    benchlands (piedmont steps) 225
    azonal inselbergs 227
    headward erosion 226
    metachronous surfaces 226
    monadnocks 227
    peneplains 224–9
    scarps 226, 227
    valley side pedimentation 227
    zonal inselbergs 227
Piedmont glaciers 268, 274
Pingos 109
Piping 115, 116
Planation 220, 222, 229
    structurally controlled planation
        229
Plastic flow 89, 92
Plate tectonics 29–32
Plateau glaciers 265, 266, 274
Plateaus 33
Playas 224
Plinthites 85
Plunging flow 144
Plutonic rocks 62
Plutonism 36
Plutonites 52
Pluvials 285
Podsols 81
Poljes 255
Polygon karstal 255
Ponors 254
Positive feedbacks 12
Postglacial uplift 33
Potash 56
Potassium feldspar 51
Potential energy 25
Pressure release exfoliation 74
Pressure release joints 74, 231
Process rate 6
Process response systems 11, 13
Process response models, *see*
    Models
Prophyries 52
Pseudogley 83
Pumice 52
Push moraines 274
Pyroclastic 53
Pyroclastic sediments 37
Pyroxene 52

Quantitative geomorphology 326

Quartz 52, 54, 78
Quartzite 58
Quaternary 43

Radial stream networks 207
Radioactive carbon isotopes 8
Radiocarbon dating 8
Radiometric dating 8
Raindrop impact crater 2
Ranker 83
Rapids 163
Recurrence intervals 46
Red earths 83
Red soils 83
Reef limestone 55
Refraction 298, 310
Regional geomorphology 4
Regional metamorphism 58
Regional relief 210
Regolith 62, 79, 81, 88
Regosol 83
Relative dating 8
Relaxation time 13, 14
Relief 17
Rendzina 83
Reservoirs 322
Residual outliers 242
Retreat moraine 274
Reynolds number 145
Rhenish Slate Mountains 83, 227,
    228
    soil catena 83
Rhine Graben (Rhine rift valley) 35,
    230, 231, 234
Rhine River 142, 164
    regulation of the Rhine 165
    terraces 186
    Rhine-Maas delta 197
Rhyolite 52
Rias 311
Ridge and valley topography 246,
    247
Riffles and pools 157
    feedbacks 157
    secondary circulation 157
Rills 115
Rip currents 300–1, 307
River terraces 183–90
    accumulation valley floor 183
    accumulation terraces 183, 188
    alluvial fans 193
    Alpine foreland 187
    climatically controlled terraces
        187
    confluence spurs 186
    correlation of terraces 189
    crustal uplift 186
    diagnostic significance 189